Green Plants: Their Origin and Diversity

GREEN PLANTS
Their Origin and Diversity

Peter R. Bell, M.A.
Emeritus Professor of Botany in
the University of London

CAMBRIDGE
UNIVERSITY PRESS

Published by the Press Syndicate of the University of Cambridge
The Pitt Building, Trumpington Street, Cambridge CB2 1RP
10 Stamford Road, Oakleigh, Victoria 3166, Australia

First published in North America by Dioscorides Press
(an imprint of Timber Press, Inc.) 1992
First published outside North America by Cambridge University Press 1992

Printed in Hong Kong

A catalogue record for this book is available from the British Library

ISBN 0 521 43875 6   paperback

CONTENTS

PREFACE

Green Plants is a thoroughly revised edition of the earlier *Diversity of Green Plants* by P. R. Bell and C. L. F. Woodcock (3rd edition, London, 1983). The continuing demand for a concise account of the algae and land plants from the point of view of their natural relationships and biology reflects the buoyant state of botanical science. Exciting advances remain a feature of all its aspects. The biophysically minded are revealing in impressive detail the electron pathways in the thylakoid membrane while paleobotanists expand significantly our knowledge of the earliest angiosperms of the Cretaceous and geneticists explore the molecular aspects of plant development. The theme of *Green Plants* is the astonishing diversity of forms which evolution has provided from the atmospheric carbon fixed by photosynthesis, the remarkable phenomenon which is basic to plant life. The treatment of the Plant Kingdom correspondingly extends from the simplest cellular organisms capable of phototrophy, the prokaryotic algae, to the complexities of the flowering plants, not omitting (so far as they are known) the essential features of the plants represented only by fossils.

The record of plant life provides a striking instance of both genetic conservation and variation. The photochemistry of the thylakoid membrane is presumably basically the same today as it was at the dawn of plant life in pre-Cambrian times, and the genetical system controlling its development likewise essentially unchanged. Variations in subsequent biochemical pathways, leading, for example, to C_3 and C_4 plants, may also be of considerable antiquity. Accompanying these stable mechanisms of phototrophy are innumerable variations in morphology, a consequence of the mutability of DNA. Natural selection has offered, and continues to offer, the principal constraint. In lush conditions, even selection, provided essential physiological and reproductive features remain unimpaired, may do little to limit diversity.

Classifications of the algae and land plants facilitate the ordered treatment of diversity. Those adopted here follow schemes in general use. The "blue-greens" (together with the Prochlorophyta) are regarded as algae. To maintain a sharp division between prokaryotic and eukaryotic organisms is to fall into the error of attributing undue weight to one character. The concept of Algae, phototrophs with a wide range of morphological, biochemical and ecological features in common, comprehends both karyotic conditions.

The preparation of the present work has involved the help, willingly given, of experts in many fields. The writer must accept responsibility for any errors remaining. In addition to the authors and publishers cited, the following kindly agreed to the reproduction of figures: The Council of the Linnean Society of London (Figs 5.7, 8.46, 9.14, 9.15); the Trustees of the British Museum (Natural History) (Figs 2.12, 3.20, 3.23, 3.24, 4.15, 4.16, 9.5); and the University of Michigan Press (Figs 3.25, 3.26, 3.27).

Nothing would have been possible without the invaluable technical assistance of John Mackey and the skilled secretarial work of Elizabeth Bell. To both my sincere gratitude.

P. R. Bell
London, 1990

___ 1 ___
GENERAL FEATURES OF THE PLANT KINGDOM

CHARACTERISTICS OF THE LIVING STATE

The living state is characterized by instability and change. A living cell consumes and releases energy by means of numerous chemical reactions, called collectively metabolism, taking place within it. Metabolism is synonymous with life. Even the apparently inert cells of seeds show some metabolism, but admittedly only a fraction of that which occurs during germination and subsequent growth. Metabolism depends upon molecular order. If this is destroyed (for example, by poisons or heat) metabolism ceases and the cell dies. In some instances, it is possible to arrest metabolism without death. With yeast and some tissue cultures, for example, this can be achieved by very rapid freezing to temperatures of the order of $-160°C$ ($-256°F$) or lower. The cells can then be preserved in liquid nitrogen ($-195°C$; $-319°F$) in an inert state. Metabolism resumes on careful thawing.

To maintain its dynamic state, a living cell must be provided with sources of energy. These are predominantly compounds of carbon. In addition, a cell requires water, since much of the metabolism takes place in the aqueous phase in the cell. Also essential are those materials necessary for the maintenance of its structure which it is unable to make for itself. Prominent among these is the nitrogen of the proteins and certain metals and other elements which, although needed only in traces, are essential components of a number of enzymes and associated molecules. Occasionally, complex organic molecules called vitamins or growth factors must also be supplied from outside.

AUTOTROPHIC AND HETEROTROPHIC NUTRITION

It is useful to divide organisms into two classes according to the manner in which their needs for organic carbon are met. Those able to utilize simple molecules with single carbon atoms are termed **autotrophs**; those requiring more complex carbon compounds rich in energy (such as sugars) are termed **heterotrophs**. Some organisms are able to switch between these alternative forms of nutrition, depending upon the environment in which they find themselves.

The assimilation of simple carbon compounds by autotrophs requires an external source of energy. This may be chemical or physical, depending upon the organism. Very many autotrophs (including the whole of the Plant Kingdom) utilize the energy of light, and are consequently known as photoautotrophs, or simply as **phototrophs**. Only the phototrophs have acquired extensive morphological diversity. Autotrophs utilizing energy from chemical sources for the assimilation of carbon are found solely among the bacteria.

Phototrophic life is made possible by unique biological molecules, chlorophyll and bacteriochlorophyll. The chemical differences between them are not profound, but their absorption spectra are distinct, as is their distribution among the phototrophs. Bacteriochlorophyll is found only in bacteria and will not be considered further here. Those organisms containing chlorophyll constitute the Plant Kingdom. So defined, the Plant Kingdom is distinct from all other organisms (including the fungi).

Chlorophyll is a complex pigment. It is green in color, and absorbs light in the blue and to a smaller extent in the red region of the spectrum. The molecule is, in part, similar to the active group of the blood pigment hemoglobin, but contains at its center magnesium in place of iron. A number of different forms are known (a, b, c, d and perhaps e), each with its characteristic absorption spectrum. Chlorophyll a, which is present in all plants, has the remarkable property of temporarily losing electrons when illuminated. Chlorophyll b, which is found in all land plants, assists in the light-harvesting process, but the functions of chlorophylls c, d, and e present in some algae, are not so well known. Chlorophyll is always accompanied by carotene and other accessory pigments (either carotenoids (xanthophylls) or phycobilins or, in a few organisms, both). The light absorbed by these pigments can be transferred to the chlorophyll.

As a result of the remarkable photochemical properties of chlorophyll a, the energy of the incident light is transformed into chemical energy. This leads to the generation in the cell of ATP, and reducing power in the form of NADPH + H$^+$ (the light reactions). These two products then bring about the reduction of atmospheric carbon dioxide in the illuminated cells, and the subsequent assimilation of the carbon in the form of carbohydrates (the dark reactions). The ability to utilize atmospheric carbon dioxide in this way (*photosynthesis*) releases the organisms concerned from the necessity of an external source of carbohydrate, and their nutritional demands are consequently relatively simple.

Two photosystems are involved in photosynthesis. The first (photosystem I) is responsible for the formation of the NADPH + H$^+$, and the second (photosystem II) with the supply of electrons to the chlorophyll of photosystem I. It is photosystem II, found only in plants (as defined in this book), which is responsible for the evolution of oxygen in photosynthesis.

THE STRUCTURE OF THE PHOTOTROPHIC CELL

Chlorophyll does not occur freely in cells, but is always associated with lipoprotein membranes. These membranes surround flattened sacs called *thylakoids*. When the membranes are seen in surface view in the electron microscope (made possible by the special technique of freeze-fracture), is clear that they bear closely packed particles (Fig. 1.1). The larger of these, about 18 nm (1 nm = $10^{-3}\mu$m) in diameter, are probably the site of the chlorophyll and carotenoids (which, like chlorophyll, are lipid soluble). The anchoring of the chlorophyll and carotenoids in a lipoprotein membrane ensures that they are held in a particular order (Fig. 1.2). Electrons can then flow along well-defined paths to the reaction center at which the radiant energy is converted into chemical. The thylakoid membrane is thus the site of the light reactions of photosynthesis, and forms the basis of plant life. In turn, the Animal Kingdom is entirely dependent upon the activity of this membrane, not only for its sustenance, but also for its respiration, since the oxygen of the atmosphere also owes its origin to photosynthesis.

Two distinct kinds of cellular organization are found among the phototrophs as a whole. In the first, termed *prokaryotic*, the cell possesses no distinct nucleus, although a region irregular in outline and of differing density occurs at the center of the cell. This is referred to as a nucleoid, and the genetic material lies therein. In the electron microscope, this region appears fibrillar rather then granular, and the fibrils indicate the site of the deoxyribonucleic acid (DNA). The protoplast of such cells is bounded by a membrane. In phototrophic cells this membrane invaginates into the cytoplasm and forms the thylakoids. Their full development depends upon light. If the cells are grown in the dark, the thylakoids disappear or become very reduced. This primordial kind of phototrophic cell is found in both the photosynthetic bacteria and in the simplest plants. The fossil record supports the view that the original phototrophs were of this prokaryotic kind.

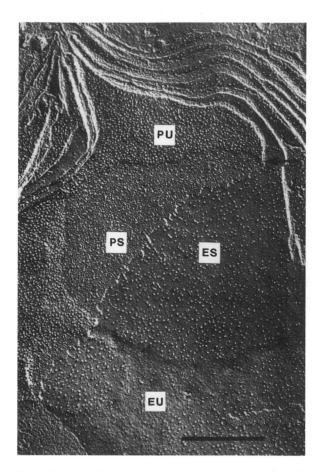

Figure 1.1. Shadowed replica of the thylakoid membranes of the chloroplast of *Euglena*
exposed by freeze-fracture. The thylakoids are either single ("unstacked") or
paired ("stacked"). Because in the conditions of freeze-fracture membranes are
pulled apart, two complementary faces (E and P) are represented in the replica.
This reveals that the particles are asymmetrically placed in the membrane (cf. Fig.
1.2). There are also differences in the frequencies of particles in stacked (S) and
unstacked (U) membranes. The arrow indicates where the membranes of two
adjacent thylakoids come together to form a stack. (From K. R. Miller and L. A.
Staehelin. 1973. *Protoplasma* 77, by permission of Springer Verlag, Vienna.) Scale
bar 0.5 μm.

Geochemical evidence of photosynthesis, and remains very suggestive of bacteria and
simple blue-green algae (Cyanophyta), come from early Archaeozoic rocks of South
Africa and Australia and are believed to be 3.3×10^9–3.5×10^9 years old (Table 1.1).
    In the cells of all other phototrophic plants, the nucleus, the photosynthetic
apparatus, and the membranes incorporating the electron transport chain of respiration
are separated from the remainder of the cytoplasm by distinct envelopes. Such cells,
termed *eukaryotic*, have evidently been capable of giving rise to much more complicated
organisms than the prokaryotic. The photosynthetic apparatus, which consists of
numerous lamellae running parallel to one another, is contained in one or more plastids.
The envelope of the plastid consists of two unit membranes, the inner of which invaginates into the central space (stroma) and generates the thylakoids. The thylakoids in the
fully differentiated plastid (chloroplast) are usually stacked. In the chloroplasts of land
plants the thylakoids are also fenestrated. Consequently, numerous small stacks, called
grana, are formed in place of a single stack, the grana being held together by stroma

lamellae (Fig. 1.3). The grana appear in the light microscope as green dots, each about 0.5 μm in diameter. Although most photosynthesis takes place in the grana, the thylakoids in the stroma also contribute.

Plastids contain both DNA and RNA, and both transcription and translation occur within them. They thus have some resemblance to phototrophic prokaryotes, but plastid biogenesis is not entirely independent of the nucleus. The information for some important components of the plastid is encoded partly in the plastid, and partly in the nuclear genomes. Also some nucleotide sequences are common to the two genomes. In eukaryotes the membranes bearing the respiratory enzymes are also segregated in a distinct organelle, the mitochondrion. Although there are structural and organizational similarities between mitochondria and plastids, in most photosynthesizing cells the mitochondria have far less internal differentiation.

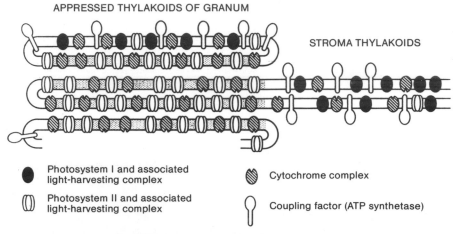

APPRESSED THYLAKOIDS OF GRANUM

STROMA THYLAKOIDS

● Photosystem I and associated light-harvesting complex

◫ Photosystem II and associated light-harvesting complex

⊘ Cytochrome complex

Coupling factor (ATP synthetase)

Figure 1.2. The molecular architecture of the thylakoid membrane. The photosystem I complexes are confined to the outer membranes of the grana and to the stroma thylakoids. The stippled regions indicate the appressed membranes of the granum. (From J. M. Anderson, W. S. Chow, and D. J. Goodchild. 1988. *Australian Journal of Plant Physiology* 15, modified.)

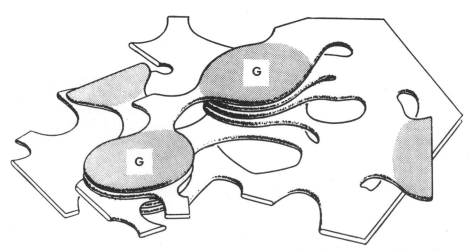

Figure 1.3. Diagram showing the arrangement of the thylakoids in the chloroplast of a higher plant. The stacked regions (grana, G) are visible as green dots in the light microscope. (From an original drawing by W. Wehrmeyer. 1964. *Planta* 63, modified.)

Table 1.1. The geological table.

Era	Period		Age (in 10^6 years)	First authentic appearance
Quarternary	Pleistocene and Recent		0–1.6	
Tertiary (or Cenozoic)	Pliocene Miocene	Upper	1.6–23.7	
	Oligocene Eocene Paleocene	Lower	23.7–66.4	
Mesozoic	Cretaceous	Upper	66.4–97.5	
		Lower	97.5–144	Angiosperms
	Jurassic	Upper	144–163	
		Middle	163–187	
		Lower (Lias)	187–208	
	Triassic		208–245	
Paleozoic	Permian	Upper	245–258	Cycad- and *Ginkgo*-like plants, *Glossopteris*
		Lower	258–286	
	Carboniferous	Upper (Pennsylvanian)	286–320	Conifers
		Lower (Mississippian)	320–360	Pteridosperms
	Devonian	Upper	360–374	Ferns, seeds
		Middle	374–387	Sphenopsids Heterospory
		Lower	387–408	Vascular plants: lycopods and thyriophytes
	Silurian		408–438	Triradiate spores
	Ordovician		438–505	
	Cambrian		505–570	
Pre-Cambrian	Proterozoic		570–2500	Calcareous algae
	Archaeozoic		2500–3000+	Fungi, algae, bacteria

EVOLUTIONARY CONSEQUENCES OF PHOTOSYNTHESIS

It seems beyond doubt from the fossil record of life, and from the biological and geological inferences that can be drawn from it, that life began in water. Although the early forms of life remain conjectural, phototrophs were probably among the first organisms to exist. Those which contained or acquired chlorophyll gave rise to the Plant Kingdom. The descendants of these early aquatic forms, which still, in the main, exploit the watery environment, are termed algae (Chapters 2, 3, and 4). They have many biochemical and structural features in common. Although some have attained morphological complexity, the unicellular forms represent the simplest plants still in existence. Some unicellular eukaryotes contain a single plastid and a single mitochondrion (e.g., *Micromonas*, Fig. 1.4).

At some stage, possibly in the Silurian period (Table 1.1) or even earlier, vegetation began to colonize the land. These early colonists, and consequently the whole of our existing land flora, almost certainly emerged from that group of aquatic plants today represented by the green algae (Chlorophyta). The Chlorophyta and the land plants (a term which means plants adapted to life on land and not merely plants growing on land) have

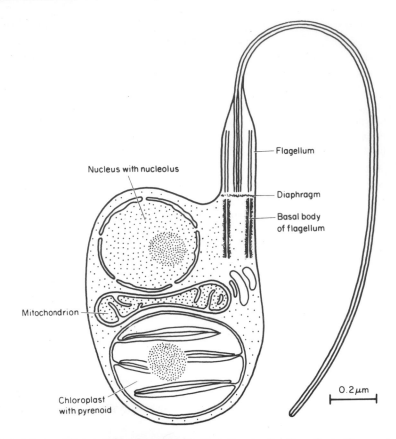

Figure 1.4. *Micromonas pusilla.* **Form and internal organization. Only the central microtubules run into the extension of the flagellum.** *Micromonas* **belongs to a small group of green algae of doubtful affinity. (Based on electron micrographs by I. Manton. 1959.** *Journal of the Marine Biological Association of the United Kingdom* **38.)**

the same photosynthetic pigments and basically the same photosynthetic apparatus. Moreover, at least one green alga (*Cladophorella*) which grows on damp mud is covered on its upper surface by a material which, judging from its resistance to acids and oxidizing agents, closely resembles cutin. This perhaps indicates the way in which the lipoidal cuticle, ubiquitous in land vegetation, originated.

Any consideration of the evolution of a photosynthesizing land flora must thus necessarily take into account the physiological features of the green algae, and how these may have been modified in the transition to terrestrial life. Recent research into algal environments is yielding much information relevant to this problem. It is commonly found, for example, that from 5 to 35% of the light striking the surface of a lake or sea is reflected, the actual amount lost depending upon the angle of incidence. The light penetrating the water is then gradually absorbed as it advances, so that up to 53% of the radiation passing the surface may be dissipated as heat in the first meter. Consequently, in warm and temperate regions, the rate of photosynthesis of submerged plants is normally controlled by the amount of light reaching them, and not by the amount of carbon dioxide in the water. We can see at once that the first colonists of land, emerging on to bare mineral surfaces, would almost certainly have had to contend with irradiances strikingly higher than those experienced by their aquatic ancestors. This would have provided opportunities for greatly increased photosynthesis.

Another discovery of recent research, also very relevant to the problem of the colonization of the land, is the surprising extent to which algae release materials derived from photosynthesis, both aliphatic molecules and phenolics, into the surrounding water. In Windermere in the English Lake District, for example, up to 35% of the total carbon fixed may be lost in this way. Losses of this order are clearly possible only from aquatic plants. As vegetation advanced from marshes, or from littoral belts subject to periodic inundation, on to relatively dry substrata, much more of the fixed carbon must have been conserved within the plant body.

The plants invading the land appear to have met the increased irradiance and aerial exposure of the land environment, not by any significant change in the structure, composition, or efficiency of the photosynthetic membrane, but by increased removal of the fixed carbon from the general metabolism. In this way, the accumulation of embarrassingly large, and possibly toxic, quantities of carbohydrate in the cells was effectively prevented. Cell walls, consisting of cellulose and hemicellulose, became thicker. Condensation products such as resin, phlobaphene, and lignin became conspicuous, and have remained so in the more primitive vascular plants. The progressive layering of cellulose microfibrils on to the growing cell walls of land plants probably tended to make the angles rounder, thus setting up the tensions which, during cell expansion, led to the appearance of air spaces at the interstices. Spaces of this kind, which ensure that the plant body is ventilated with saturated air, are not a feature of algae. Lignin, of which the phenols of algal cells may have been a precursor, is a product of cells undergoing programed death. It is laid down within cell walls and fills the spaces initially occupied by water, thus both sealing and strengthening the wall. Tracheidlike cells may have originated from the death of the elongated cells normally formed at the center of axes. Initially, they were probably in discontinuous patches (as in the gametophyte of *Psilotum*) and at that stage lacked a conducting function.

Natural selection ensured that those forms survived in which the various destinations of the fixed carbon were not disadvantageous to the growth and reproduction of the plant as a whole. Lignified tissue, for example, led to the evolution of xylem, providing both a skeleton supporting the plant in space and an effective system for the transport of water and solutes. Massive plant bodies, which seem to have appeared relatively early in the evolution of the land flora, also made possible the confinement of photosynthesis to specialized regions, such as leaves and fronds. The amount of assimilation per unit mass of the plant was thereby reduced. Simultaneously, the increase in the amount of living, but non-photosynthesizing, tissue naturally increased the call on metabolizable assimilates. Both factors ensured that the multicellular colonists of the land remained in balance with their environment without interference with the fundamental features of photosynthesis.

Cutin and sporopollenin, condensation products of a fatty nature, also take on essential roles in land plants. In the form of the cuticle, covering all cell-air interfaces, and with the assistance of the regulatory stomata, cutin makes possible the maintenance of the saturated atmosphere within the plant body. As a consequence, plants were able to migrate to areas of low humidity. Sporopollenin forms a protective coat on the spores of land plants and, in some instances, lines membranes within plants separating reproductive regions from the surrounding somatic tissue.

In the course of evolution, many complex and bizarre forms of growth have appeared in land plants, but the material from which they are fashioned has remained predominantly carbon, extracted from the atmosphere. This diversity can be related to the tetravalent nature of carbon and to the great range of compounds that can be formed from it. Had not the photosynthetic fixation of this versatile element arisen on the earth's surface, plant life, and that of animals which depend on it, would have been impossible. Indeed, it is difficult to conceive of any alternative form of life appearing in its absence.

THE MOBILITY OF PLANTS

Although the earliest plants probably soon acquired motility of the kind seen today in *Chlamydomonas* (Chapter 3), this appears to have been rapidly lost in the evolution of higher forms (although often retained in their gametes). The immobility of large and firmly anchored plants is naturally a disadvantage, not shared by the higher animals, at times of natural catastrophe, such as volcanic eruption or fire. Plants, however, very frequently possess a remarkable mobility, or at least a ready transportability by agencies such as wind and water, in their reproductive bodies. Fern spores, for example, have been caught in aeroplane traps in quantity at 1500 m (5000 ft.) and even higher, and the hairy spikelets of the grasses *Paspalum urvillei* and *Andropogon bicornis* have been encountered at 1200 m (4000 ft.) above Panama. Lakes, seas, and the coats and feet of animals also play their part in distributing plants. The immobility of the individual is thus frequently compensated for by the mobility of the species, and devasted areas and new land surfaces become colonized with amazing rapidity and effectiveness.

Some plants (e.g., *Glechoma*) produce stolons which appear to explore the neighboring ground. Since the plantlets becoming established on richer areas will come to dominate the stand, this behavior has been fancifully referred to as "foraging."

LIFE CYCLES

Although developmental cycles are known in the prokaryotic Cyanophyta, a well-defined cycle involving meiotic segregation of the genetic material and its subsequent recombination by sexual fusion is found only in eukaryotic phototrophs. That part of the cycle in which the nucleus contains a single set of chromosomes, is termed **haploid**, and the complementary part of the cycle in which two sets are present **diploid**. The cycle is seen at its simplest in the unicellular algae of aquatic environments (Chapters 2, 3, and 4), where haploid individuals in certain circumstances behave as gametes and fuse, so forming a **zygote**. The zygote, which contains a diploid nucleus, either undergoes meiosis at once, or only after some delay, in which case the diploid condition can be thought of as having an independent existence. Either the haploid or the diploid phase, or both, may be multicellular. The multicellular plant is called a **gametophyte** if it produces gametes directly, and a **sporophyte** if it produces, following meiosis, individual cells (called spores or meiospores) which either behave as gametes immediately or develop into gametophytes. Each phase may also multiply itself asexually. These various possibilities are summarized in Figure 1.5.

A life cycle is thus basically a nuclear cycle, and it is not necessarily accompanied by any morphological change. In the algae *Ulva* and *Dictyota*, for example, the gametophyte and sporophyte are superficially indistinguishable, and it is necessary to observe the manner of reproduction to identify the phase of the cycle to which any individual belongs. Such a life cycle is termed **isomorphic** (or homologous). Frequently, however, the two phases of the cycle have different morphologies, one often being less conspicuous than the other, and sometimes parasitic upon it. These cycles are termed **heteromorphic** (or antithetic). Although the algae show both isomorphic and heteromorphic life cycles, those of land plants are exclusively heteromorphic. Occasionally there may be a morphological cycle without a corresponding nuclear cycle, as in the apogamous ferns, but this is regarded as a derived condition.

Gametes are always uninucleate and, when motile, usually naked cells. In the simplest form of sexual reproduction, termed **isogamy**, the two gametes involved in fusion are free cells and morphologically identical. Nevertheless, detailed investigations continue to show that gametes from the same parent rarely fuse. Some measure of self-

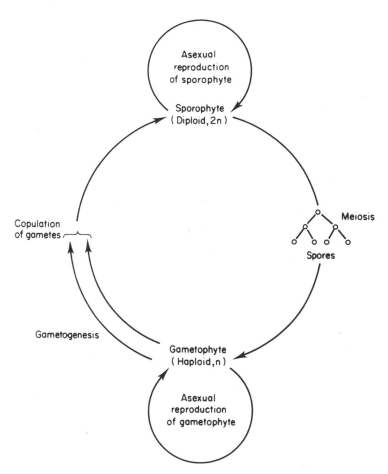

**Figure 1.5. The life cycle of autotrophic plants generalized. The large circle represents sexual
reproduction. Only relatively few species display all the reproductive potentiali-
ties shown.**

incompatibility, and hence physiological differentiation between the parents, appears to
be the general rule.

Isogamy was probably the most ancient condition, and this appears to have been
succeeded by **anisogamy**. Here the gametes, although still free cells, are morphologically
dissimilar, but usually differ in little more than size. The larger, which may also be less
mobile, is called the female. The extreme form of anisogamy is **oogamy**, in which the
female gamete, now called an egg cell or ovum, is large, nonmotile, and filled with food
materials. The egg cell may either float freely in water, as in the alga *Fucus*, or be retained
in a chamber, as in some algae and all land plants. The chamber bears various names
according to the group of plants being considered. Since the progression from isogamy is
accompanied in many algal groups by an increase in somatic complexity, it seems very
probable that this morphological progression is also a phylogenetic one.

In several instances of sexual reproduction it has been shown that one or both
gametes produce traces of chemical substances, termed pheromones (or gamones), which
cause the appropriate gametes to approach each other. The chemistry of these
pheromones varies widely. In some algae they are peptides. In the ferns the male gametes
are attracted to the opened egg chambers by a pheromone which may be malic acid. This
substance is known to have a striking chemotactic effect *in vitro*.

LIFE CYCLES OF TRANSMIGRANT FORMS

The transition to a terrestrial environment clearly presented a number of problems in relation to sexual reproduction. Although all land plants are oogamous and are presumably derived from oogamous algae, fluid was still necessary in the initial land plants to allow the motile male gametes to reach the stationary female. This problem appears to have been met first by the egg becoming enclosed in a flask-shaped chamber, the *archegonium*, in the neck of which the male gametes accumulate, and second by the male gamete becoming an efficiently motile cell. The male gametes of the lower archegoniate plants (Chapters 5, 6, and 7), termed *spermatozoids* (or antherozoids), are remarkable cytological objects. Each is furnished with two or more highly active flagella, and both the cell and nucleus have an elongated snakelike form, well suited for penetration of the archegonial neck. Dependence upon water is thus reduced to the necessity for a thin film in the region of the sex organs at the time of maturity of the gametes.

The archegonium is common to all the lower land plants, but its origin remains tantalizingly obscure. It may have appeared immediately before the colonization of the land, possibly as a consequence of morphogenetic tendencies seen today in association with the eggs of some Charophyceae and certain red algae.

If so, these antecedents of the transmigrants are no longer represented among living algae. Whatever the exact time of the evolution of the archegonium, however, there are no compelling reasons for regarding it as having been evolved more than once. Since the fossil record indicates that the most primitive forms of land plants were probably all archegoniate, it follows that the migration of plants on to land was also very likely a unique event.

If the transmigrant forms were archegoniate, what was the nature of their life cycles? This is largely a matter for conjecture. However, as will be seen in later chapters, except for one approach to isomorphy (in the living Psilopsida), the lower archegoniate plants possess markedly heteromorphic life cycles in which the conspicuous phase is either the gametophyte (Bryophyta), or the sporophyte (Lycophyta, Sphenophyta, Filicophyta). The transmigrants possibly had an intermediate position, which more or less isomorphic cycles. Nevertheless, heteromorphic cycles were probably very soon developed as terrestrial vegetation diversified and exploited particular features of the new environment. The cycle in which the sporophyte was the most highly developed phase clearly had the greater evolutionary potential, since, with the exception of the Bryophyta, it is characteristic of all existing terrestrial vegetation.

SEXUAL REPRODUCTION IN LATER TERRESTRIAL VEGETATION

An important step in the evolution of sexual reproduction on land was undoubtedly the emergence in the archegoniate plants of heterospory. This involves the production of spores of two sizes, the larger giving rise to a wholly female gametophyte and the smaller to a male. In homosporous archegoniate plants, considerable growth of the gametophyte is often required before it acquires the ability to produce egg cells. In the primitive heterosporous plants, however, the small female gametophyte formed on germination of the megaspore produces one or a few archegonia in a very short time. The microspore also develops rapidly, and spermatozoids are soon liberated from the diminutive male gametophyte. Heterosporous reproduction is thus coupled with a reduction of the time spent in the gametophytic phase. The shortening of the life cycle increased the rate at which new forms could appear and hence promoted evolution.

In higher archegoniate plants (Chapter 8) we see how sexual reproduction becomes increasingly independent of water. These archegoniates are exclusively heterosporous, but the megaspore is retained and germinates within a specialized sporangium called an ovule. In some forms (e.g., *Cycas*, *Ginkgo*), fertilization is still effected by flagellate male gametes, but the only fluid necessary is a small drop, immediately above the archegonia, into which the gametes are released. Other archegoniate plants escape even from this requirement. The male gametophyte is filamentous, and, as a consequence of its growing towards the female gametophyte, it liberates the male gametes (which now lack any specialized means of locomotion and are probably moved passively) directly into an archegonium. In a few allied plants (e.g., *Gnetum*) modifications of the female gametophyte result in the disappearance of the archegonium. Ultimately we arrive at the embryo sac and the finely ordered cytology that is characteristic of the sexual reproduction of the flowering plants. Comparative morphology and the fossil record indicate that the morphological sequence we have considered here also represents the evolutionary development of sexual reproduction in land plants. Compared with the cytological elegance of fertilization in an angiosperm, the clumsy spermatozoid of *Cycas* is thus not only barbarous, but also primitive.

CLASSIFICATION OF THE CHLOROPHYLLOUS PHOTOTROPHS

A classification provides the shelving on which knowledge of the Plant Kingdom can be arranged in an orderly fashion. The ideal is a classification which arranges plants according to their level of organization and in their natural alliances. In every classification there is an element of subjectivity. Consequently, as knowledge expands, judgments need to be modified. The primary classification followed in this book is shown in Table 1.2. The classifications if the subkingdoms will be found subsequently at the beginnings of Chapters 2, 5, and 6. The aim throughout is to present a general view of the principal kinds of organization encountered in the Plant Kingdom. Although the approach is systematic, a purely systematic treatment is not attempted and would be inappropriate. The lower plants receive proportionately greater attention. The fossil evidence indicates that they retain features present at crucial stages in plant evolution. Familiarity with them is essential for an understanding of today's diversity.

Table 1.2. The Plant Kingdom, phototrophs containing chlorophyll and evolving oxygen during photosynthesis.

Subkingdom	Description	Chapter(s)
Algae	Predominantly plants of aquatic environments, or persistently damp situations exposed to saturated atmospheres. Unicellular, colonial, or multicellular, the multicellular forms lacking a well-developed vascular system. Reproductive mechanisms relatively unspecialized. Complex and thickened walls associated only with resting cells.	2, 3, 4
Bryophyta	Terrestrial or epiphytic, some aquatic. The sporophytic phase normally determinate and partly dependent upon the gametophyte. Multicellular, external surfaces covered with a cuticle, but that of the gametophyte relatively permeable. Vascular systems, if present, not highly differentiated. Sexual reproduction dependent on presence of water. Spores with exine but only in a few groups heavily thickened and ornamented.	5
Tracheophyta	Almost entirely confined to land, rarely completing their life cycle in a submerged state. The gametophytic phase relatively small or rudimentary, the sporophyte not dependent upon it. Sporophyte often of indefinite growth, regularly provided with a cuticle, normally impermeable, and almost always with stomata and internal air spaces. Well-defined vascular systems consisting of xylem and phloem. Reproductive regions often with elaborate morphology. Spores usually with a well-developed and acetolysis-resistant wall. Fusion of male and female gametes only in lower forms dependent on extraneous water.	6, 7, 8, 9

__ 2 __
THE SUBKINGDOM ALGAE
Part 1

BIOLOGICAL FEATURES OF ALGAE

The simplest phototroph imaginable is a single cell floating in a liquid medium, synthesizing its own sugar, and reproducing at intervals by binary fission. Such organisms do, in fact, exist in both fresh and salt waters. Examples are provided by the cyanophyte *Synechococcus* and the minute marine *Micromonas* (Fig. 1.4).

These organisms are examples of algae, the group of plants showing the greatest diversity of any major division of the Plant Kingdom. They range from minute, free-floating, unicellular forms to large plants, exclusively marine, several meters in length. Many of the smaller algae form a component of **plankton**, the communities of minute plants and animals which float at or near the surface of fresh waters and oceans. Algae are responsible for a large part of the photosynthesis in the biosphere, the productivity of some coastal communities in warm seas exceeding that of the tropical rain forest. Much of the carbon so fixed enters the food chain of the aquatic heterotrophs.

Despite the enormous range in size, the algae remain comparatively simple in organization. In the smaller multicellular species (e.g., *Merismopedia*, Fig. 2.4; *Pediastrum*, Fig. 3.8) the cells resemble each other in appearance and function, and they can be regarded as forming little more than an aggregate of independent units. In the larger, however, there is morphological and cellular differentiation, although usually less extensive than that in most land plants. The few heterotrophic forms, mostly small, are regarded as derived.

Many algae are fully immersed and firmly attached to the substratum. Together with a few vascular plants, these constitute the **benthos** and contrast with the floating plankton. The attachment may be by a disklike holdfast, which forms a firm union with the surface of a rock or stone, or branched, penetrating soft material such as muds. Branched, rootlike attachments (as in the Charales) may participate in the absorption of minerals, but any resemblance to the root system of higher plants is distant.

The largest algae are found only in the sea. The restriction of these forms to a marine environment is perhaps accounted for by the relative impermanence of inland waters in geological time and the consequent limiting of the opportunity for the evolution of similar complexity in these situations. Although marine algae are sometimes able to withstand inundation in fresh water (e.g., *Fucus*), and occasionally may even become adapted to permanently low salinity (e.g., *Ulva* and *Enteromorpha*), they do not normally survive indefinitely or grow in these conditions. Presumably fresh waters are unable to supply minerals at a rate adequate for their metabolism. A large alga in European seas is *Laminaria* (Fig. 4.18), some species of which may reach 4 m (13 ft. 4 in.) in length. Off the west coast of North America are found the gigantic *Nereocystis* and *Macrocystis*, with thalli commonly exceeding 50 m (165 ft.). Maintaining the integrity of a thallus of this size raises substantial mechanical problems. Although the sea provides considerable supporting upthrust, currents and turbulence cause more sustained tensions and pressures than similar movements in a gaseous medium. The toughness and hard rubbery resistance to any kind of distortion found in the larger algae are thus necessary qualities for survival in the oceans. These attributes arise principally from the general properties of the cell walls and surface, not from any specialized strengthening elements.

As would be expected of a group exploiting the aquatic habitat, algae have a number of distinctive biochemical characteristics. Many, for example, accumulate fats and oils rather than starch, and others, polyhydric alcohols. The microfibrils of the cell walls of the eukaryotic algae contain, in several instances, the polysaccharides mannan and xylan in addition to cellulose. The nitrogenous polysaccharide chitin is found as an outer layer of the wall in *Cladophora prolifera* and possibly in *Oedogonium*. Pectin, a polymer based on galacturonic acid, is a common component of algal cell walls, sometimes forming a distinct outer sheath (e.g., *Scenedesmus*, Fig. 3.7). Colloids such as fucin and fucoidin, unknown outside the algae, occur in the amorphous matrices of the walls of brown algae. Alginic acid, which occurs in quantity in the middle lamellae and primary walls of several brown algae, is extracted commercially and finds a wide range of uses as an emulsifier in industry. Complex mucilaginous polysaccharides rich in galactan sulfates are characteristic of the red algae. Dimethyl sulfur compounds assist osmoregulation in marine phytoplankton. Gaseous derivatives of these escape and contribute to the sulfur content of the atmosphere.

Many unicellular algae, occurring both singly and in colonies, and the unicellular reproductive cells of more complex algae are motile. In all prokaryotic forms, and in some eukaryotic (e.g., diatoms), movement is probably brought about by directed jets of mucilage. The motility of many eukaryotes, however, depends upon the presence of flagella. An unexpected and remarkable discovery of electron microscopy was that all flagella produced by eukaryotic organisms have a common basic structure, providing a characteristic picture in transverse section (Fig. 2.1). Nine pairs of microtubules, each pair oriented tangentially, are equally spaced around the periphery of the flagellum. Although the microtubules of each pair are similar in diameter (18–25 nm), they differ in profile. Viewed from the base of the flagellum outwards, the microtubule on the right (the "A" tubule) usually appears circular in outline, whereas that of the "B" tubule on the left is not completely so. The portion of the "B" tubule shared withe the "A" tubule commonly follows the curvature of the latter. In addition to the peripheral microtubules, two free microtubules usually lie symmetrically at the center. These are often slightly wider than the peripheral tubules. Usually, but not always in plant flagella, two short arms can be made out on the "A" tubule. These consist of a special protein, dynein, an ATPase.

The microtubular system of the flagellum constitutes the **axoneme**. The microtubules of axonemes appear to be quite similar to others in the cell, but they are not so sensitive to colchicine. In some instances, flagellogenesis may even continue in the presence of this antimicrotubular drug. Movement of the axoneme is probably caused by the paired microtubules sliding over one another. The mechanism is not, however, entirely understood. More detailed information will probably come from the study of mutants in which the structure of flagella is in some way defective.

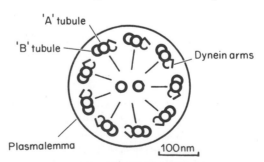

Figure 2.1. Diagram of transverse section of eukaryotic flagellum viewed from the base. The "spokes" radiating from the center to the peripheral doublets can usually be made out. Sometimes an ill-defined sheath is present around the central pair.

Formerly, two classes of flagella were recognized: "whiplash" considered to be smooth, and "Flimmer" furnished with rows of minute hairs (**mastigonemes**). It now seems doubtful whether algal flagella are ever entirely smooth, but appendages are certainly much more conspicuous in some groups than others. Appendages other than hairs are also known. The single flagellum of *Micromonas*, for example, is covered with minute scales, and that of the related *Pyramimonas* with minute scales of two distinct kinds. The electron microscope has shown that, in many instances, these scales are assembled in Golgi bodies and transported to the surface in vesicles.

The nature of the surface, and other features of the flagella, such as their number, arrangement, and method and kind of insertion, provide considerable assistance in identifying the relationships of the eukaryotic algae. The Rhodophyta are outstanding in having no flagellate forms. A few unicellular forms belonging to the Haptophyta have, in addition to two flagella, a third flagellumlike organ. The structure of this is much simpler and less regular than that of the "9 + 2" flagellum.

There is no evidence that the major groups of algae have any close relationship with each other. Nevertheless, there are sufficient morphological, physiological, and ecological similarities between these plants to make the term "alga" a useful one. Study of the structure and reproduction of the algae reveals a number of ways in which these simple phototrophs have increased their morphological and reproductive complexity. We shall in the main be concerned with the illustration and discussion of these trends, and we shall not attempt a complete taxonomic or morphological survey of any group.

GENERAL CLASSIFICATION OF THE ALGAE

The general classification of the algae followed in this work, based upon the nature of the chlorophylls present in the photosynthetic membranes, is shown in Table 2.1.

ALGAE CONTAINING CHLOROPHYLL a

Algae in which the chlorophyll is wholly or predominantly chlorophyll a embrace both prokaryotic and eukaryotic forms. These algae are probably the closest among the living algae to the original cellular phototrophs. They are, therefore, of particular interest in relation to the origin of plant life.

Prokaryotic Forms

The prokaryotic forms which contain solely chlorophyll a comprise the well-characterized division Cyanophyta. Many exist as single cells and are representative of the simplest phototrophs. In others, the cells are aggregated either loosely or into groups, each with a distinctive morphology (e.g., *Gloeotrichia*, Fig. 2.8). Filamentous forms also occur, and a few attain a multiseriate and pseudoparenchymatous level of organization (*Stigonema*). Collectively the cyanophytes are often known as the "blue-green" algae.

CYANOPHYTA

Habitat	Water, swamps, soil, occasionally endolithic.
Pigments	Chlorophyll a; β-carotene; myxoxanthin, zeaxanthin; biliproteins (allophycocyanin, phycocyanin, and phycoerythrin).
Food reserves	Cyanophycean starch (similar to glycogen), polyphosphate granules (volutin), cyanophycin (a polymer of arginine and aspartic acid).

Cell wall components	Murein, hemiculluloses.
Reproduction	Asexual. Genetic recombination observed, but mechanism unknown.
Growth forms	Unicellular, cellular aggregates, filamentous, a few pseudoparenchymatous.

Cytology. The cells, rarely exceeding 10 μm in diameter, are not unlike those of bacteria. Indeed, despite the presence of chlorophyll a, they are often classified with the bacteria ("Cyanobacteria") instead of with the algae. The resemblance is enhanced by the presence of the peptidoglucan murein in the cell walls, their being subject to attack by viruses similar to the bacteriophages, and the bearing by some of filamentous appendages (fimbriae) otherwise found only in Gram-negative bacteria.

Under the light microscope, partly because of the small size of the cells, the photosynthetic pigments appear dispersed in the cytoplasm. The electron microscope reveals that this is an erroneous impression and that the cells contain the same kind of photosynthetic membrane as that present in higher plants. The thylakoids, which probably arise as, and may remain, invaginations of the plasmalemma, form loose stacks parallel to the longitudinal axis of the cell, or a number of concentric whorls. Less commonly they form a three-dimensional reticulum, and, in older cells, aggregate into a crystalline body about 0.3 μm in diameter which closely resembles the prolamellar bodies of etioplasts of higher plants. Blue-green algae from hot springs have yielded the first pure preparations of the chlorophyll a reaction center at which the light energy is transformed into chemical (Photosystem I).

Also seen in the electron microscope are small particles about 30 nm in diameter, lying between the thylakoids. These are the site of the biliprotein pigments (**phycobilisomes**). They trap the light energy which is utilized in Photosystem II, that part of photosynthesis which leads to the evolution of oxygen. Elsewhere in the cell, clusters of minute cylinders containing gas (gas vesicles) may be found. These vesicles, bounded by a sheet consisting solely of protein, regulate buoyancy. They collapse if the cells are subjected to sudden mechanical shock. Vacuoles similar to those of eukaryotic cells appear to be absent.

In many filamentous species the chains of cells are interrupted by occasional, conspicuously larger cells called **heterocysts** (Fig. 2.2). There are often indications of regularity in their spacing, and the relative simplicity of the system may facilitate the identification of the factors leading to their initiation. The heterocysts contain

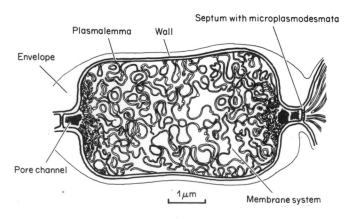

Figure 2.2. *Anabaena.* **Longitudinal section of a heterocyst. (After Fay, in Carr and Whitton. 1973.** *The Biology of Blue-Green Algae.* **Blackwell Scientific Publication, Oxford.)**

Table 2.1. Classification of subkingdom Algae.

Division	Class/Subclass	Order

Algae containing wholly or predominantly chlorophyll a

CYANOPHYTA	Cyanophyceae	Chroococcales
		Nostocales
		Stigonematales
RHODOPHYTA	Rhodophyceae	
	Bangiophycidae	
	Florideophycidae	

Algae containing chlorophylls a and b

PROCHLOROPHYTA		
CHLORACHNIOPHYTA		
CHLOROPHYTA	Prasinophyceae	Pedinomonadales
		Pyraminodales
		Pterospermatales
	Chlorophyceae	Volvocales
		Tetrasporales
		Chlorococcales
		Chlorosphaerales
		Chaetophorales
		Oedogoniales
		Sphaeropleales
	Ulvophyceae	Ulotrichales
		Ulvales
		Prasiolales
		Cladophorales
		Siphonocladales ⎫
		Codiales ⎪
		Caulerpales ⎬ Siphonales
		Dichotomosiphonales ⎪
		Dasyclydales ⎭
	Charophyceae	Klebsormidiales
		Mesotaeniales
		Desmidiales
		Zygnematales
		Coleochaetales
		Charales
	Pleurastrophyceae	Tetraselmidales
		Pleurastrales
		Trentepohliales
EUGLENOPHYTA	Euglenophyceae	Euglenales
		Eutreptiales

Algae containing chlorophylls a and c

CHRYSOPHYTA		
XANTHOPHYTA		
HAPTOPHYTA		
(Prymnesiophyta)		
DINOPHYTA	Desmophyceae	
(Pyrrophyta)		
	Dinophyceae	
BACILLARIOPHYTA		
CRYPTOPHYTA		
PHAEOPHYTA	Phaeophyceae	Ectocarpales
		Sphacelariales
		Cutleriales
		Laminariales
		Fucales
		Dictyotales

chlorophyll a and some other pigments, but are altogether paler and, by contrast, may appear empty. Electron microscopy reveals that they contain an elaborate and often reticulate membrane system. The wall of a heterocyst is conspicuously thickened except for a pore at one or both poles. Its protoplast is there separated from those of its neighbors solely by a thin septum, in which fine channels (microplasmodesmata) can sometimes be discerned.

Heterocysts have attracted considerable attention since, in those species possessing them, they are the sites of fixation of atmospheric nitrogen. The enzyme responsible, nitrogenase, is irreversibly inactivated in aerobic conditions. It is significant that the heterocysts lack photosystem II and therefore generate no intracellular oxygen. Indeed, their frequently elaborate membrane systems may indicate considerable respiratory activity. This would, in turn, generate an anaerobic environment within the cells, which the conspicuously thickened walls (Fig. 2.2) may help to maintain. Of particular note is that those species without heterocysts, which are able to fix nitrogen, can do so only at very low oxygen tensions. In a few instances heterocysts have been observed to regain pigmentation and then to germinate. On these grounds some have considered them to be vestigial reproductive cells.

The cell walls of the Cyanophyta contain a layer of the peptidoglucan murein adjacent to the plasmalemma. The walls are thus chemically and structurally similar to those of Gram-negative bacteria, and their formation is similarly disorganized by penicillin. The presence of murein also accounts for the dissolution of the wall, as those of bacteria, by lysozyme. The outer part of the wall is commonly a distinct layered sheath, sometimes pigmented, and frequently consisting largely of gelatinous or slimy hemicelluloses. The creeping movements of some species (reaching up to 4 μm s^{-1}) are caused by localized excretion of mucilage through fine pores, about 4 nm in diameter, in the wall. Some strains of other species are capable of more rapid motion (up to 25 μm s^{-1}). No motile apparatus has been detected and the cause of the motion is unknown.

The gelatinous material of the outer wall frequently holds cell colonies together (Figs. 2.3 and 2.4). The morphological integrity of the more strikingly filamentous forms often depends upon the cells being held in linear sequence by the toughness of the sheath. The continuity of the sheath is broken at points of branching either "false," characteristic of *Tolypothrix* (Fig. 2.5), or "true," as seen in the filamentous *Mastigocladus* and the pseudoparenchymatous *Stigonema* (Fig. 2.6).

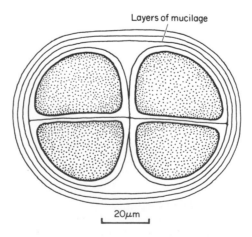

Figure 2.3. *Chroococcus.* **Colony held together by successive sheets of mucliage.**

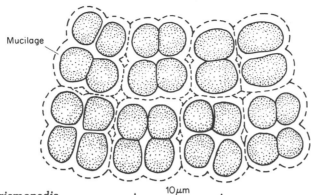

Figure 2.4. *Merismopedia.*

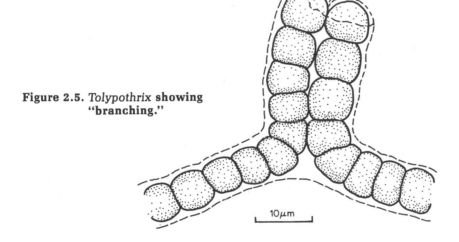

Figure 2.5. *Tolypothrix* **showing "branching."**

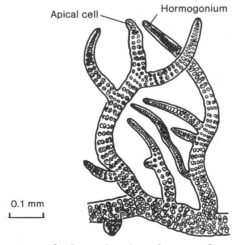

Figure 2.6. *Stigonema.* **Part of plant showing the pseudoparenchymatous thallus. A segmenting apical cell can be distinguished at the tips of the branches. Hormogonia consisting of two or a few cells are formed at the ends of some of the branches. (After Frémy, redrawn from Fritsch. 1945.** *The Structure and Reproduction of the Algae, 2.* **Cambridge University Press.)**

Distribution. The Cyanophyta are widely distributed, occurring in both soil and water, though rarely at a pH below 4. Some are marine (e.g., *Trichodesmium*, one of the organisms responsible for the color of the Red Sea). Others are found under such extreme conditions as snowfields and hot springs (where they can survive temperatures of up to 85°C/185°F), and beneath the eroded surfaces of rocks in Antarctic deserts. Many occur as slimes on rocks, damp soils, and tree trunks, and as scums on stagnant water. Some live as endosymbionts in the protoplasts of colorless flagellates and amoebae (*Glaucocystis* is often regarded as a composite organism of this kind), and others as colonies in the tissues of higher plants (e.g., the presence of *Anabaena* in the fern *Azolla* and the gymnosperm *Cycas*). Cyanophyta are also components of some lichens. These symbiotic associations are probably related to the ability of many species both to photosynthesize and fix atmospheric nitrogen, with the subsequent release of metabolites to the host. The fertility of many swampy tropical soils (e.g., rice *padi*) is dependent upon nitrogen fixation by Cyanophyta. Conversely, in temperate regions they often contaminate fresh water, particularly where these become enriched by mineral nutrients from agricultural drainage (**eutrophication**). Some cyanophytes secrete calcium carbonate and others silica, and so build up laminated rocks (**stromatolites**). Living examples are found in Western Australia and in North America, and fossil stromatolites occur as far back as the pre-Cambrian.

Reproduction. Although genetic recombination has been reported, the observed reproduction of the Cyanophyta is entirely asexual, in its simplest form involving nothing more than cell division. This, correlated with the absence of distinct nuclei, is not associated with the formation of any recognizable mitotic figure or chromosomes. In some species, mainly those that are unicellular, the cell enlarges, and the protoplast meanwhile divides to form many naked daughter cells termed **baeocytes.** When these are eventually released they develop a cell wall and become new individuals. In filamentous species, reproduction is frequently by simple fragmentation. This is rarely indiscriminate, but involves the filament becoming interrupted at intervals by dead cells (**necridia**). The necridia ultimately rupture and the intervening fragments of filament, some 5–15 cells long and now termed **hormogonia** (Fig. 2.7), are liberated. They often show motility before developing into a new individual. Many, but not all, Cyanophyta have been observed to produce **akinetes** in unfavorable conditions. These develop from single cells. There is considerable increase in size and an accumulation of cyanophycin and polysaccharide. Photosynthetic activity declines and ultimately the envelope becomes massive. Akinetes probably play an important role in the preservation of species. They appear capable of surviving prolonged deposition in aseptic mud, and also desiccation. Germination takes place rapidly in favorable conditions. Reserve materials are catabolized, thylakoids are regenerated, and cell division begins. Local dissolution of the envelope releases the new individual.

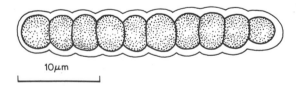

10μm

Figure 2.7. *Nostoc.* **Hormogonium.**

A form of alternations of generations is found in some Cyanophyta. In *Nostoc*, for example, a heterocystous cycle can be made to alternate with an akinete-producing cycle, light being essential for the switch to the akinete cycle.

Representative species. The Cyanophyta contain three principal orders: the **Chroococcales**, mainly single or aggregated spherical cells, and the filamentous **Nostocales** and **Stigonematales**.

Although containing the simplest blue-green algae, some of the **Chroococcales** are nevertheless colonial forms with a regular and conspicuous symmetry. In *Chroococcus* (Fig. 2.3) single cells are occasionally seen, but more usually, and always in the similar *Gloeocapsa*, the cells remain held together after division in a mucilaginous matrix. These aggregates of indefinite size and shape are referred to as *palmelloid* forms. In *Merismopedia* (Fig. 2.4) the cells are arranged in regular rows to form a plate, and in *Coelosphaerium* a hollow sphere. Other geometric arrangements are characteristic of further genera in this order. Although all cells in these colonial forms appear to be of similar status and function, there is evidence of polarity, since in vegetative reproduction divisions appear to take place more readily in certain directions than others.

The **Nostocales** are filamentous. The trichomes (the chain of cells as distinct from the sheath) range in width from 20 μm to 1 μm. Sometimes the width diminishes progressively along a filament so that it ends in a fine point (e.g., *Gloeotrichia*, Fig. 2.8).

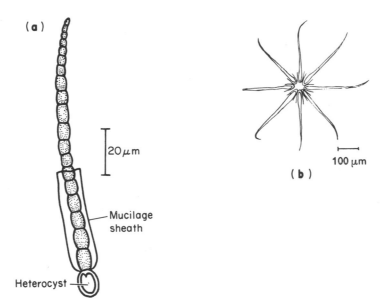

Figure 2.8. *Gloeotrichia.* **(a) Single trichome with basal heterocyst. (b) Star-shaped cluster of trichomes.**

The Nostocales fall into species in which heterocysts have never been observed, and those in which they are often conspicuous. Examples of the former are *Oscillatoria* and *Spirulina*. The filaments of *Oscillatoria* are straight, and the cells often furnished with numerous gas vesicles. Individual filaments make oscillatory movements caused by regions of alternating curvature being propagated along the filament. The source of these movements is unknown. *Spirulina* is a form with a characteristic spiral filament. It forms mats in soda lakes in Central Africa and is compressed into cakes used as a human food.

Nostoc (Fig. 2.9) is a prominent heterocystous form. *Nostoc commune* is common on damp soil. Divisions occur sporadically along the filaments resulting in spherical masses of intertwined threads embedded in gelatinous mucilage, the whole often reaching 2 cm (0.75 in.) or more in diameter. *Anabaena* is a frequent component of water blooms. *Anabaena flos-aquae* contains a powerful neuro-muscular poison known to have caused the death of animals drinking from affected ponds and lakes.

Clustering of filaments is characteristic of several Nostocales. *Anabaena*, for example, grown in an agitated medium, is uniformly dispersed, but if the shaking is stopped the filaments soon group themselves into a few tight bunches. This is probably a

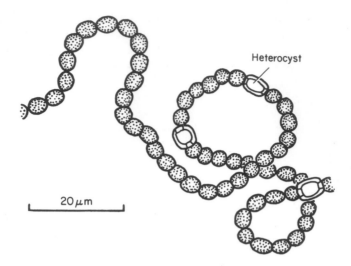

Figure 2.9. *Nostoc.* **Habit.**

consequence of the chance cohesion of filaments, followed by the gliding of the filaments along each other within the coalescent mucilage. Sometimes the clustering leads to forms with a distinct morphology. In *Gloeotrichia*, for example, the trichomes radiate from a central plate formed by a coalescence of the larger basal cells (Fig. 2.8).

The order **Stigonematales** contains the most elaborate Cyanophyta. They are generally heterocystous. Some, such as those submerged on rocks, show **heterotrichy,** the production of two kinds of filaments, prostrate and erect. Many members of the order show true branching, the branches being initiated by an oblique division of a cell in the parent filament. In some species of *Stigonema* the prostrate trichomes become pluriseriate and approach a parenchymatous condition (Fig. 2.6). The organization of these forms is no less complex than that of many filamentous eukaryotic algae.

Eukaryotic Forms

The eukaryotic forms provided solely with chlorophyll a fall into a circumscribed group called, since many are bright pink in color, the Rhodophyta ("red algae"). The pigmentation results from the presence of the biliproteins (phycobilins) phycoerythrin and phycocyanin. The chlorophyll a is accompanied in a few members by a small amount of an additional chlorophyll (d).

The chloroplasts are discoid or lobed, and simple in structure. The thylakoids usually lie parallel, but they are not stacked, and, in some forms, the biliproteins are present as phycobilisome particles attached to the outside of each pair of thylakoid membranes. These features of the chloroplast of the Rhodophyta are also those of the photosynthetic region of the cells of the Cyanophyta. Floridean starch, which appears outside the chloroplasts, is chemically similar to the amylopectin of higher plant starches.

RHODOPHYTA

Habitat	Aquatic (mainly marine).
Pigments	Chlorophylls a, (rarely) d; α- and β-carotene; lutein, zeaxanthin; biliproteins (phycoerythrin, phycocyanin).
Food reserves	Floridean starch, compounds of sugars and glycerols.

Cell wall components	Cellulose, hemicelluloses, sulfated polysaccharides.
Reproduction	Asexual and sexual (oogamous).
Growth forms	Unicellular, filamentous, pseudoparenchymatous.
Flagella	None.

General features of the red algae. Although some species frequent rock pools, most Rhodophyta live in the deep waters of warm seas and are most commonly seen when washed up on beaches. The ability to live in the ocean depth depends on the presence of the biliproteins. Light reaching these regions lies principally at the middle of the visible spectrum, and this coincides with the maximum absorption of phycoerythrin. The energy is immediately transmitted to the chlorophyll which, at these wavelengths, experiences little direct excitation. The bright green color of some freshwater red algae probably results from the photodestruction of phycoerythrin in higher irradiances. An organism of this kind is the unicellular *Cyanidium* which on biochemical and structural features is probably correctly placed with the red algae. It thrives in waters issuing from volcanic springs which are often hot to the touch and may be as acid as pH 2. Some species of tropical seas develop a calcareous exoskeleton and contribute to the formation of coral reefs. Miocene limestones (e.g., of Minorca) are sometimes almost entirely of red algal origin. Weathering of the limestone reveals the remains of the original plants. These were more or less spherical and about 5 cm (2 in.) in diameter. The hard ball-like remains are known as "rhodoliths".

A number of species, often showing reduced pigmentation, may be true parasites on parenchymatous brown algae and larger members of their own division.

A few red algae are of economic importance. *Porphyra* (Fig. 2.10), for example, has long been used in Europe ("laver bread") and in the Far East ("nori") as a foodstuff. *Chondrus crispus* ("carrageen") and *Rhodymenia* ("dulse") are similarly used to lesser extent in Europe. *Gelidium*, particularly from the Pacific, is the principal source of agar. The walls of the red algae in general are notably mucilaginous.

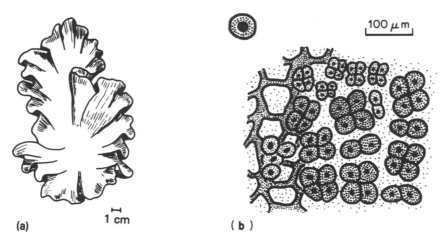

(a)   1 cm   (b)

Figure 2.10. *Porphyra*. **(a) Habit. (b) Transverse section of portion of thallus producing clusters of monospores. ([b] After Esser. 1976.** *Kryptogamen*. **Springer, Berlin.)**

Subclasses of red algae. The Rhodophyta fall into two subclasses: The Bangiophycidae and the Florideophycidae. Unicellular forms are found only in the Bangiophycidae, and this group also displays simpler sexual reproduction. There are also cytological differences, one of the most conspicuous being the presence of well-formed pit connections between the cells in many Florideophycidae. There are no motile forms.

Even in sexual reproduction, which is uniformly oogamous, the dispersal of the male gamete is entirely passive.

Morphology and reproduction of the Bangiophycidae. Apart from the unicellular forms, the simplest Bangiophycidae, mostly epiphytes of other marine algae, are heterotrichous with intercalary growth. The marine *Porphyra* has a sheetlike thallus resembling the chlorophyte *Ulva*, but for the most part only one cell thick. It develops from a filament in which the cells divide transversely and longitudinally, the divisions being confined to one plane. The cell walls consist of two parts: an inner consisting of xylan in a microfibrillar form and an outer of mannan. Cellulose is absent.

Asexual reproduction of the Bangiophycidae is by means of *monospores*. These are formed by simple division of vegetative cells (Fig. 2.10b) and are liberated from the surface or margins of the thallus. Monospores show bipolar development and regenerate the parent plant.

Details of sexual reproduction in the Bangiophycidae are known for only a few species. The simplest form is shown by *Rhodochaete*, a small filamentous marine epiphyte. Some of the cells produce single male gametes (*spermatia*). Elsewhere in the same thallus, a cell otherwise indistinguishable from a vegetative cell acts as an oogonium (usually in the Rhodophyta termed a *carpogonium*) containing a single egg cell. The coming together of the spermatium and carpogonium appears to depend entirely upon water currents. Nothing is known about the details of fertilization. The zygote is said to divide once, and one of the cells to be liberated. These give rise to diploid plants resembling the haploid gametophytes. They reproduce by monospores, some of which develop into diploid plants and others into haploid. The site of meiosis is uncertain.

More is known about sexual reproduction in *Porphyra*. Here a group of cells in the thallus undergoes repeated divisions, each cell yielding 64 or 128 compartments. Each of these is liberated as a spermatium (Fig. 2.11a). The carpogonia differentiate from vegetative cells and each contains a single egg cell (Fig. 2.11b). The drifting spermatia are trapped and retained by the mucilage surrounding the carpogonia. Fertilization takes place by the protoplast of the spermatium first putting forth a narrow process which penetrates, possibly by means of localized lysis, the membrane of the carpogonium (Fig. 2.11c). The body of the spermatium then passes entirely into the female cell.

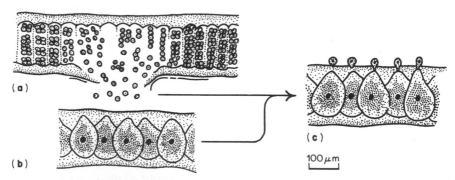

Figure 2.11. *Porphyra.* **(a) Portion of thallus producing spermatia. (b) Thallus with oogonia. (c) Spermatia beginning to penetrate oogonia. (All after Esser. 1976.** *Kryptogamen.* **Springer, Berlin.)**

After fertilization, the zygote divides and releases several carpospores. The germination of these is unipolar. They give rise to a small filamentous diploid plant most commonly found growing on the inside of oyster and mussel shells. This phase, initially placed in a separate genus *Conchocelis*, may reproduce itself by monospores, but other spores (*conchospores*) produced in special sporangia, regenerate the more conspicuous gametophyte. Meiosis is believed to occur in the conchosporangium.

The life cycle of the freshwater *Bangia* is similar to that of *Porphyra*, but there are frequent aberrations in both genera. Apomixis, for example, is well known. Although the life cycle may appear normal, fertilization is omitted. Also, in some forms, the *Conchocelis* phase may be entirely lacking.

Morphology of the Florideophycidae. The Florideophycidae are basically filamentous in construction. Apical growth is well represented. Representative of the simpler Florideophycidae is *Batrachospermum*, one of the few Rhodophyta of fresh water. The vegetative organization, an axis bearing whorls of branches, is very similar to that of the green alga *Draparnaldiopsis*. The filaments are enveloped in copious mucilage; this together with the dark pigmentation of the cells causes colonies of the alga *in situ* in ponds and streams superficially to resemble masses of frog-spawn. In *Nemalion* (Fig. 2.12a), common in the intertidal zones of North Temperate shores, the center of the axis is occupied by several filaments ascending in parallel and adhering to each other. These are surrounded by an investment of short and freely branched laterals enveloped in mucilage.

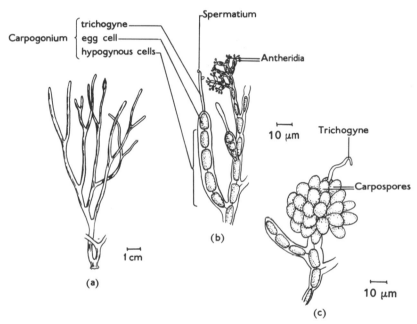

Figure 2.12. *Nemalion multifidum.* **(a) Habit. (b) Fertile lateral branch. (c) Formation of carpospores after fertilization. (After Newton. 1937.** *A Handbook of the British Seaweeds.* **British Museum Publications.)**

A more complex form of multiaxial construction is seen in *Polysiphonia*, common in littoral vegetation throughout the world. The thallus, which consists of a central axis bearing freely branching laterals (Fig. 2.13a), shows well-defined apical growth. This results from a dome-shaped apical cell surmounting a central column of cells which are recognizable throughout the thallus. This axial column is surrounded from well below the summit by numerous columns of other cells produced by oblique divisions in the apical region. The structure thus becomes pseudoparenchymatous and, in parts, especially at the nodes, hardly distinguishable from a truly parenchymatous condition. The manner in which in many multiaxial forms the parallel columns of cells diverge at the apex gives the impression of a fountain (and hence the term "fountainlike growth").

Reproduction of the Florideophycidae. Asexual reproduction in the Florideophycidae takes place by a variety of spores, including monospores, all non-motile and in many instances known only in conditions of pure culture. A curious feature

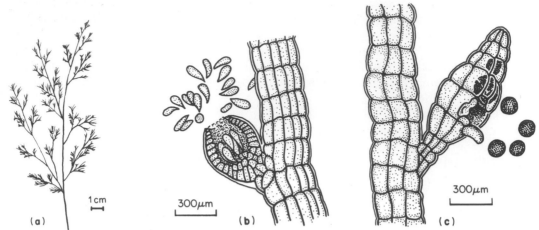

Figure 2.13. *Polysiphonia.* **(a) Habit. (b) Mature cystocarp discharging carpospores. (c) Sporophyte with a tetrasporangiate branch from which some of the tetraspores are escaping.**

of the epiphytic *Centroceras* is the development of lateral branches of four to five cells which are regularly abscised, and serve as a means of vegetative reproduction.

Even in the simpler Florideophycidae, sexual reproduction displays a number of features not encountered in other divisions of the algae. The male and female gametangia, in many species produced on separate plants, arise on specialized side branches. In *Nemalion* (Fig. 2.12b), for example, the antheridia are budded off from mother cells lying at the tips of short tufted branches; as many as four or five antheridia may come from one mother cell. Each antheridium liberates a single spermatium through an apical slit. The carpogonium, surmounted by a long tubular, hairlike process (the **trichogyne**) terminates a short carpogonial branch. When a spermatium makes contact with a trichogyne above an unfertilized egg, it becomes attached, and the intervening cell walls break down, allowing the contents of the spermatium to pass into the carpogonium. After fertilization a mucilaginous plug is secreted at the base of the trichogyne. The zygote germinates almost immediately and becomes surrounded by a tuft of short-branching filaments, often with brownish pigmentation. Carpospores are cut off from the ends of these filaments (Fig. 2.12c). In *Batrachospermum*, the carpospores of some species germinate to form a small heterotrichous *Pseudochantransia*-stage, which may reproduce itself by monospores. Meiosis occurs sporadically in cells towards the tips of the erect branches. Some of the cells containing haploid nuclei grow into the characteristic gametophytic plant.

A more elaborate cycle is seen in the Mediterranean *Liagora*, the thallus of which is calcified. Here the carpospores again develop into diploid heterotrichous plants which reproduce themselves by monospores. Occasionally **tetrasporangia** are produced in which meiosis occurs, each tetrasporangium producing four tetraspores. These give rise to filamentous haploid plants indistinguishable from the filamentous diploids from which they arose. These are a protonemal stage of the gametophyte. Ultimately buds are formed on the protonemal branches from which the adult *Liagora* thalli develop. *Liagora* thus shows heteromorphy in both the gametophytic and sporophytic parts of the life cycle.

In an advanced member of the Florideophycidae, such as *Polysiphonia*, the initiation of sexual reproduction is much as in simpler forms. Antheridia are produced in the middle region of short filaments. When mature, the filaments become club-shaped with the antheridia, almost colorless and with refractive walls, densely packed at the surface.

Development of the female filament is more complex. Although initially a single file of cells, all but the tip becomes multicellular. One of the peripheral cells in the central region gives rise to a carpogonial branch terminating in a carpogonium surmounted by a trichogyne. The basal cells divide sparingly giving rise to a few auxiliary cells all lying close to the carpogonium. During the development of this carpogonial branch, adjacent peripheral cells divide and form an urn-shaped sheath which grows up and ultimately encloses the carpogonial branch. This structure, from which only the trichogyne projects, is termed the **pericarp**.

Following fertilization, the trichogyne degenerates and the zygote makes a pitlike connection with an adjacent auxiliary cell. The diploid nucleus then migrates into this cell and its original haploid nucleus apparently degenerates. The diploid nucleus divides mitotically and a small branching, filamentous **carposporophyte** is produced. Simultaneously, evidently stimulated by fertilization, the pericarp grows into the mature **cystocarp** which completely encloses the carposporophyte. Ultimately, the terminal cells of the carposporophyte become carposporangia. Each liberates a single densely pigmented carpospore. These escape through the apical orifice (ostiole) of the cystocarp (Fig. 2.13b).

The diploid carpospores germinate directly and give rise to plants quite similar to the gametophyte. Segments towards apices of the branches become fertile. A peripheral cell undergoes a number of divisions leading to a structure not unlike a carpogonial branch. In this branch, one or more of the cells give rise to tetrasporangia. A tetrasporangium is initially uninucleate, but the nucleus undergoes meiosis and four tetraspores are formed. These escape by rupture of the sporangium (Fig 2.13c) and germinate to form normal gametophytic plants. Unlike *Liagora*, *Polysiphonia* thus shows heteromorphy only in the sporophytic phase.

In *Griffithsia*, vegetative cells of male and female plants have been fused experimentally (parasexual fusion). The resulting binucleate heterokaryons, analogous to those of ascomycete and basidiomycete fungi, generate plants with diploid (tetrasporangiate) morphology. Actual nuclear fusion is therefore not essential for the new form of growth. Functional tetrasporangia have also been reported following parasexual fusion.

Relationships of the Rhodophyta. The absence of all but a few unicellular representatives of the Rhodophyta makes the assessment of the wider relationships of the division difficult. Motile forms are generally considered to be absent. Some have regarded the puzzling organism *Glaucocystis*, found in acid pools and often interpreted as a composite cell incorporating a cyanophyte as a symbiont, as representative of a group close to the Rhodophyta (see also Chapter 3).

The similarities, both biochemical and structural, between the chloroplasts of the Rhodophyta and the photosynthetic cells of the Cyanophyta are certainly striking. In both, for example, the biliproteins are present as phycobilisomes lying between the thylakoids. Furthermore, both the Rhodophyta and Cyanophyta lack flagella. It is, indeed, conceivable that the Rhodophyta arose from a prokaryotic source resembling the Cyanophyta. The transition from the prokaryotic to the eukaryotic condition could have taken place either by the internal photosynthetic and respiratory membranes of a cyanophyte becoming enveloped by bounding membranes, or by a unicellular cyanophyte invading an already existing heterotrophic eukaryote and ceasing to be recognizable as an individual. Both routes are conceivable, and both are likely to remain matters of speculation into the foreseeable future.

There is an undoubted similarity between the reproductive process in *Polysiphonia* and that in ascomycete fungi, but, if this resemblance is anything more than coincidental, it seems more likely that the fungi, being heterotrophic, are the derived forms.

The calcareous coral-forming Rhodophyta (e.g., *Lithothamnion*) have a fossil record that extends back to the Cretaceous. All these species belong to the more advanced Florideophycidae.

___ 3 ___
THE SUBKINGDOM ALGAE
Part 2

ALGAE CONTAINING CHLOROPHYLLS a and b

In proceeding from the "chlorophyll a" to the "a + b" algae a striking difference is seen in the arrangement of the photosynthetic thylakoids. Thylakoids whose membranes contain only chlorophyll a tend to be clearly separate from each other. In the presence of chlorophyll b, the apposed faces of the thylakoids are closely appressed, either generally or regionally. The stacking of the thylakoids may be in pairs or in greater numbers.

Prokaryotic Forms

The prokaryotic algae containing chlorophylls a and b are placed in the Prochlorophyta. So far, only a few examples are known and the classification is clearly tentative.

PROCHLOROPHYTA

Habitat	Symbiotic, freshwater, marine.
Pigments	Chlorophylls a, b; β-carotene (α- in one form); zeaxanthin; phycobilins in one form.
Food reserves	Starch (where known).
Cell wall components	Probably cyanophytelike.
Reproduction	Presumably asexual.
Growth forms	Unicellular, filamentous
Flagella	None.

Of the three prochlorophytes discovered, *Prochloron* is the most studied. It is a unicellular extracellular symbiont of ascidians, although a free-living form has now been discovered. The cells are 10–20 μm in diameter and divide by binary fission. The thylakoids are more or less concentric and closely stacked. The ribosomes are similar in size to those of the Chlorophyta, but the ribosomal RNA is more like that of the Cyanophyta.

Prochlorothrix is free-living, filamentous, and planktonic, first discovered in shallow eutrophic pools in Holland. It is capable of being grown in pure culture in mineral medium. The filaments reach a width of 0.7 μm and consist of elongated cells without a conspicuous sheath. The thylakoids run parallel to the walls and lie close together.

A third prochlorophyte, also free-living, has been discovered as a prominent component of picoplankton (the plankton which will pass through a 2 μm mesh). The cells measure only 0.6–0.8 μm in diameter, and are thus smaller than the coccoid Cyanophyta. The thylakoids are concentric and closely stacked. The principal photosynthetic pigment in addition to chlorophyll b is a divinyl chlorophyll a-like compound. This organism is so far unnamed.

Eukaryotic Forms

The cells of all "a + b" algae other than the Prochlorophyta are eukaryotic and, in respect of the photosynthetic mechanism, have much in common with higher plants.

CHLORACHNIOPHYTA

Habitat	Warm marine; planktonic.
Pigments	Chlorophylls a, b. Other thylakoid pigments await identification. Biliproteins absent.
Food reserves	Possibly a β-1,3 glucan, outside plastids.
Cell wall components	None, except around cysts; chemical nature not known.
Reproduction	By fission, or zoospores following cyst formation.
Growth forms	Amoeboid; cysts formed in depleted medium.

The division Chlorachinophyta contains a single genus, *Chlorachnion*. It forms floating colonies of bright green cells, each about 10 μm in diameter, entirely naked and connected by reticulopodia. The chloroplasts are regularly bilobed with a prominent projecting central pyrenoid, and surrounded by one or two sheets of endoplasmic reticulum.

Spherical, walled cysts are formed in aging cultures. They germinate in fresh medium to give rise directly to the amoeboid stage, or to zoospores, which bear a single lateral flagellum furnished with fine hairs. They sometimes appear to be produced in tetrads, but it is not known whether meiosis is involved. On settling, the zoospores regenerate the amoeboid stage.

Chlorachnion has, so far, been found only in a few warm marine habitats. The conspicuous manner in which its chloroplasts are ensheathed in endoplasmic reticulum suggests that it may have arisen in the past from a heterotrophic amoeboid organism which incorporated chloroplasts from another source.

CHLOROPHYTA

Habitat	Aquatic (mainly freshwater), terrestrial in moist situations, a few epiphytic.
Pigments	Chlorophylls a, b; β-carotene (α-carotene less prominent); lutein, violaxanthin, neoxanthin; other xanthophylls less widely distributed.
Food reserves	Starch, rarely inulin, oils, and fats.
Cell wall components	Cellulose, crystalline glycoproteins, various hemicelluloses; occasionally sporopollenin.
Reproduction	Asexual and sexual (isogamy, anisogamy, and oogamy).
Growth forms	Flagellate, coccoid, filamentous, rarely foliaceous or siphonaceous.
Flagella	2 or 4, occasionally numerous, hairs or scales inconspicuous.

The Chlorophyta (green algae) in respect of metabolism, photosynthetic pigments, and ultrastructure show much in common with the vascular plants and the bryophytes. We shall here consider the more common representatives and other species of interest contained within the five classes shown in Table 2.1. The classification of the Chlorophyta, initially depending largely upon the degree of development of the thallus and, where present, the nature of the sexual reproduction, now rests increasingly upon ultrastructure and comparative biochemistry. The shape of the chloroplast (chromatophore) remains, however, a useful and fairly readily observable feature. In some genera they are characteristically large and are present singly or in very small numbers in each cell.

Prasinophyceae

This class consists principally of unicellular planktonic organisms. The flagellate representatives include *Micromonas* (**Pedinomonadales**) (Fig. 1.4) and *Pyramimonas* (**Pyramimonadales**). *Micromonas*, abundant in the sea, is naked. The single axoneme is unusual in that the peripheral microtubules do not ascend above the lower portion of the flagellum. Reproduction is by fission. Electron micrographs reveal that the mitochondrion and plastid divide at the same time as the cell. *Pyramimonas* has four flagella set in an apical pit. Two kinds of scales are found on the flagella; on the body of the organism they are intermixed with yet a third kind. The scales reach the surface through a canal opening into the apical pit near the insertions of the flagella. *Halosphaera*, whose bright green spherical cells are frequent in plankton, is nonmotile in the vegetative state. The **Pterospermatales** are also nonmotile and characteristically thick-walled. Despite the abundance of many of the prasinophytes, the life cycles are imperfectly known. No form of sexual reproduction has been detected.

Unicellular forms recalling prasinophyte algae have been recorded from Proterozoic deposits in China and the South Urals. *Tasmanites*, a spore-like fossil of the Devonian and subsequently, is believed to have been the resting cyst of a prasinophyte. The wall contains sporopollenin.

Chlorophyceae

The Chlorophyceae fall into a number of well-defined orders. The **Volvocales**, for example, range from unicellular to a unique multicellular form (*Volvox*), but a simple biflagellate cell is a structural element common to the order.

Unicellular forms of the Volvocales. Representative of the unicellular forms of the Volvocales is *Chlamydomonas*, numbering some 400 species, mostly freshwater. The cells, which rarely exceed 30 μm in major diameter, contain a single chloroplast, usually basin-shaped, and one or more mitochondria (Fig. 3.1). Towards one side of the chloroplast, in which the thylakoids show loose stacking, forming irregular grana, lies a conspicuous *pyrenoid*. This proteinaceous body, a common feature of algal chloroplasts, is the site of the enzyme which incorporates CO_2 into ribulose biphosphate, the initial step

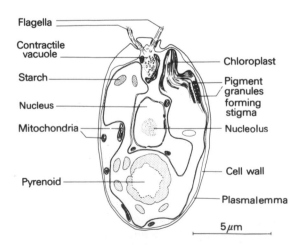

Figure 3.1. *Chlamydomonas reinhardtii.* **Longitudinal section of cell. (From electron micrographs by Ursula W. Goodenough.)**

in the assimilation of atmospheric carbon. A granular *stigma* ("eyespot"), associated with a bovinelike rhodopsin and carotenoid pigments, also lies within the chloroplast. Although possibly photosensitive, this property is not confined to the stigma, since phototaxis persists in mutants in which it is lacking. Each cell also contains two pulsating vacuoles which discharge their contents at short intervals and so play an important role in the osmoregulation of the cell. The two flagella are inserted into narrow pits at the apex of the cell. The emergent portion is finely hairy

The cell wall of *Chlamydomonas* lacks cellulose and consists of a number of layers, some of which form a crystalline lattice made up of glycoprotein sub-units. Among the sugars involved are mannose, arabinose, and galactose. Forms related to *Chlamydomonas* have a similar wall, but with differences in detail.

Cultures of *Chlamydomonas* can be raised so that the cells divide synchronously, facilitating the study of metabolic changes during the cell cycle. *Chlamydomonas* has also figured prominently in research into chloroplast genetics, and an insight into the control of wall formation has been gained from wall-less mutants. Knowledge of the nature and regulation of flagellar movement has also come from *Chlamydomonas*. The whole flagellar apparatus, its function unimpaired, can be isolated and studied *in vitro*. The breast stroke motion of the flagella (Fig. 3.2), which is uniplanar, can be reversed by increasing the concentration of calcium ions in the medium.

Dissection has also shown that the natural beat of the flagellum adjacent to the eyespot is lower than that of its fellow. The beat is nevertheless synchronized in the intact cell. The basal bodies of the flagella are associated with RNA, but the presence of DNA is disputed.

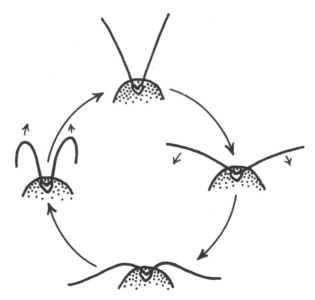

Figure 3.2. The beating cycle of the paired flagella of *Chlamydomonas*.

Reproduction of unicellular forms. Asexual reproduction of *Chlamydomonas* is by fission. The parent flagella are resorbed and the protoplast divides in a manner resembling the cleavage division of animal cells to form two to eight daughter cells (**aplanospores**). These secrete cell walls and acquire flagella before being liberated, but on an agar surface they may continue to lack flagella and form palmelloid colonies. Sexual reproduction involves the fusion of gametes. These are structurally similar to vegetative cells. but smaller. Gametogenesis can be induced by culturing in nutrient-(particularly nitrogen-) deficient medium. Fusion of morphologically similar gametes (isogamy), the simplest method of combining nuclear information, is found in some species of *Chlamydomonas*, but other species show anisogamy, and a few oogamy. Aggregation of gametes is promoted by agglutins, complexes of membrane and glycoprotein shed from the flagella of activated cells. A medium in which the + strain has been grown will cause aggregation (isoagglutination) of the − strain, and vice versa. When the cultures of the + and − strains are mixed, there is first animated cell movement, and then pairing of opposed strains (Fig. 3.3). The first contact is between the tips of the flagella. These then extend laterally and come to lie side by side, the cell walls in some species being simultaneously shed. As the cells become accurately aligned, a protoplasmic bridge forms between the apposed apical papillae, and the flagella simultaneously cease to pair. Fusion is completed in 15–20 minutes, leading to a zygote with four flagella. Recognition and adhesion of gametes of opposed mating type, and actual fusion of the cells are under separate genetic control. Experiments shown that (+) mutants show defects in adhesion and (−) mutants in fusion.

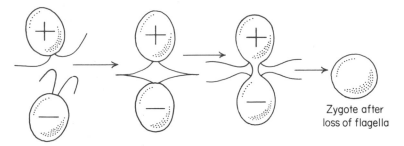

Zygote after
loss of flagella

Figure 3.3. Mating and zygote formation in isogamous *Chlamydomonas*.

Zygotes continue to swim, but show negative phototaxis and eventually settle, although in one species (*C. variabilis*) the swimming phase lasts long enough for the zygotes to have been formerly described as a separate organism. Settling of the zygotes is accompanied by loss of flagella and the secretion of a thickened wall.

Reproduction is anisogamous in *C. braunii*, the − strain producing four macrogametes and the + strain eight microgametes. The oogamous state is approached in *C. coccifera* in which the cells of one strain produce a single macrogamete and those of the other 16 or 32 microgametes. It is reasonable to envisage anisogamy having evolved from isogamy, and experiments with isogamous strains have shown how this may have come about. A large number of mitoses before gametogenesis results in smaller gametes. Controlling the number of mitoses in cultures of different mating types can thus lead to gametes differing in size. Anisogamy (and ultimately oogamy) could have resulted from differences of this kind having become genetically fixed and linked with mating type.

Meiosis occurs on germination of the zygote and four new individuals are produced. There is 2 : 2 segregation of mating type, indicating Mendelian inheritance of this feature. The chloroplast DNA of one parent, however, appears to be absent from the zygote and is probably depolymerized following syngamy.

Although representative of the simplest eukaryotic green plants, *Chlamydomonas* clearly possesses features of structural and functional complexity far more impressive than those found in the prokaryotic algae. It would appear that the attainment of the nucleate state represents an advance in organization without which others, such as flagellate motility and sexual reproduction, are impossible.

An alga similar to *Chlamydomonas*, but lacking a cell wall, is *Dunaliella*. In conditions of extreme salinity (e.g., in salt pans) *Dunaliella* accumulates glycerol to such a high concentration that the cells are able to resist desiccation by exosmosis. In these conditions the cells often become pigmented bright orange. On the basis of its structure, *Polytoma*, a heterotrophic form, is also probably related to *Chlamydomonas*.

Morphology and reproduction of motile colonies. The Volvocales also contain motile colonies composed of identical cells, each similar in morphology to *Chlamydomonas*. *Gonium* (Fig. 3.4), for example, has a flat plate of 4 or 16 cells (depending on the species), which are regularly arranged and held together in a tough mucilaginous matrix. The 16 cone-shaped cells of *Pandorina* (Fig. 3.5), however, are arranged in a sphere. In both asexual reproduction (in which a single cell gives rise to a new spherical colony) and gametogenesis, all the cells of these simple colonies become involved simultaneously.

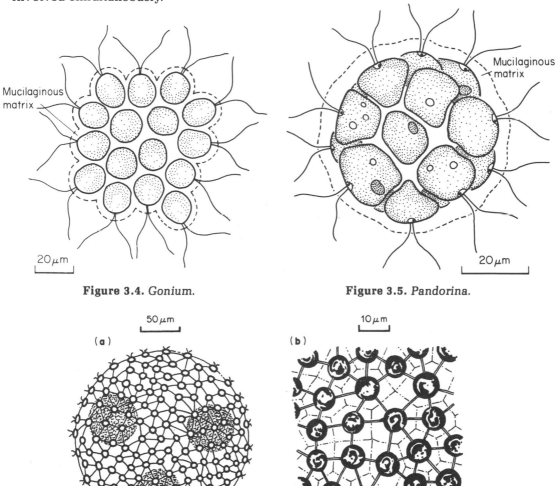

Figure 3.4. *Gonium.*

Figure 3.5. *Pandorina.*

Figure 3.6. *Volvox.* **(a) Asexual reproduction, three daughters lying within the cavity of the parent. (b) Detail of the connections between individual cells.**

In *Volvox* itself (Fig. 3.6) there is evidence of both coordination of activity and division of labor between the constituent cells. The organism takes the form of a hollow sphere, the number of cells forming its surface always being a power of 2, indicating synchronized division during growth. The cells, which individually resemble *Chlamydomonas*, are held together by mucilage. the refractive boundaries between the sheaths of individual cells often give a hexagonal pattern to the surface. The center of the sphere is filled with less viscous mucilage. The cells remain connected by cytoplasmic strands, a feature of significance in relation to the coordination of their activity. The flagella, for example, beat in unison producing a steady rolling motion. Despite the radial symmetry, the organism has an anterior and posterior. In the anterior region, the stigmata are more conspicuous than elsewhere. Some species may reach diameters of almost 1 mm (0.04 in.).

The reproduction of Volvox is correspondingly complex. Asexual reproduction takes place by the enlargement and repeated longitudinal division of a vegetative cell so that eventually a new sphere of cells is formed bulging into the hollow center. Initially, the cells of this daughter individual are oriented inversely in relation to those of the mother, but, at the completion of cell division, the new sphere invaginates and the orientation of the cells consequently reverses. The daughter is simultaneously liberated into the cavity, and may even itself generate a daughter before it is released by the death of the original mother. On release, the daughter organism expands to its mature volume.

Sexual reproduction in Volvox, in which both monoecious and dioecious species occur, is truly oogamous. The gametangia may develop from any cell or, when the distinction of anterior and posterior is well marked, be confined to the posterior region. The mature oogonia are large flask-shaped cells without flagella. Each is developed from a single vegetative cell and remains in position up to and during fertilization. The male organ, the antheridium, similarly develops from a vegetative cell. Successive divisions lead to a bowl-shaped mass of biflagellate spermatozoids (antherozoids). After release, the aggregate migrates to an oogonium and there breaks up into individual gametes. The details of fertilization are little known. Subsequently, the zygote forms an oospore, which in some species acquires a thickened wall and is able to survive unfavorable conditions. Meiosis occurs at germination, but, since only one flagellate swarmer is released, three of the meiotic products evidently fail to survive. Successive divisions of the swarmer lead to a small spherical individual, but after repeated asexual reproduction the mature-size characteristic of the species is ultimately regained.

In some dioecious species, individuals entering the sexual phase and becoming female can induce others still vegetative to become male, a consequence of a "male-inducing substance," polypeptide in nature, secreted by the female. This is only one example of such correlative systems in Volvox, and the situation in the genus as a whole is evidently complex.

The transition from colony to multicellular individual in the Volvocales. Volvox differs from simpler Volvocales in possessing somatic cells which play no part in reproduction. These cells consequently perish when an individual ruptures to release a daughter, or when its integrity is destroyed by the liberation of numbers of oospores. Because of the evident organization in the thallus, and the presence of somatic cells, the term colony is inappropriately applied to Volvox. This organism has reached the level of a multicellular individual. In the Volvocales as a whole, the transition from colonial forms to Volvox is a gradual one, and although the status of Volvox is unambiguous, that of many of the intermediates is less clearly defined. The term coenobium is often used for a colony of cells in which there is some degree of coordination.

The Tetrasporales. The Tetrasporales are a small order of simple algae consisting of Chlamydomonas-like cells embedded in mucilage. The cells either lack flagella or possess flagella which are either nonmotile (e.g., Tetraspora) or move only feebly in the mucilaginous matrix. Motility is usually present in the reproductive stages (zoospores and gametes). The zoospores of some species have four flagella.

The Chlorococcales. The Chlorococcales are a heterogeneous order of lowly green algae in which motility is confined to zoospores and gametes. Although there are common features in reproduction, the Chlorococcales show much diversity, and interrelationships appear distant. Most members of the order occur in fresh waters, a few live in the oceans and moist places on land, and others are endophytic in the intercellular spaces of higher plants, symbionts with lower animals, or constituents of lichens. *Chlorococcum* and *Chlorella* are unicellular, the latter occurring both free in soil and water, and as the endosymbiont in the coelenterate *Hydra*. *Chlorella*, one of the plants with which Priestley and Ingenhousz first demonstrated photosynthesis in 1779, has been used extensively in studies of algal metabolism. The thin cell wall contains sporopollenin, possibly enhancing its water-repellent properties. Another unicellular form, *Botryococcus*, is commonly planktonic and sometimes forms water blooms. Up to 75% of its dry weight may consist of hydrocarbons, and the dried remains of blooms resemble sheets of crude rubber. *Oocystis*, similar to *Chlorella*, has figured prominently in studies of the manner in which the layers of cellulose microfibrils are deposited successively in the developing cell wall. *Prototheca*, a parasite of fish, is also placed in the Chlorococcales. Although heterotrophic and colorless, the cells resemble those of *Chlorella*, and the plastids accumulate starch.

 Scenedesmus (Fig. 3.7) and *Pediastrum* (Fig. 3.8) are representative of the simpler coenobial forms. Sporopollenin has been detected in the wall of *Scenedesmus* (as in *Chlorella*) and lignin in the spines. These carotenoid and phenolic polymers may also be present in cell walls elsewhere in the order.

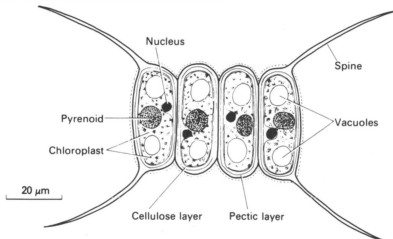

Figure 3.7. *Scenedesmus quadricauda.* **Section of coenobium in the plane of the spines, based on photo- and electron micrographs. The middle layer of the three-layered wall is micellar in structure, but its chemical nature is not certain. The spines consist of bundles of micelles emerging from the middle layer.**

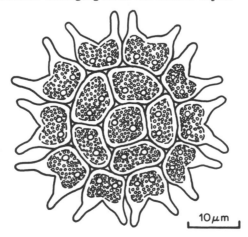

Figure 3.8. *Pediastrum.*

Hydrodictyon is altogether remarkable. The multinucleate cells, each with a reticulate chloroplast containing numerous pyrenoids, form a hollow cylindrical net with closed ends (Fig. 3.9). Individual cells may reach 5–10 mm (0.2–0.4 in.) in length, and the whole coenobium extends to 20 cm (8 in.) or more. At the other extreme is the diminutive *Protosiphon*. The thallus consists of a single sphere, about 0.3 mm (0.01 in.) in diameter, anchored to the substratum by a colorless rhizoid which may reach a length of 1 mm (0.04 in.) (Fig. 3.10). The sphere contains several nuclei and a reticulate chloroplast with several pyrenoids. *Protosiphon* is frequent on damp mud and walls, and some forms can withstand extremes of heat and salinity, as in desert soils. Pure cultures have been shown to produce bacteriostatic substances.

The curious unicellular *Glaucocystis* has also been placed by some in this order (see also Chapter 2).

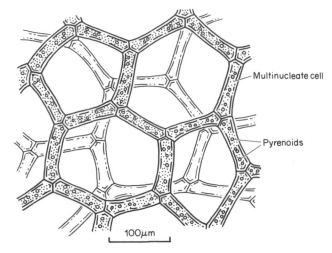

Figure 3.9. *Hydrodictyon.* **Portion of the cylindrical reticulum of a growing individual showing the multinucleate cells with conspicuous pyrenoids. (After Esser. 1976.** *Kryptogamen.* **Springer, Berlin.)**

Reproduction in the Chlorococcales. Asexual reproduction by simple fission does not occur in the Chlorococcales. In *Chlorella*, the nucleus undergoes two mitoses, and the cytoplasm is divided between the four daughter nuclei. A cell wall forms around each of these nonmotile aplanospores. They are then liberated from the mother cell and grow to the mature size.

In the coenobial forms each cell produces aplanospores or zoospores, depending upon species. In *Scenedesmus*, the aplanospores aggregate before, or immediately after, release to form a new coenobium. In *Hydrodictyon*, the multinucleate cells become transformed into a mass of biflagellate zoospores. These are retained, shed their flagella, and then form a new coenobium within the mother cell. This ordered reassembly is probably controlled by microtubules since, in the presence of colchicine, the daughter coenobium is deformed. Following rupture of the mother cell, the young coenobium is free to expand to its mature size.

Sexual reproduction is isogamous or anisogamous, the biflagellate gametes conjugating laterally and not at their flagellate poles. Meiosis takes place as the zygote germinates, yielding four zoospores. Ultimately the zoospores settle, lose their flagella, and develop into conspicuous polyhedral cells known as "polyeders." These become multinucleate and germinate to release a rudimentary coenobium. This then enlarges to its mature size. In *Hydrodictyon*, the young coenobium is two-layered. The individual cells enlarge and becomes multinucleate, and the coenobium acquires its typical cylindrical form.

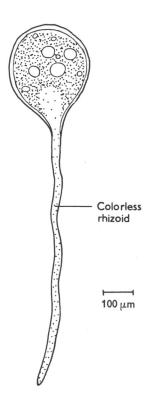

Colorless
rhizoid

100 μm

Figure 3.10. *Protosiphon botryoides.* **(After Klebs, from Fritsch. 1935.** *The Structure and Reproduction of the Algae, 1.* **Cambridge University Press.)**

Reproduction in *Protosiphon* is little specialized. Simple asexual reproduction takes place by budding, but biflagellate swarmers are often produced after flooding. These may behave as either zoospores or gametes. There is no evidence for the existence of mating types since gametes from the same plant may fuse. Germination of the zygote, probably the occasion of meiosis, may be immediate or delayed.

The Chlorosphaerales. The Chlorosphaerales are a small order of simple algae consisting of *Chlorella*-like cells. The cells occur in groups or short threads. *Chlorosphaera antarctica* occurs on snow. Its pigmented resting stages contribute to the "red snows" of antarctic regions. Biflagellate zoospores are produced by some species.

The Chaetophorales. A distinct advance in the organization of a filamentous thallus is found in the Chaetophorales where the thallus is typically composed of both prostrate and erect components, and is consequently termed heterotrichous. The prostrate system is typically a flat plate attached to the substratum, from which arises the erect branching system bearing the reproductive organs and often characteristic hairs. Although elements of an aerial and a prostrate system are almost always detectable, many genera show greater development of one component than of the other, sometimes almost to its exclusion. The maintenance of the dominance of one system has been shown in some species to depend upon environmental factors. In *Stigeoclonium* (Fig. 3.11), for example, magnesium depresses the growth of the erect system, while an excess of nitrogen inhibits that of the prostrate component.

Environmental factors, as well as affecting the quantitative relationships of the two systems, also influence the extent to which the erect system produces hairs and branches. Hairs may be concerned with the absorption of mineral nutrients. When these are deficient, the production of hairs is conspicuously increased.

Correlated with their morphological plasticity, the Chaetophorales show a remark-able range of habitats extending from the littoral zones of seas and lakes to damp soil (e.g., *Fritschiella*), tree trunks (e.g., *Pleurococcus*), and particularly in the tropics and sub-tropics to the surfaces of leaves. Sexual reproduction is correspondingly diverse, both isogamy and oogamy being represented. The life cycle shows little development of the zygotic phase.

Stigeoclonium is representative of those members of the Chaetophorales in which both the prostrate and erect components of the thallus are easily recognizable, and each is more or less well developed. The species are commonly attached to submerged stones or woodwork, or are epiphytes on the leaves of aquatic angiosperms. They differ widely in external morphology. In general, the erect system ends in long, thin, hyaline hairs, and is less branched than the prostrate system (Fig. 3.11). The latter often forms a pseudo-parenchymatous sheet as a consequence of the close packing of the branches, but detailed investigation of its structure in natural habitats is rendered difficult by the tenacity with which it adheres to the substratum. The vegetative cells contain a peripheral girdle-shaped chloroplast with one or more pyrenoids.

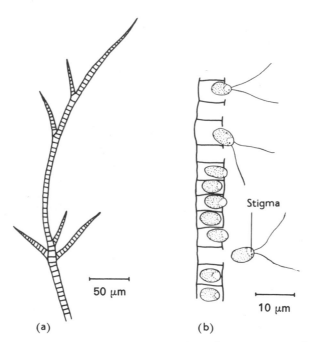

Figure 3.11. *Stigeoclonium tenue.* **(a) Terminal portion of erect system. (b) Release of gametes.** **(After West and Fritsch. 1927.** *A Treatise on the British Freshwater Algae.* **Cambridge University Press.)**

Possibly allied to *Stigeoclonium* is the terrestrial alga *Fritschiella* in which the prostrate system, buried in damp mud, produces nodules of cells which serve as perennat-ing organs.

Draparnaldia and its close relation *Draparnaldiopsis* are two aquatic members of the order in which the erect system is dominant. The upright axes of *Draparnaldia* consist of large barrel-shaped cells, from which arise the highly branched whorls of laterals with much smaller cells (Fig. 3.12). Frequently, these laterals, many of which terminate in hairs, are so profuse that the axis is quite obscured. The chloroplasts of the lateral branches are notably better developed than those of the axes, indicating some localiza-tion of function within the upright system. *Draparnaldiopsis* is similar, but the axis con-

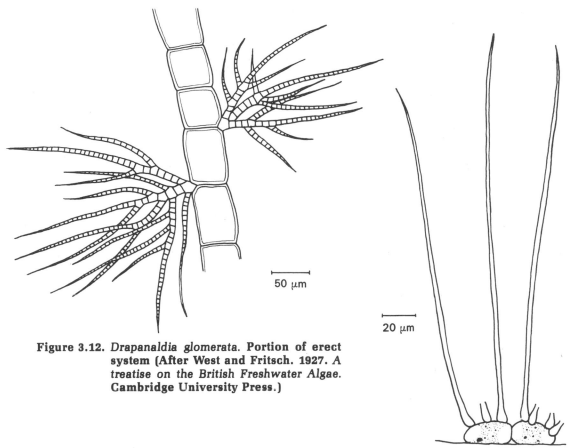

Figure 3.12. *Drapanaldia glomerata*. **Portion of erect
system (After West and Fritsch. 1927. *A
treatise on the British Freshwater Algae.*
Cambridge University Press.)**

Figure 3.13. *Aphanochaete polychaete*. **(After West and Fritsch. 1927.
A Treatise on the British Freshwater Algae. Cambridge
University Press.)**

sists of long and short cells, and the whorls of fine laterals, usually originating in four
tufts, arise only from the short cells. The tufts consist of radiating long and short
branches. Other branches, arising particularly from the bases of the longer laterals, turn
down, branch freely, and invest the internode in a cortical sheath, often thicker than the
axis itself. In both genera, the prostrate system is vestigial, and it is represented by a holdfast, the function of which is assisted by the outgrowth of rhizoids from adjacent cells of
the main axis.

Aphanochaete (Fig. 3.13) contains a number of epiphytes common on aquatic
angiosperms. The genus is representative of those members of the Chaetophorales in
which the prostrate component is dominant. It is composed of cells much like those of
Stigeoclonium, but each bears, on its dorsal surface, one or more hair-like cells which
soon lose their protoplasmic contents. They then become brittle and are readily broken
off.

Pleurococcus (*Desmococcus*) may also belong here. The cells contain a single
parietal chloroplast and resemble those of the Chaetophorales. They occur singly, or in
small groups or complanate aggregates, and occasionally in damp situations in short filaments. *Pleurococcus* is the main component of the friable incrustation (often mixed with
other algae and lichens) which forms on the shaded side of walls and tree-trunks. It is
probably the commonest green alga.

Reproduction and Relationships of the Chaetophorales. In the asexual reproduction of *Stigeoclonium*, the production of zoospores is confined to the erect system, and in

that of *Draparnaldia* and *Draparnaldiopsis* to the lateral branches. Zoospores are usually quadriflagellate, but smaller biflagellate forms are also present in some genera.

The sexual cycles of the Chaetophorales are not well known. Both isogamy and anisogamy have been reported, and *Aphanochaete* is oogamous. Motile gametes have been described as both biflagellate and quadriflagellate. Zygotes are commonly bright orange, and meiosis is believed to occur on germination.

Pleurococcus has no known method of reproduction other than simple cell division.

The morphological diversity of the Chaetophorales, and the restricted knowledge about their life cycles, make it difficult to assess their relationships with other green algae. Isolated specialized forms, such as *Draparnaldia* and *Aphanochaete*, are a conspicuous feature of the order. Although the cells of the Chaetophorales usually contain only a single basin-shaped chloroplast, their growth forms perhaps foreshadow later developments in plant evolution. The heterotrichous habit, for example, well developed in some genera, is also a feature of the gametophytes of many bryophytes (see Chapter 5).

The Oedogoniales. The Oedogoniales are a well-defined order, possessing several unique features. Some have attached so much weight to these as to consider removing the order from the Chlorophyta. The Oedogoniales are not, however, anomalous in such fundamentals as wall structure and pigmentation, and they are probably better regarded as a small group that has diverged from the main line of evolution, the intermediate stages being no longer represented. The Oedogoniales comprise only three genera: *Oedogonium*, *Oedocladium*, and *Bulbochaete*. Of these, *Oedogonium* is by far the most common and best known.

The thallus in the Oedogoniales. The thallus of *Oedogonium* is an unbranched filament (Fig. 3.14). When young, the filaments are attached by a basal holdfast, but, unless the water they inhabit is flowing, the mature condition is free-floating. The individual cells have a single, large nucleus and a reticulate chloroplast. Pyrenoids are frequent at the interstices of the reticulum. Cell division is normally intercalary, but the way in which the new cell walls are produced is highly peculiar (Fig. 3.15). The first indication that a cell is about to divide is the formation of a ring of wall material towards the upper end of the cell, just below the septum. The nucleus then divides and a septum forms between the daughter nuclei, but this septum remains free at the periphery. during the divisions of the nucleus, the ring in the upper part of the cell becomes larger and crescent-shaped in vertical section. Eventually, the cell wall breaks transversely at this level, and the ring is

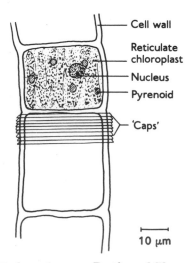

Figure 3.14. *Oedogonium* sp. Portion of filament.

drawn out longitudinally to form a cylinder of new wall material. Meanwhile the septum between the nuclei moves up the cell, reaches the bottom of the newly formed cylinder, and then fuses peripherally with the longitudinal walls. The wall of the lower cell is thus largely that of the mother cell, while that of the upper, except for a conspicuous cap of original wall at the anterior end, is wholly new. The presence of caps, which if the daughter cell goes on dividing may be several in number, at the anterior ends of the cell is a feature diagnostic of the Oedogoniales.

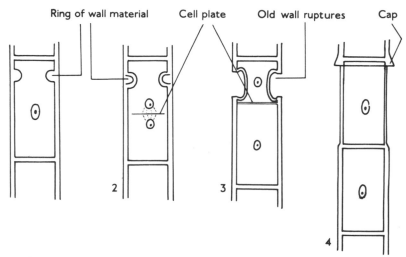

Figure 3.15. *Oedogonium*. **Stages in cell division. Diagrammatic.**

Reproduction in the Oedogoniales. Both asexual and sexual reproduction occur in *Oedogonium*. Asexual reproduction is by means of zoospores which, like others of the Oedogoniales, are unique in possessing an apical ring of many short flagella. They arise singly in cells at various sites along the filament, while in the sporangium the zoospore is surrounded by a mucilage sheath giving the wall a characteristic two-layered appearance. The zoospore is released by rupture of the sporangium, thereby causing a break in the filament. It emerges still surrounded by mucilage, but this is soon shed and the zoospore becomes almost spherical (Fig. 3.16a). After a brief motile phase, the zoospore settles and irregular rhizoidlike outgrowths from the flagellate end attach it to the substratum. A new filament then develops.

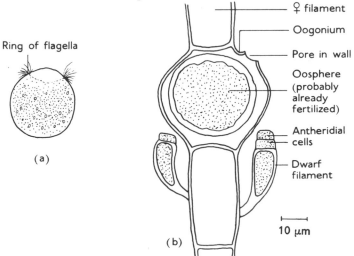

Figure 3.16. *Oedogonium* sp. **(a) Zoospore. (b) Oogonium with adjacent dwarf males.**

Both monoecious and dioecious species of *Oedogonium* are known. Large, almost spherical, oogonia are produced within the filament from the upper cell of a vegetative division. The lower cell may divide again, and another oogonium develop, but more usually it remains as the supporting cell. A single, dense oosphere is formed in each oogonium (Fig. 3.16b). The oosphere shrinks away from the cell wall and develops a colorless receptive spot as it matures. When it is fully mature, a pore appears at the anterior end of the oogonium adjacent to the receptive spot of the oosphere. Meanwhile, antheridia are produced by division of a vegetative cell into several disk-shaped portions, each of which produces two (in some species, four) pale green male gametes. These are similar, both in morphology and method of liberation, to the zoospores. After fertilization, the zygote forms an oospore, often reddish in color and regularly surrounded by a thickened wall.

In some dioecious species, the male and female filaments differ morphologically, and the male may even be reduced to a minute individual epiphytic upon the female. In these nannandrous species, the dwarf males arise from the female plants by way of a special propagule called an androspore. This is produced in a manner similar to that of the male gametes, except that only one androspore is produced per cell. Resembling zoospores, except for their smaller size and yellowish color, the androspores are attracted chemotactically to the oogonium and its supporting cell. Here they settle and give rise to small filaments (Fig. 3.16b). After a few cells have been produced, antheridia are cut off, and reproduction proceeds as in the isomorphic (macrandrous) species.

Although oospores are sometimes able to germinate immediately, a long resting period seems generally to be necessary. There is evidence that chilling hastens germination. When growth is eventually resumed, the normal course is for four haploid protoplasts to be extruded, each of which develops a crown of flagella, swims away, and settles to produce a new plant. In experimental conditions, unchilled oospores have been germinated to give rise directly to a single filament with very large cells. Oogonia and antheridia are produced, but fertilization has not been observed. It seems likely that these giant forms are produced without meiosis, the nuclei consequently containing twice the normal number of chromosomes. They provide an example in the algae of the aposporic production of a gametophyte.

In laboratory cultures, the sexual phase can be induced by increasing the carbon dioxide concentration of the medium. With dioecious species it is also necessary to intermix male filaments with female to stimulate the production of oogonia. The secretion of hormonelike substances into the medium seems indicated (cf. **Volvox**).

Related to *Oedogonium* is *Bulbochaete*. *Bulbochaete* resemble *Oedogonium* in essentials, but the filaments are branched and terminate in hairs. In *Oedocladium* the filaments are both branched and form a heterotrichous system.

Relationships of the Oedogoniales. The affinities of the Oedogoniales are obscure. The hairs of *Bulbochaete* and the heterotrichous system of *Oedocladium* recall features of the Chaetophorales, and perhaps indicate a distant relationship. Another feature in common is the peripheral girdle-shaped chloroplast in each cell. The small gametelike zoospores of some Chaetophorales may indicate the origin of nannandry in some Oedogoniales. Multiflagellate zoospores occur elsewhere only in some siphonaceous algae (e.g., *Derbesia*), but a close relationship seems unlikely here.

The Sphaeropleales. The small order Sphaeropleales is also included in the Chlorophyceae. It includes the distinctive genus *Microspora*. The filaments, free-floating at maturity, are encountered in freshwater lakes and marshes. The cells contain a reticulate chloroplast lacking pyrenoids. The cell walls disjoin in characteristic H-shaped pieces when the filaments fragment (Fig. 3.17). Asexual reproduction (other than by fragmentation) is by bi- or quadriflagellate zoospores. Sexual reproduction is isogamous, the gametes being biflagellate. The site of meiosis in the cycle appears not to have been determined.

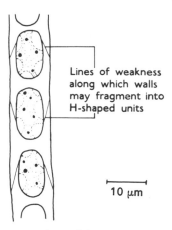

Lines of weakness
along which walls
may fragment into
H-shaped units

10 μm

Figure 3.17. *Microspora pachyderma.* **Portion of filament. (After West and Fritsch. 1927.** *A Treatise on the British Freshwater Algae.* **Cambridge University Press.)**

Ulvophyceae

The orders grouped into this class show a great diversity of habit ranging from the filamentous to the siphonaceous.

The Ulotrichales. Among the filamentous forms are the Ulotrichales. *Ulothrix* itself has a simple filament lacking branches. The genus has species in both fresh and saline waters. When mature, the filaments form loose, free-floating bundles. All the cells are of equal status, except for the basal attachment cell usually present in young filaments. The vegetative cells, which are often wider than long, are uninucleate and have a single peripheral chloroplast (Fig. 3.18a). Cell divisions are sporadic and intercalary. Fragmentation of the filament is common, but this is caused principally by accidental breakage; simultaneous dissociation of the filament into segments has rarely been observed.

In the asexual reproduction of *Ulothrix* 1 to 32 zoospores (the number depending upon the species) are produced in each cell by division of the protoplast. The mature zoospores are pear-shaped and quadriflagellate (macrozoospores) (Fig. 3.18b above). In some species, smaller biflagellate zoospores (microzoospores) are also produced, inter-

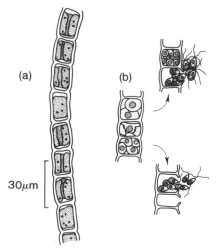

(a)   (b)

30μm

Figure 3.18. *Ulothrix zonata.* **(a) Vegetative filament showing the girdle-shaped chloroplast. (b)** *Above,* **production of quadriflagellate zoospores;** *below,* **production of gametes. Diagrammatic.**

mediate in size between macrozoospores and gametes. The zoospores are liberated through a pore in the wall of the parent cell, each zoospore surrounded by a mucilaginous sheath. The free zoospore closely resembles a unicellular member of the Volvocales, devoid of its cell wall. After settling, the zoospore attaches itself by the posterior end (to which the stigma has now shifted), and grows out laterally, producing a holdfast cell on one side, and new vegetative cells on the other. Sometimes, after the initial division in the parent cell in a filament, aplanospores with resistant walls are produced instead of zoospores.

In sexual reproduction, gametogenesis resembles the production of zoospores in asexual reproduction. The gametes, however, are uniformly biflagellate, and 8, 16, 32, or 64 (the number again depending upon the species) are produced in each gametangium (Fig. 3.18b below). *Ulothrix* is physiologically heterothallic, gametes from the same filament being unable to unite. The zygote is mobile for a short while, but then secretes a resistant wall and enters a resting period during which it may become attached and form a small unicellular plant. Germination commences with meiosis followed by one or two mitoses and the formation of zoospores or aplanospores, the mating types again segregating. The diploid state is thus represented only by the zygotic cell and has no prolonged existence. According to some accounts, the biflagellate swarmers are exclusively gametic.

Monostroma is also currently placed in the Ulotrichales. Although beginning life as a filament, lateral divisions also occur so that the mature plant has a flat, leaflike thallus, uniformly one cell thick, attached to the substratum by a holdfast. Some species are notable for being able to tolerate great changes of salinity in, for example, estuarine conditions. The life cycles are not fully known. In some species, the mature plant appears to be gametophytic and in others sporophytic. The gametes are biflagellate and the zoospores quadriflagellate. Anisogamy is present in some species.

The Ulvales. The Ulvales contain forms very similar to *Monostroma*. *Ulva*, for example, has a foliaceous thallus, but, in this instance, two cells thick (Fig. 3.19). Each cell, as in the Ulotrichales, contains a single chloroplast. At cell division each chloroplast

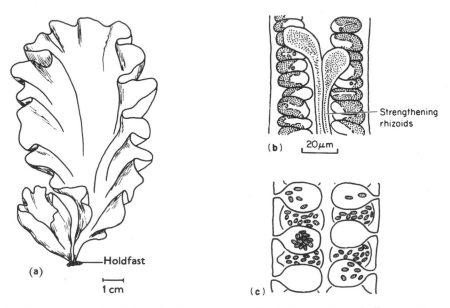

Figure 3.19. *Ulva lactuca.* **(a) Habit. (b) Transverse section of vegetative thallus. (c) Transverse section of thallus producing zoospores. ([a], [c] After Newton. 1973.** *A Handbook of the British Seaweeds.* **British Museum. [b] After Thuret, from Fritsch. 1935.** *The Structure and Reproduction of the Algae,* **1. Cambridge University Press.)**

divides at the same time as the nucleus. *Ulva* was the first multicellular plant in which it was possible to follow unambiguous chloroplast division with the electron microscope.

Young plants of *Ulva* always begin their development as simple filaments. division in principally two dimensions results in a flattened expanse of tissue expanding from a narrow stalk and holdfast rhizoids which grow between the two cell layers, strengthening the thallus (Fig. 3.19b). This occurs particularly in the stalk. Mutants are known which remain filamentous, the cells evidently no longer acquiring the ability to divide in more than one direction.

The marine intertidal zone is the characteristic habitat of both *Ulva* and of the closely related *Enteromorpha* which has a peculiar tubular thallus (Fig. 3.20). Both species, like *Monostroma*, can tolerate wide variations in salinity and may be found far up tidal estuaries.

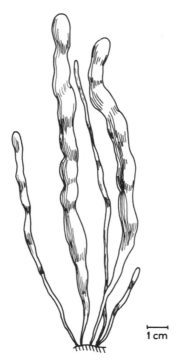

Figure 3.20. *Enteromorpha intestinalis.* **The thallus consists of hollow thin-walled cylinders. (After Newton. 1937.** *A Handbook of British Seaweeds.* **British Museum.)**

The asexual reproduction of *Ulva* resembles that of *Ulothrix*, most of the vegetative cells taking part (Fig. 3.19c). After zoospores have been discharged, the parent thallus remains as a bleached framework of empty cells.

Sexual reproduction is again similar to that of *Ulothrix*, and *Ulva* is also physiologically heterothallic. The zygote, however, does not undergo reduction division on germination, but grows instead into a diploid thallus identical with that of the haploid plant. The production of zoospores is accompanied by meiosis (although there are some irregularities), equal numbers of both mating strains being produced. *Enteromorpha* differs from *Ulva* in being anisogamous.

There are no intermediate forms in the Ulvales indicating how an isomorphic life cycle, such as that of *Ulva*, might have originated. It is possible that a relatively simple mutation prevented meiosis at the zygotic stage and that the inhibition remained effective until after many cell generations. Since there is no evidence that chromosome number itself determines the form of growth in any group of plants, this delaying of mitosis would have allowed the development of a diploid thallus closely resembling that of the haploid.

In the evolution of *Ulva*, the emergence of a distinct sporophyte was probably accompanied by the progressive elaboration of the vegetative structure of both phases of growth.

The Prasiolales. *Prasiola*, representative of an order usually placed with the Ulvales, is a familiar plant on tidal rocks and shore lines, particularly where the substrate is rich in nitrogen (for example, from bird droppings or seal excrement). The thallus resembles a small *Ulva* in form and attachment, but is only one cell thick. The nonsexual plant reproduces itself by spherical spores produced from the distal part of the thallus. In the sexual plant, the distal portion of the thallus is the site of meiosis. The haploid cells so produced then undergo several mitoses, building up groups of haploid cells within the diploid thallus. Each group of cells ultimately differentiates into gametes, either biflagellate males or nonflagellate females. A gametophytic phase thus exists for some time within the confines of the sporophyte.

The fusion of a male and female gamete takes place progressively. For a time, one male flagellum remains extruded and the zygote is motile. Ultimately, there is total absorption and a walled zygote is formed. Little is known about germination and the early development of the sporophyte. Sexual reproduction tends to predominate in the plants lower in the intertidal belt and asexual reproduction in the higher.

The Cladophorales. The remainder of the Ulvophyceae are characterized by multinucleate cells. The simpler are placed in the order Cladophorales. All are filamentous. *Cladophora* itself has some 160 species, some marine and others freshwater. The filaments show true branching, buds developing towards the anterior end of the elongated, cylindrical vegetative cells (Fig. 3.21). The bundle of filaments is usually attached below a rhizoidlike cell to a firm surface in the substratum. The internal structure of the cells is complex, each cell being multinucleate and having a single reticulate chloroplast with many pyrenoids. The cell wall comprises three layers, the inner of cellulose, a central layer of hemicellulose, and finally an outer coating, containing up to 70% protein and possibly chitinous, which gives the alga its characteristic crisp feeling. It also provides a surface which is rapidly colonized by small epiphytes, notably diatoms. Division of the protoplast, which does not regularly follow nuclear division, is accompanied by the formation of transverse septa. These develop from the margin towards the center and

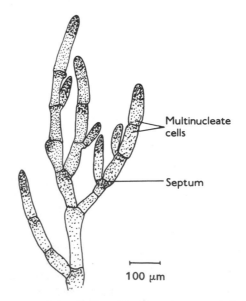

Multinucleate cells

Septum

100 μm

Figure 3.21. *Cladophora* **sp. Upper part of the thallus showing characteristic branching and the greater density of the cytoplasm towards the apical ends of the cells.**

gradually acquire a complicated lamellate structure at the periphery, which may impart some flexibility to the older filaments. Growth of the filaments is apical in *Cladophora*, but intercalary growth is general elsewhere.

A curious feature exhibited by some species of *Cladophora* and certain other genera is the tendency, when subjected to a gentle rolling motion in water, to aggregate into cushions or spheres. These are termed "aegagropilous" species because of the fancied resemblance of these aggregates to the balls of wool found in goats' stomachs. *Cladophorella*, a plant of damp places, is of interest because its upper cells secrete what appears to be a true cuticle, otherwise unknown in the algae.

Both the asexual and sexual reproduction of Cladophorales resemble that of *Ulothrix*. The zoospores and gametes are produced in non-specialized cells, and copulation is usually isogamous, rarely anisogamous. Fusion of the gametes is lateral. An isomorphic life cycle has been demonstrated in some species, but in others the cycle is heteromorphic. Before these were understood, the gametophytic and sporophytic phases, being unrecognized, were assigned to separate genera.

Vegetative reproduction by fragmentation of branches also occurs, and seems to be the sole means of reproduction of the aegagropilous species. In some species, survival through unfavorable periods is afforded by the formation of thick-walled resting spores packed with food reserves.

The Siphonales. The more complex Ulvophyceae with multinucleate cells are all siphonaceous. In current classifications the siphonaceous algae are distributed among several orders (see Table 2.1), but they can be conveniently considered together ("Siphonales"). The thalli normally contain multinucleate protoplasts and are consequently termed *coenocytic*. Dividing walls are largely absent until the formation of the reproductive organs. The range in form of the thallus is considerable; it may consist of a simple, unbranched tube or a complex mass of interwoven filaments. Those species with complex thalli are invariably marine, and the thalli are often mechanically strengthened by superficial deposits of calcium carbonate. There are numerous chloroplasts, and in some species they have particularly tough envelopes. They can remain functional following ingestion by marine invertebrates. The siphonaceous algae also differ from the rest of the Chlorophyta in containing both α- and β-carotene, instead of β-carotene alone, and at least one additional xanthophyll (siphonaxanthin). Almost all are confined to tropical or warm seas.

Dichotomosiphon (**Dichotomosiphonales**) is filamentous with a superficial resemblance to *Vaucheria*. It is aquatic and has been found at depths of up to 15 m (50 ft.) in fresh water. *Valonia* (**Siphonocladales**) is marine and consists solely of a bladderlike cell reaching a few centimeters in diameter. It often bears clusters of daughter vesicles, and short basal branches from a rhizoidlike holdfast. *Valonia* has been extensively used in experiments on wall structure, permeability, and absorption of electrolytes.

Acetabularia (**Dasyclydales**) (Fig. 3.22) has a mushroom-shaped thallus when mature, the cap reaching a diameter of about 1 cm (0.4 in.) The whole delicate plant is stabilized by a shell of calcium carbonate. The central axis produces whorls of deciduous branches during growth. Unlike most of the siphonaceous algae, *Acetabularia* is uninucleate throughout its vegetative development, the nucleus, large and conspicuous, remaining at the base of the stalk. If the stalk is removed from the nucleate portion, it remains capable of some growth and the apex may even begin to form a cap. Nevertheless, the nucleus is clearly essential for continued growth and morphogenesis. Decapitation and grafting experiments have shown that the nucleus produces a sequence of "morphogenetic substances" (probably messenger ribonucleic acids) which ascend into the stalk and determine the kind of growth which occurs at the top. Grafts made between the stalk of one species and the nucleate portion of another have revealed that the cytoplasm of the first species can affect the expression of the genetic information contained in the nucleus of the second. The fructosan inulin has been reported as a reserve product in *Acetabularia*. The chloroplasts can be isolated comparatively easily. They were the first with which it was possible to demonstrate normal photosynthesis *in vitro*.

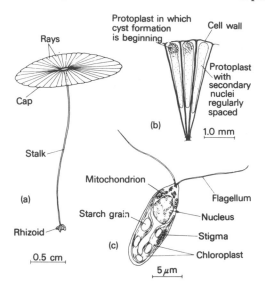

Figure 3.22. *Acetabularia mediterranea.* **(a) Cell with mature cap. (b) Detail of cap during cyst formation. (c) Longitudinal section of gamete.**

The **Codiales** show a considerable advance in vegetative organization. The thallus of *Bryopsis* (Fig. 3.23), for example, possesses a main axis, from which branches arise pinnately. These also branch, leading to a complanate bipinnate, and in some species tripinnate, condition. At the insertion of each branch the cell walls are constricted and conspicuously thickened. This feature undoubtedly has mechanical significance, since a simple branched tube lacking septa and with walls of uniform thickness would become structurally unstable beyond a certain point. In the vegetative condition the cytoplasm is spread as an even layer containing numerous nuclei and chloroplasts. The frondlike thallus of *Bryopsis* rises more or less vertically and is anchored to the substratum by rhizoids produced from a small prostrate filament. Fragments of the thallus will regenerate as new plants, but rhizoid outgrowths are formed only from the morphologically lower end. This is an example of polarity, widespread in the Plant Kingdom, but particularly open to investigation in *Bryopsis* because of its occurrence in a relatively simple unicellular system. *Derbesia* is a plant of warmer seas. Here there is a distinct "rhizome" attached to the substratum by lobed holdfasts and producing from its upper surface branching tubular threads.

Codium (Fig. 3.24) has a wider distribution and extends into cooler seas. It is representative of the most elaborate vegetative organization found in the siphonaceous algae. The thallus is made up of closely packed, interwoven filaments (hyphae), although the outward morphology varies widely with species. In *C. tomentosum* the thallus is a system of dichotomously branched axes, each about 0.5 cm (0.2 in.) in diameter, anchored at the base, Other species are flattened, forming a cushion or plate, or spherical. In all species, a weft of branched filaments gives rise to a continuous covering of elongated vesicles at the exterior.

Caulerpa (**Caulerpales**) also has a large creeping "rhizome" from which both rhizoids and upright "fronds" arise (Fig. 3.25). The "fronds" show a great variation in shape when mature, different species having been named from their distinctive forms (Fig. 3.26). Internally, all species of *Caulerpa* show ingrowths of the cell wall forming

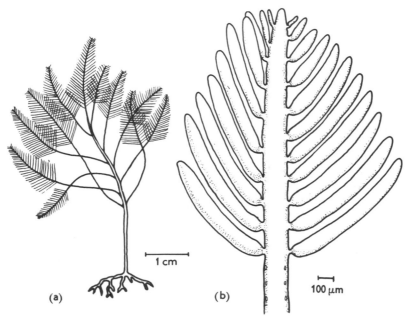

Figure 3.23. *Bryopsis plumosa.* **(a) Habit. (b) Terminal portion of branch. (After Newton. 1937.** *A Handbook of the British Seaweeds.* **British Museum.)**

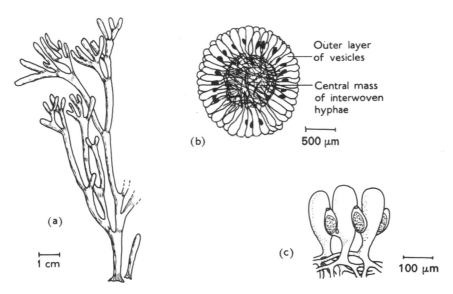

Figure 3.24. *Codium tomentosum.* **(a) Habit. (b) Transverse section of thallus. (c) Gametangia. (All after Newton. 1937.** *A Handbook of the British Seaweeds.* **British Museum.)**

throughout the thicker portion of the thallus, a web of interconnecting bars. This adds to the mechanical stability of the thallus, and the increased surface area of the protoplast may facilitate the passage of minerals. The fibrillar polysaccharide of the wall is not cellulose but a β- 1,3 xylan, a polymer of a pentose sugar. The walls also contain callose, another β- 1,3 glucan, also found in the sieve tubes and reproductive structures of land plants. *Halimeda* (Fig. 3.27) is a related tropical alga with a complex thallus of variable morphology. Calcium carbonate is deposited in the side walls of the outer vesicles. These deposits survive after the death of the plant and yield one of the several forms of coral. The

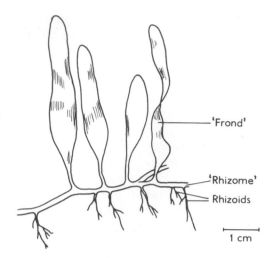

Figure 3.25. *Caulerpa prolifera.* Portion of plant. (After Taylor. 1960. *Marine Algae of the Eastern Tropical and Subtropical Coasts of the Americas.* University of Michigan Press.)

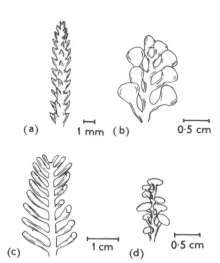

Figure 3.26. *Caulerpa.* Types of "fronds." (a) *C. cupressoides.* (b) *C. racemosa* var. *macrophysa.* (c) *C. asmeadii.* (d) *C. peltata.* (All after Taylor. 1960. *Marine Algae of the Eastern Tropical and Subtropical Coasts of the Americas.* University of Michigan Press.)

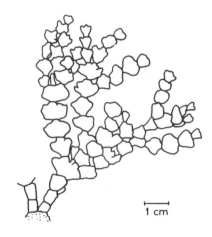

Figure 3.27. *Halimeda simulans.* Portion of thallus. (After Taylor. 1960. *Marine Algae of the Eastern Tropical and Subtropical Coasts of the Americas.* University of Michigan Press.)

fossil record of lime-secreting siphonaceous algae goes back as far as the Upper Silurian (see Table 1.1).

Although *Derbesia* produces multiflagellate zoospores (resembling those of *Oedogonium*), all other siphonaceous algae appear to lack specialized means of asexual reproduction. In *Valonia*, for example, vegetative multiplication takes place by the sporadic detachment of the daughter vesicles. In the larger forms, segments, branches, or parts of the "rhizome" become detached and establish themselves as new plants. In *Codium* and *Caulerpa*, this seems to be the principal method of reproduction.

The sexual cycles of many Siphonales are still obscure. In general, it appears that reproduction is isogamous or anisogamous, and oogamy absent. In species of *Bryopsis*, terminal pinnae, normally the smallest, become transformed into gametangia and are cut off from the rest of the thallus by a septum. The gametangia are conspicuously opaque as a consequence of the considerable multiplication of plastids. The gametes are liberated by dissolution of the apex of the gametangia. They are biflagellate, but there is striking anisogamy and, since each plant usually produces gametes of only one size, a clear trend to dioecy. The larger gametes also contain a distinctly green plastid, whereas the smaller gametes are yellowish. After fusion, the zygote develops immediately into a new plant. Sexual reproduction in *Codium* and *Caulerpa* appears to be basically similar to that in *Bryopsis*. Meiosis has been detected in gametogenesis, so the thallus in these forms is probably normally diploid.

Sexual reproduction in *Acetabularia* has many curious features. During vegetative growth the chromosomes in the single nucleus undergo continued endoreplication. When growth is completed, this basal nucleus then divides into several thousand secondary nuclei. These ascend into the cap which eventually becomes cleft into uninucleate compartments. Each compartment then becomes a cyst from which biflagellate gametes are ultimately liberated (Fig. 3.22b and c). Although the gametes are morphologically identical, their pairing behavior indicates the presence of + and − mating strains, a situation recalling that already encountered in *Chlamydomonas*. Despite much research, it is still not absolutely certain when meiosis occurs, but it is either preceding the formation of the cysts or during gametogenesis. The zygote therefore is diploid. There is no resting period, and the formation of a new plant begins immediately.

The existence of heteromorphic life cycles has been confirmed in *Derbesia*. Zoospores from some species of *Derbesia* have been observed to grow into vesicular plants reaching about 1 cm (0.4 in.) in diameter formerly placed in the genus *Halicystis*. These are now known to be gametophytic. Zygotes formed from the fusion of biflagellate gametes have yielded *Derbesia*. Other species of *Halicystis* may be gametophytes of *Bryopsis* whose sexual cycles have not yet been fully investigated. There are many indications that surprising discoveries may yet be made about the life cycles of the siphonaceous algae.

The relationships of the siphonaceous algae are obscure, but the origin may have been in a form resembling the present Chlorococcales, an order in which tendencies towards the coenocytic habit are evident. An ancestor may have been a simple coenocyte resembling *Protosiphon*.

Charophyceae

This class contains unicellular, filamentous, and heterotrichous forms. Of particular interest are metabolic and structural features which resemble those found in higher plants. In the non-mitochondrial part of respiration, for example, the enzyme glycolate oxidase, characteristic of land plants, replaces glycolate dehydrogenase present elsewhere in the Chlorophyta. During cell division in the Charophyceae, the nuclei tend to move well apart in telophase and the spindle to persist. In some members, the new transverse wall forms from coalescing vesicles at the equator in the manner in which the phragmoplast functions in higher plants. This contrasts with the situation in the Chlorophyceae. A significant feature in the reproductive cells of some species is an array of minute plates and microtubules at the root of the flagella in zoospores and gametes, very similar to the **multilayered structure** (MLS) found in archegoniate spermatozoids. Although these features may have arisen more than once in the course of evolution, they have drawn particular attention to the Charophyceae in relation to the question of the kind of algae from which land plants may have arisen.

The Klebsormidiales. Among the simpler filamentous forms is *Klebsormidium*, most species of which occur in the soil. The cells are not unlike those of *Ulothrix*. Asexual reproduction occurs by means of biflagellate zoospores. Sexual reproduction is isogamous.

The Mesotaeniales and Desmidiales. The unicellular members of the class fall into the orders Mesotaeniales and Desmidiales. The Mesotaeniales contain the so-called "saccoderm desmids," the cell walls of which consist of only one portion of (so far as known) uniform age (cf. the "placoderm desmids"). The cells contain two platelike chloroplasts symmetrically placed on each side of the nucleus (Fig. 3.28). The cell walls are without ornamentation and lack pores. The cell sap is sometimes pigmented, and these desmids (mostly *Ancylonema*) are, in part, responsible for the "red snow" seen on Alpine glaciers and in the Arctic. Saccoderm desmids also occur in bog pools.

Asexual reproduction is by a simple division, several cells often being held together within a mucilage sheath. Sexual reproduction involves conjugation of two cells. Meiosis and division of the zygote into four daughter cells take place after a resting period. Mendelian segregation of mating type has been demonstrated in pure cultures.

Most of the desmids are referable to the order Desmidiales (the placoderm desmids). They are typically unicellular (although sometimes the cells may be aggregated in chains) and are conspicuous in the phytoplankton of oligotrophic meres. The cells have complex shapes and a precise and striking symmetry. In many forms the cell is divided into two halves, one being the mirror image of the other, connected by a narrow central portion, the isthmus. The nucleus usually lies in the isthmus, and one or more chloroplasts are present in each half cell. Even in those forms in which the central constriction is less conspicuous (e.g., *Cosmarium*, Fig. 3.29) or absent (e.g., *Closterium*), the cells still display exact bilateral symmetry. In *Closterium*, the cells are narrowed towards the poles and slightly curved. There is a small vacuole at each pole. Crystals of barium sulfate are commonly found in the vacuoles even when the concentration of barium ions without is very low.

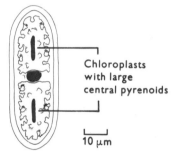

Chloroplasts
with large
central pyrenoids

10 μm

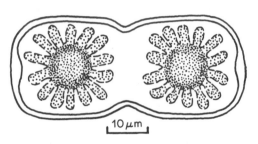

10 μm

Figure 3.28. *Cylindrocystis brebissonii.* **The two prominent stellate chloroplasts are symetrically placed in the cell, and the nucleus lies between them.**

Figure 3.29. *Cosmarium diplosporum.* **(After West and Fritsch. 1927.** *A Treatise on the British Freshwater Algae.* **Cambridge University Press.)**

The cell wall is made up of an inner cellulose layer and an outer layer of variable composition, frequently containing iron compounds or silica. It is particularly this outer layer, often patterned with spines and other protuberances, which provides the specific characteristics and puts the desmids among the most beautiful of microscopic objects. The cells are frequently surrounded by an investment of mucilage secreted by minute pores in the wall. It is probably the localized secretion of such mucilage which enables the cells to perform slow movements.

Asexual reproduction of the placoderm desmids is by fission. The details of the process are not yet fully understood, but in outline it is as follows. After nuclear division, a ring of wall material develops at the center of the isthmus and grows inwards to form a septum. When complete, the daughter cells separate. Each then regenerates the missing half and the symmetry is restored (Fig. 3.30). This results in two adult organisms in each of which half of the thallus is inherited from the previous generation and the other half newly synthesized. This is an example of semi-conservative replication of an organism recalling that of the molecule of double-stranded DNA.

Sexual reproduction involves conjugation of cells. A protuberance may be formed from one cell and directed towards the other, but in many species the walls break down at the isthmus and the protoplasts emerge and fuse. Following fusion, a thick-walled zygospore is formed. Meiosis and germination occur after a resting period. As in the saccoderm desmids, segregation of mating types has been demonstrated.

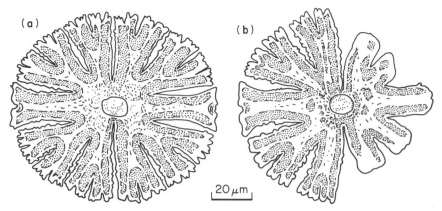

(a)   (b)

20 μm

Figure 3.30. *Micrasterias.* **(a) Mature individual preparatory to division. (b) Following division, the beginning of regeneration of new half cell.**

The Zygnematales. The thalli of the order Zygnematales consist of unbranched and free-floating filaments, although rhizoidlike outgrowth may attach the basal cell to the substratum in the young state. Almost all are found in fresh water.

The cells are often markedly elongate, having an axile ratio of 5:1 or more. The nucleus lies near the center suspended by cytoplasmic strands, and the conspicuous chloroplast is in the form of a flat plate (e.g., *Mougeotia*), a helical band at the periphery of the cell (e.g., *Spirogyra*, Fig. 3.31), or it consists of two stellate portions (e.g., *Zygnema*). Numerous pyrenoids occur in each form of chloroplast. In strong light, the chloroplast of *Mougeotia* turns so that its narrow edge is presented to the source. Experiments with polarized light have shown that the stimulus to this movement comes from an array of photoreceptive molecules at the periphery of the cytoplasm. The proteinaceous pigment phytochrome is also involved.

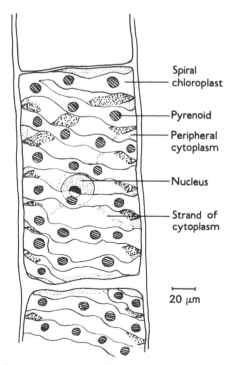

Spiral chloroplast

Pyrenoid

Peripheral cytoplasm

Nucleus

Strand of cytoplasm

20 μm

Figure 3.31. *Spirogyra* **sp. Cytology of vegetative cell.**

Growth of the filament is intercalary, all the cells except the basal cell being capable of division. The dividing wall grows in from the margin, but the central region of the septum is completed by the formation of a phragmoplast.

Asexual reproduction is solely by fragmentation of the filament, and is preceded by changes in the transverse walls. In some species, the middle lamella of the transverse wall becomes gelatinous and a turgor difference arises between the cells. This causes one cell to bulge into its neighbor and the junction between the cells becomes strained to breaking point. In other species, a ring or collar of wall material forms on each side of the septum and, in a manner not yet understood, leads to separation. The comparative ease with which the cells separate has led some to regard the Zygnematales as basically unicellular.

Sexual reproduction involves conjugation, occurring normally towards the end of the growing season, but factors such as nitrogen deprivation, pH, and altitude (presumably because of its relationship with mean temperature) have all been found to influence it. Before conjugation, the two participating filaments must lie parallel and come close together (Fig. 3.32a). The necessary adjusting movements are probably brought about by the localized secretion of mucilage. Since solitary filaments show no directed movement, the first approach of one filament to another, presumably of opposite mating type, must involve some form of chemotaxis. Once alignment has taken place, further mucilage secretion holds the filaments in position, and opposed cells begin to form protuberances directed towards each other. The production of protuberances is not necessarily simultaneous, and a cell may be touched by the protuberance of its partner before it has itself become active. The production of a similar structure is then immediately induced. The growth of the protuberances pushes the filaments apart at the site of conjugation. If an unmated protuberance touches a cell which has already formed a connection, there is no further response (Fig. 3.32c). Eventually, growth of the protuberances ceases, the appressed distal walls dissolve, and a conjugation tube is formed between the

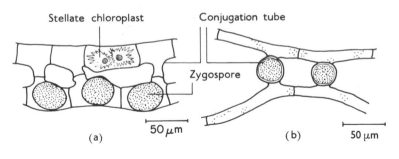

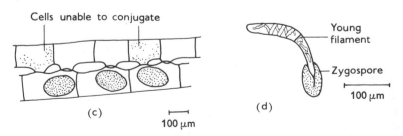

Figure 3.32. **Sexual reproduction in the Conjugales. (a)** *Zygnema stellinium.* **(b)** *Mougeotia parvula.* **(c)** *Spirogyra setiformis.* **(d)** *Spirogyra velata.* **Germination of zygospore. (After West and Fritsch. 1927.** *A Treatise on the British Freshwater Algae.* **Cambridge University Press.)**

two cells. With the formation of the tube complex, physiological changes take place within the protoplasts. There is an increase in starch and a decrease in permeability and the osmotic pressure of the cell sap. Before completion of the conjugation tube, no sexual differentiation of the two protoplasts is apparent, but as soon as communication is established one protoplast withdraws from its wall and passes slowly through the tube to fuse with its partner. Since all the protoplasts in one filament behave in the same way (Fig 3.32c), it is legitimate to regard one filament as male and one as female. After fusion, the chloroplast of the male gamete is resorbed, and the zygote contracts.

The process of conjugation is subject to a number of variations. In *Spirogyra*, for example, adjacent cells of the same filament may sometimes conjugate. In *Zygnema* and *Mougeotia* the zygote is often formed in the conjugation tube (Fig. 3.32b). Sexual differentiation of a morphological nature is seen in *Sirogonium*. Here the male gametangium, regularly associated with a sterile cell, is notably smaller than the female. In all instances the zygote forms a spore (zygospore) with a thick wall containing sporopollenin. After a period of dormancy, the spore germinates and gives rise to a single filament (Fig. 3.32d). In those species investigated, meiosis has been shown to occur before germination, but three nuclei degenerate. The survival of only one meiotic product prevents a direct analysis of Mendelian segregation.

There are evident affinities in relation to sexual reproduction between the Mesotaeniales, Desmidiales, and Zygnematales. These three orders have often been grouped together as the Conjugales.

The Coleochaetales. The Coleochaetales are a small order of which *Coleochaete* is fully representative (Fig. 3.33). The genus contains many freshwater epiphytes. The habit is basically heterotrichous, but the prostrate elements is usually dominant. An exception is *C. pulvinata* is which the erect branches form a conspicuous hemispherical cushion. Sheathed bristles are a characteristic feature of the thallus of *Coleochaete*.

The prostrate portion of some species of *Coleochaete* (e.g., *C. scutata*, often epiphytic on leaves of the aquatic angiosperm *Naias*) becomes parenchymatous. A phragmoplast has been observed at cell division and the dividing walls are penetrated by plasmodesmata.

Coleochaete is also outstanding in certain features of its oogamous sexual reproduction. The cells which differentiate into oogonia terminate short branches, although they may ultimately appear lateral because of continued growth from the penultimate vegetative cell. The oogonium develops a long neck, termed a trichogyne (Fig. 3.33b), which opens at the tip when the egg is ready for fertilization. The antheridia arise in the fila-

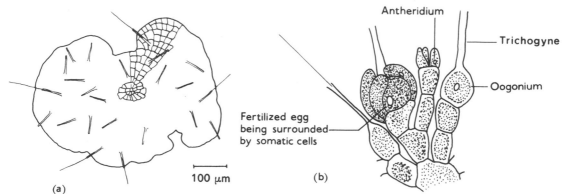

Figure 3.33. *Coleochaete.* **(a)** *Coleochaete scutata.* **Surface view. (b)** *Coleochaete pulvinata.* **Sexual reproductive structures. ([a] After West and Fritsch. 1927.** *A Treatise on the British Freshwater Algae.* **Cambridge University Press; [b] after Pringsheim, from Fritsch. 1935.** *The Structure and Reproduction of the Algae, 1.* **Cambridge University Press.)**

ments, often just above the oogonium, and each produces a single colorless spermatozoid. On liberation, they are attracted, presumably chemotactically, to the receptive
spot at the tip of the trichogyne and fertilization follows.

As a consequence of fertilization, but in a manner still unknown, the vegetative cells
adjacent to the oogonium are stimulated into growth and envelop the zygote in a continuous parenchymatous sheath. Symplastic connections between the zygote and the
covering cells are lacking, but the common wall may develop irregular projections suggestive of the thickened and labyrinthine boundary found between the gametophyte and
embryo sporophyte in archegoniate plants. The sheath eventually dies, but its cells contribute to the wall which forms around the resting zygote. This wall is acetolysis-resistant
and probably contains sporopollenin. In the prostrate species, the zygotes are retained in
the parent thallus.

After a resting period, the zygote germinates and the nucleus undergoes meiosis.
This is followed by several mitoses giving rise to a plate of 16 to 32 wedge-shaped cells.
The original wall of the zygote eventually bursts and each cell liberates a biflagellate
swarmer. Asexual reproduction of the parent thallus also takes place by the liberation of
single zoospores.

The multilayered structure beneath the basal bodies of the two flagella is well developed in *Coleochaete*. It consists of a flat ribbon of about 100 microtubules overlying an
array of narrow plates, each about 50 nm deep and with a central discontinuity (Fig. 3.34).
The alignment of the microtubules is transverse to that of the plates. One end of the ribbon
continues without curvature into the body of the cell (Fig 3.34, right). Although minute,
the multilayered structure is very distinctive. That of *Coleochaete* is closely similar to the
corresponding structure in bryophyte spermatozoids (Chapter 5).

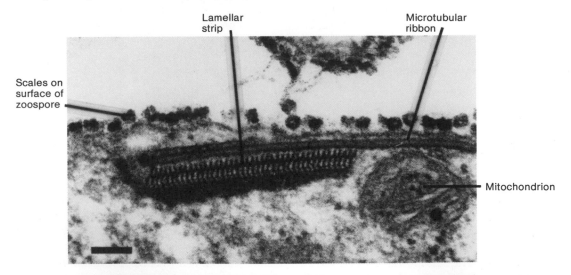

Figure 3.34. *Coleochaete, zoospore.* **Section of the multilayered structure (MLS) at the anterior
of the cell. The section is parallel to the alignment of the microtubules of the
microtubular ribbon (spline). The plates of the lamellar strip are perpendicular to
the microtubules and are consequently cut transversely. The anterior edge of the
MLS lies beneath the basal bodies of the flagella. The microtubular ribbon continues to the posterior of the cell and is presumably an important structural element. (Electron micropraph by H. J. Sluiman. 1983. Reproduced from**
Protoplasma **115, by permission of Springer Verlag, Vienna.) Scale bar 0.1 μm.**

The Charales. In the Charales, the Chlorophyta reach the highest level of differentiation encountered in the division, including many striking features unrepresented elsewhere in the Plant Kingdom. Nevertheless, they remain basically similar to other Chlorophyta in pigmentation, metabolism and the limited nature of their anatomy. There are only six living genera, the remainder being fossil. The living *Chara* and *Nitella* are common in base-rich waters, particularly those which are not too fast-moving and with a muddy substratum to provide anchorage. A few Charales are found in brackish waters (e.g., *Lamprothamnium*). Many species develop an exoskeleton of calcium carbonate (hence the name "stoneworts"). This is clearly an ancient feature since charalean exoskeletons, which are readily preserved as fossils, can be traced back as far as the Devonian.

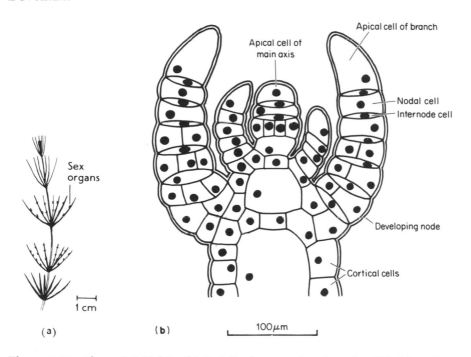

Figure 3.35. *Chara.* **(a) Habit. (b) Detail of apex of main axis. ([b] After Esser. 1976.** *Kryptogamen.* **Springer, Berlin.)**

The thallus of the Charales always possesses a clear main axis, growing from a dome-shaped apical cell (Fig. 3.35). Segments are cut off at regular intervals and in a single column from the flat base of this cell, and each segment immediately divides again by a transverse wall. The upper cell of this division becomes a nodal initial, and the lower, which does not divide again, an internode cell. The nodal initials remain meristematic and give rise to whorls of branches, mostly of limited growth. The mature plant thus has a morphology reminiscent of *Draparnaldiopsis*. The internode cells often achieve remarkable lengths (up to 10 cm [4 in.] or more in *Nitella*), and the cytoplasm commonly shows vigorous streaming, rates of up to 100 μm per second being common. The streaming persists in protoplasts extracted enzymically from internodal cells, although these adopt a spherical form. The chloroplasts, which are discoid and lack pyrenoids, are, however, little affected since they lie in an outer, less active layer (ectoplasm). The streaming is associated with actin microfilaments, about 50 μm of actin, accompanied by myosin, being present in a single cell. The nucleus often fragments in mature cells.

The large size of the internode cells of *Nitella* makes them ideal for studying permeability. It is even possible to cut off the ends and wash out the contents, leaving the plasmalemma intact. The biophysical properties of this membrane can then be investigated without interference from the cytoplasm.

In *Nitella*, the internodes consist only of the single internode cell, but in *Chara* this becomes surrounded by a cortex formed of rows of cells growing down from the lower cells of the node.

Unlike most benthic algae, the Charales are usually anchored to a soft substrate. Attachment is by colorless branched rhizoids which issue from the lower subterranean nodes. The rhizoids are negatively gravitropic and contain aggregated crystals of barium sulfate. These may act as statoliths and, by deflecting the flow of Golgi vesicles containing wall materials to one side of the cell, cause differential growth, and hence curvature.

Vegetative propagation, the only method of asexual reproduction of the Charales, is by the formation of tuberous outgrowths on subterranean nodes or rhizoids. These become filled with starch and can withstand unfavorable conditions. The nodal outgrowths often have a stellate symmetry ("starch stars").

Sexual reproduction is oogamous, and highly specialized. The male and female reproductive structures, which consist of antheridia and oogonia surrounded by envelopes of sterile tissue, are sufficiently large to be seen with the naked eye. In both monoecious and dioecious species, the antheridia develop just below the insertions of the lateral appendages (Fig. 3.36). A complicated sequence of divisions leads to the mature antheridium.

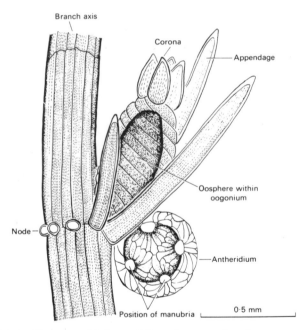

Figure 3.36. *Chara* **sp. Male and female reproductive organs.**

The eight peripheral cells of the original primordium form the wall of the antheridium. Attached to the inner surface of the wall are the **manubria**. These rod-shaped receptacles give rise to numerous threads of spermatogenous cells (Fig. 3.37) which fill the cavity of the antheridium. Each spermatogenous cell produces one spermatozoid. A multilayered structure is present during differentiation. This resembles the multilayered structure of the motile cells of *Coleochaete*, but only 20–30 microtubules are present in the upper layer. The mature spermatozoids, with the long narrow nucleus (Fig. 3.37c), resemble those of the simpler archegoniate plants. The wall of the mature antheridium is often bright red.

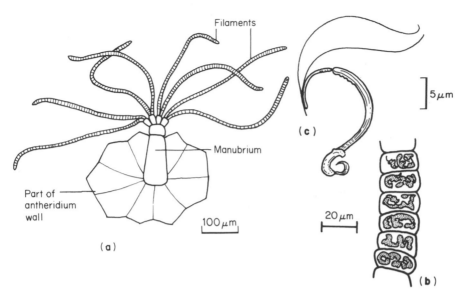

Figure 3.37. *Chara.* **(a) Antheridial filaments. (b) Differentiating spermatozoids in cells of filament. (c) Biflagellate spermatozoid.**

Oogonia, when mature, are surrounded by a spiral of elongated vegetative cells, at the tip of which short columns of rounded cells come together as a corona (Fig. 3.36). Within this sheath is a single egg cell which accumulates food reserves and becomes largely opaque. Spermatozoids penetrate slits which appear in the coronal region, and fertilization occurs at a small clear area near the apex of the egg, the so-called "receptive spot." After fertilization, a cellulose membrane is secreted by the zygote. This, together with the wall of the oogonium and the inner walls of the spiral cells, which become thickened and indurated, serves to enclose the zygote in an almost impervious jacket, often reinforced externally with calcium carbonate. Red light promotes germination, indicating that it is controlled by a phytochrome system.

Germination, which occurs after a resting period, is almost certainly accompanied by reduction division, but only one of the tetrad of nuclei so produced remains intact; the remainder degenerate. Division of the intact nucleus leads to the production of two cells, which, as the membranes of the zygote break open at the apex, yield from the one a rhizoid and from the other an erect green filament. This filament is the protonema, a stage of development peculiar, in the Chlorophyta, to the Charales. Cell divisions occur in the protonema, and it differentiates into nodes and internodes. The first node gives rise to rhizoids and additional protonema, and the second to a whorl of laterals, one of them developing into a normal main axis, and the others into yet more protonema. The mature plants thus arise at lateral branches of protonemal filaments.

At least one species of *Chara* has been shown to be parthenogenetic, but aberrations of the life cycle have been little investigated.

The relationships of the Charales are largely speculative. The fossils show that they are an extremely ancient type of algal organization, even fossilized zygotes (*Gyrogonites*) being recognizable in rocks as old as the Devonian. The small discoid chloroplasts are more like those of the higher plants than of Chlorophyta generally, but there is no evidence that terrestrial plants came directly from *Chara*-like ancestors. The Charales share with the desmids and *Spirogyra* the presence of barium sulfate crystals in the cells, and whorled branching and an archegoniumlike oogonium are represented in *Draparnaldia* and *Coleochaete* respectively. Although there may be distant relationships, the Charales are best regarded as a highly specialized, and in certain ecological situations highly successful, order of aquatic plants, long separated from the main trends of algal evolution. They are possibly now represented by fewer forms than earlier in geological time.

Pleurastrophyceae

The small class Pleurastrophyceae contains flagellate, coccoid, and filamentous forms, mostly freshwater and occasionally terrestrial. The basal bodies of the flagella of the motile stages show characteristically a counterclockwise orientation, in contrast to the clockwise orientation found in the Chlorophyceae. The most notable genus currently included here is *Trentepohlia* (**Trentepohliales**), worldwide in distribution and common in warm, moist situations. The heterotrichous thallus grows on rocks, bark, and shaded leaves, as well as on soil and humus, and is often conspicuous for its orange rather than green appearance. The cell walls are thick and often layered, and this feature probably accounts for the ability of the alga to resist desiccation. When growth is inhibited, as in conditions of drought, there is a marked increase in the pigmentation of the cells, and concurrently an accumulation of fat, in which the red pigment (haematochrome) is dissolved. Some species of *Trentepohlia* are components of lichens.

Asexual reproduction of *Trentepohlia* is by biflagellate zoospores and takes place only in damp conditions. Sporangia may be sessile or stalked. The stalked sporangia are often distributed by wind, the liberation of the zoospores taking place in damp conditions. Purely vegetative propagation also occurs by dispersal of fragments of the prostrate system.

Isogamous sexual reproduction has been observed in some species. A multilayered structure has been detected in the motile cells but it is not identical with that occurring in the Charophyceae.

Evolution within the Chlorophyta. It is reasonable to regard the unicellular forms as relicts of the most primitive green algae. Diversification can be envisaged as being accompanied by a progression towards colonial, coenobial, and filamentous organization, this taking place in a number of parallel lines resulting in the major classes recognized today. In Table 3.1, the classes are placed according to their relative complexity. The

Table 3.1. The living green algae arranged according to the level of complexity attained in their respective classes, and their relative position in the evolution of "a + b" plants. The relationships between the classes appear too remote to indicate affinities. A class which has reached a thalloid level of organization may nevertheless retain flagellate forms (e.g., *Chlamydomonas* within the Chlorophyceae).

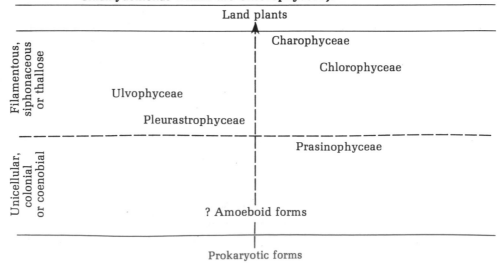

eukaryotic forms presumably arose from the prokaryotic, as discussed in relation to the Rhodophyta. Although *Prochloron* immediately excites interest as a possible ancestor of the "a + b" plastid, there are substantial differences in the carotenoids. A direct relationship seems unlikely, and the name Prochlorophyta now seems unfortunate.

There are, however, compelling reasons for regarding the Charophyceae as being close to the algae from which the land plants arose. This is discussed further in Chapter 5.

EUGLENOPHYTA

Habitat	Freshwater, a few marine.
Pigments	Chlorophylls a, b; β-carotene; diadinoxanthin, other xanthophylls less prominent.
Food reserve	Paramylum.
Cell wall components	No cell wall, but a pellicle containing a helically arranged structural protein present in many.
Reproduction	Asexual, sexual doubtful, but if present isogamous.
Growth form	Predominantly unicellular, flagellate.
Flagella	2, but only 1 (rarely 2) emerging from the gullet, Flimmer, anterior.

The Euglenophyta have been assigned to both the Plant and Animal Kingdoms, since the division contains both green autotrophic and colorless heterotrophic forms. Many of the autotrophic Euglenophyta are also able to thrive in the dark if supplied with suitable metabolites, so they can be regarded as facultative heterotrophs. Races free of chloroplasts can be raised by growing cultures at 35°C (95°F), when cell division proceeds faster than chloroplast fission so that some cells eventually lack them altogether. Similar colorless races can also be produced by low concentrations of streptomycin, a drug which inhibits chloroplast replication. It seems probable that naturally heterotrophic species have evolved by spontaneous loss of plastids from autotrophic forms.

Although heterotrophic nutrition, a feature of animals, is common in the Euglenophyta, the evidence points to the ability to ingest organic food materials as being, in living forms, a secondary feature. Heterotrophic nutrition, both facultative and obligate, is, in fact, known in many other algae. Examples are provided by the Cyanophyta, and by *Polytoma* (Volvocales) and *Prototheca* (Chlorococcales) in the Chlorophyta. The Euglenophyta are nevertheless outstanding in the extent to which they have developed this facility. Similarly a cell wall, absent in the Euglenophyta and animal cells generally, is by no means always present in other flagellate unicellular algae, It is lacking, for example in *Dunaliella*, otherwise, similar to *Chlamydomonas*, and in *Micromonas*. Flagellate zoospores and gametes are also commonly naked.

The node of nutrition and presence or absence of a cell wall thus appear to be relatively plastic features at the flagellate level of organization. It should be noted, however, that although there is evidence that photosynthetic activity has been lost in the course of evolution, resulting in heterotrophic forms, there is no indication that it has been spontaneously acquired. Indeed, it is difficult to envisage how a heterotrophic form lacking plastids could generate a photosynthetic apparatus *ab initio*.

Euglena, representative of the **Euglenales**, is normally autotrophic, the cells containing several discoid or bandlike chloroplasts (Fig 3.38). Each has a compound envelope of three membranes, the outer of which is continuous with the endoplasmic reticulum.

Some species abound in fresh water rich in organic material, such as seepage from dung-hills and farmyards, while others occur on damp mud by rivers, salt-marshes, and similar places.

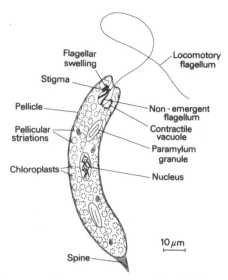

Figure 3.38. *Euglena spirogyra.* **Reconstruction of cell based on photo- and electron micrographs. (From G. F. Leedale, B. J. O. Meeuse, and E. G. Pringsheim. 1965.** *Archiv für Mikrobiologie* **50.)**

The cell is retained in a characteristic spindlelike shape by a rigid outer layer of cytoplasm termed a pellicle. Electron microscopy has shown that the pellicle contains a bandlike proteinaceous component wound in several helices around the cell. This structural protein lies against the plasmalemma and is associated with microtubules. In those species in which the pellicle is flexible, a flowing peristaltic movement is commonly observed, particularly if flagellar movement is constrained or the flagella are shed. The causes of this "metabolic" movement are obscure. The chloroplasts lie towards the periphery of the less viscous cytoplasm, and the lamellae characteristically consist of three appressed thylakoids. A pyrenoid is present, but the storage product paramylum (a β-1,3 glucan) is formed outside the chloroplast. The paramylum granules are highly crystalline and bounded by a single membrane. The nucleus lies in or near the posterior half of the cell, and chromosomes remain contracted throughout interphase, a feature also seen in the dinoflagellates. At the anterior end of the cell is a small invagination, the gullet, which in heterotrophic forms may serve for the ingestion of food. A system of vacuoles discharges at intervals into the gullet and provides for osmoregulation. There are two flagella, only one of which, normally Flimmer (Fig. 3.39), emerges through the mouth of the gullet. The base of this flagellum bears a thickening, believed to be a photoreceptor, close to the stigma in the adjacent cytoplasm.

Possibly, belonging to the heterotrophic Euglenophyta is *Scytomonas*, a common intestinal parasite. Its cytology closely resembles that of *Euglena*, and paramylum and fat occur as food reserves.

Binary fission is the common method of reproduction of *Euglena*. By the end of nuclear division, the locomotor apparatus is replicated and cleavage of the whole cell proceeds from the anterior to the posterior end. Nuclear division resembling meiosis has been reported, but well-established instances of sexual reproduction in the Euglenophyta are lacking.

Although predominantly flagellate, a few Euglenophyta are encapsulated and form dendroid colonies. These can be regarded as the result of developmental trends from the flagellate state which parallel those in other algal groups, but which have reached only a rudimentary level of morphological complexity. The **Eutreptiales** differ from the Euglenales in being biflagellate.

The relationships of the Euglenophyta with other algae are obscure. In the presence of chlorophylls a and b they resemble the Chlorophyta, but the carotenoid composition of the plastids is significantly different. The ability of *Euglena* to grow heterotrophically in the absence of plastids is unique among the algae. This would be in line with the view that *Euglena* had originated from a flagellated heterotroph which had taken in plastids from another source, thereby becoming phototrophic. Any relationship with the Chlorophyta would then be remote and possibly indirect. This would be in agreement with the substantial biochemical and metabolic differences between the Chlorophyta and Euglenophyta, and the regular presence of an extra membrane around the plastids of the Euglenophyta.

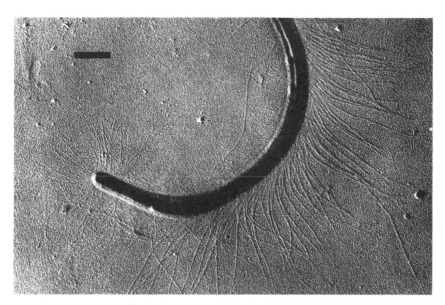

Figure 3.39. The tip of the flagellum of *Euglena gracilis* **showing the ''Flimmer'' hairs. Shadowed preparation. (Electron micrograph by G. F. Leedale. © Biophoto Associates.) Scale bar 1 μm.**

4

THE SUBKINGDOM ALGAE
Part 3

ALGAE CONTAINING CHLOROPHYLLS a AND c

Although chlorophyll a is always present in the algae considered in this chapter, the amount of chlorophyll c is sometimes small. Chlorophyll b is always absent.

The "a + c" algae show a number of organizational trends resembling those seen in the Chlorophyta. There are also features not represented in living Chlorophyta, but possibly present at some stage in their evolutionary history. The chlorophyll c-containing algae are sometimes referred to as the chromophytes

CHRYSOPHYTA

Habitat	Aquatic (mainly freshwater).
Pigments	Chlorophylls a, c; β-carotene; fucoxanthin
Food reserves	Fat, chrysolaminarin (leucosin).
Cell wall components	Cellulose, hemicelluloses, often with siliceous scales.
Reproduction	Asexual, occasionally sexual.
Growth forms	Flagellate, coccoid, colonial, rarely filamentous; amoeboid stages in some forms.
Flagella	Two unequal (one Flimmer, usually the longer), or one (Flimmer), anterior; in some uniflagellate forms a second present as a stump detectable only with the electron microscope.

Although the Chrysophyta are strikingly different in color (the cells usually contain two golden-brown chloroplasts), the members of this division show many growth forms resembling those found in the Chlorophyta. The highest level of organization attained, however, is only the simple filament, suggesting that evolution, although parallel, has also been very much slower than in the green algae.

The Chrysophyta are of considerable interest from the physiological and biochemical points of view, and they are often important components of plankton.

The flagellate habit is represented by a number of species, often referred to as the chrysomonads, some of which are able to alter their shapes and method of locomotion. *Ochromonas*, for example, is commonly pear-shaped, with two unequal flagella (one up to six times the length of the other) emerging from a depression at the anterior end (Fig. 4.1a). Sometimes, however, it may lose this shape and produce narrow outgrowths (*rhizopodia*), the flagella being lost. This amoeboid phase may also pass over temporarily into a thick-walled cyst. *Ochromonas* is also nutritionally versatile and at least one species is capable of feeding on unicellular blue-green algae.

In the common planktonic *Mallomonas* (Fig. 4.1b), the pectinaceous wall is covered with imbricating siliceous scales, some of which bear delicate hinged needles, often as long as the cell itself, possibly assisting flotation. The curious Silicoflagellineae (**Dictyochales**) are also placed with the Chrysophyta. These are uniflagellate planktonic organisms furnished with an intracellular siliceous skeleton. The cell has little or no wall,

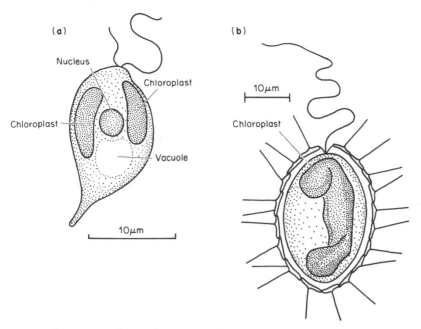

Figure 4.1. (a) *Ochromonas.* (b) *Mallomonas.* **After Scagel, Bandoni, Rouse, Schofield, Stein, and Taylor. 1966.** *An Evolutionary Survey of the Plant Kingdom.* **Wadsworth, Belmont.)**

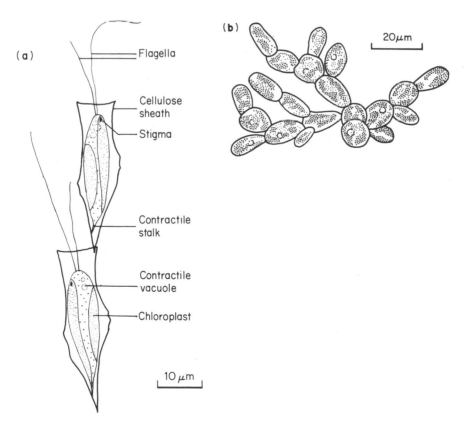

Figure 4.2. (a) *Dinobryon cylindricum.* **(After West and Fritsch. 1927.** *A Treatise on the British Freshwater Algae.* **Cambridge University Press.) (b)** *Phaeothamnion.*

and produces rhizopodia. Although the silicoflagellates are now rare, their skeletons are well known as fossils from the Tertiary onwards, indicating former abundance.

The longer flagellum in *Ochromonas* is furnished with mastigonemes which themselves bear minute filaments. Electron microscopy has shown that, after beginning their assembly in the perinuclear space, the mastigonemes are transferred to Golgi vesicles, where the filaments are added. They then pass, still in vesicles, to the site of emergence of the flagellum and appear in two rows on the flagellar surface.

A number of palmelloid and coccoid forms are commonly encountered in the field. *Synura* forms a free-floating colony in the plankton of drinking-water reservoirs. The individual cells are similar to *Mallomonas*, but are held together in a gelatinous mucilage. *Hydrurus*, which forms brownish layers on rocks in alpine streams, is similar but the mucilaginous matrix branches irregularly. In *Dinobryon* (Fig. 4.2a), frequent in both fresh water and the sea, the cells are elongate, and each is enclosed in a cellulose sheath. The sheaths are commonly connected head to tail.

The only filamentous form of note is *Phaeothamnion* (Fig 4.2b), found on rocks in flowing streams. The club-shaped cells form branching filaments which readily disarticulate when disturbed.

The reproduction of the Chrysophyta is predominantly asexual, longitudinal fission being characteristic of the unicellular flagellates. In *Dinobryon*, one product of the division moves out of the sheath and forms a new sheath attached to the rim of the old, accounting for the characteristic form of the colony. Copulation of flagellated cells has been observed in some instances, but little is known of life cycles.

XANTHOPHYTA

Habitat	Aquatic (freshwater), damp soil.
Pigments	Chlorophylls a, c, (e); β-carotene; heteroxanthin, diadinoxanthin, diatoxanthin, vaucheriaxanthin.
Food reserves	Fats, chrysolaminarin.
Cell wall components	Cellulose, hemicellulose.
Reproduction	Asexual and sexual, in some forms oogamous.
Growth forms	Flagellate, coccoid, colonial, filamentous, siphonaceous.
Flagella	Typically two, strikingly unequal, the larger usually Flimmer.

In this division, the amount of chlorophyll c is again often small. Chlorophyll e, reported to occur in some members, is now regarded as possibly a breakdown product of chlorophylls a or c. The absence of fucoxanthin from the plastids causes the Xanthophyta to be yellow-green in appearance compared with the Chrysophyta. Despite the relatively few representatives, the division again shows a series of growth forms paralleling those of the Chlorophyta. In the Xanthophyta, the organization attained exceeds that in the Chrysophyta. *Vaucheria*, for example, has a well-established siphonaceous habit and complex oogamy.

The inequality of the flagella and their insertion to one side of the apex are distinguishing features of this division, accounting for the earlier name Heterokontae. The cell walls in many forms show a tendency to fall into two halves. The chloroplasts, in contrast to those of the Chlorophyta, will often turn blue in dilute hydrochloric acid.

Representative of the simplest flagellate forms is *Heterochloris*, and of the coccoid *Chlorobotrys* (Fig. 4.3a), common in bog and fen pools. *Chlorobotrys* has a spherical cell resembling *Chlorella*. Several cells are usually held together in mucilage forming a free-floating colony.

Of the filamentous forms, *Tribonema* (Fig. 4.3b) is the most notable. The short, unbranched threads are encountered in fresh water and on damp earth, usually in base-rich situations. The cell wall is clearly made up of two overlapping halves. Consequently, the filament, when disrupted, breaks up into a number of pieces H-shaped in optical section, closely resembling the situation in the green alga *Microspora*. It is the H-shaped piece rather than the complete cell which is the basic unit of the filament so far as the wall is concerned. At cell division, instead of the usual septum being formed, a new H-shaped piece is produced at the center of the parent cell. Internally, the cell organelles are disposed in much the same way as they are in *Microspora*, except that several lenticular plastids occupy the peripheral cytoplasm, instead of a single bandlike plastid. Although the cells normally have a single nucleus, there is evidence in some species of multinucleate cells, perhaps indicating a transition to a coenocytic condition.

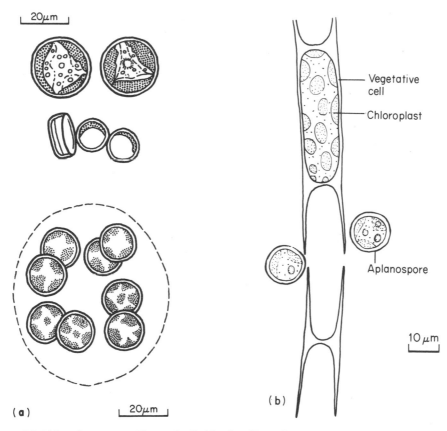

Figure 4.3. (a) *Chlorobotrys* sp. *Above,* **individual cells and germinating cyst;** *below,* **aggregate of cells held together in mucilage. (b)** *Tribonema bombicinum.* **(After West and Fritsch. 1927.** *A Treatise on the British Freshwater Algae.* **Cambridge University Press.)**

The simplest of the siphonaceous forms is *Botrydium* (Fig. 4.4), superficially similar to *Protosiphon*, but differing in its forked rhizoid and its ability to divide vegetatively. The multinucleate cells, reaching up to 2 mm (0.08 in.) in diameter and containing numerous disk-shaped chloroplasts, are frequently encountered on damp mud. *Vaucheria* (Fig. 4.5) is filamentous and branched, the single cell containing numerous nuclei and chloroplasts. The chloroplasts move in relation to light, the greatest exposure occurring at low irradiances. The action spectrum indicates that the photoreceptor is probably a

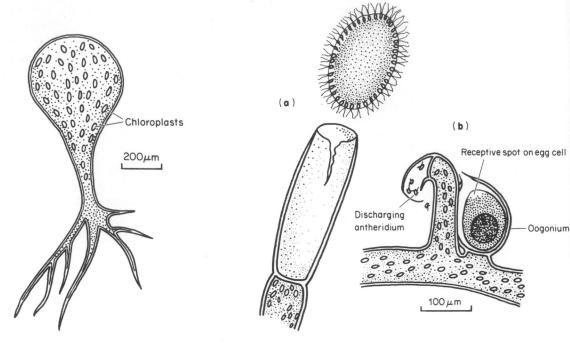

Figure 4.4. *Boytrydium granulatum.*

Figure 4.5. *Vaucheria.* **Reproductive organs. (a) Release of zoospore. (b) Gametangia.**

flavoprotein. Various species are found in fresh water and on damp earth, some becoming greenhouse pests. In terrestrial situations, colorless rhizoids attach the thallus to the substratum. Some species become encrusted with calcium carbonate and contribute to the formation of tufa.

Asexual reproduction in *Botrydium* is by heterokont zoospores generated and released when the parent plant is flooded. In dry conditions, aplanospores may be produced, or even cysts, from the rhizoidal regions. *Tribonema* reproduces vegetatively by fragmentation of the filament and asexually by the production of biflagellate zoospores. These are produced singly or in pairs in the parent cells and released by separation of the two halves. In place of zoospores, the cells of the filament may sometimes give rise to thick-walled aplanospores. As these germinate, the protoplast enlarges and causes the wall to fall into two pieces revealing its structural similarity to the walls of the filament. Asexual reproduction in *Vaucheria* occurs frequently, and involves the formation of a peculiar zoospore (Fig. 4.5). The apical region of a filament, rich in oil droplets and plastids, is cut off by a transverse septum. The many nuclei within this segment migrate to the periphery of the protoplast, and opposite each nucleus emerges a pair of flagella. The tip of the filament becomes gelatinous, and a multiflagellate zoospore, which can be regarded as a mass of unseparated uninucleate zoospores, escapes. On coming to rest, the flagella disappear, a wall is secreted, and the zoospore begins to germinate. Two or three filaments emerge from the spore, one of which usually acts as a holdfast. In some species, particularly the terrestrial, aplanospores lacking flagella are produced in a similar manner. The spore is liberated by breakdown of the sporangial wall, and germination may begin before liberation is complete.

Sexual reproduction involving isogamy has been reported in *Tribonema* and *Botrydium*, but reaches its fullest expression in the oogamous *Vaucheria* (Fig. 4.5). The onset of the reproductive phase is indicated by the formation of antheridia. The tip of a lateral branch, in which many nuclei but few plastids accumulate, is cut off by a transverse septum. The cytoplasm becomes apportioned between the nuclei, and each uninucleate protoplast then differentiates into a biflagellate gamete. Meanwhile the female organ develops, in many species next to the male on the same filament. Although the oogonium is initially multinucleate at maturity, it contains a single uninucleate egg cell. When the egg is mature, a short beak develops asymmetrically at the apex and the egg becomes accessible to the male gametes. Only one penetrates the egg cell, and its nucleus comes to lie by that of the egg, not fusing with it until it has swollen to an approximately equal volume. The zygote becomes surrounded by a highly impervious wall and may remain dormant for several months. Germination, which results in the formation of a new filament, is preceded by meiosis.

Since, in most species, fusion takes place between gametes from the same filament, sexual reproduction in *Vaucheria* provides for little more than the interpolation of a resting stage in the life cycle. A few species, however, are dioecious, allowing the possibility of genetic recombination.

If, as seems probable, the Xanthophyta are the result of a line of evolution from some ancestral motile form, parallel to that of the Chlorophyta, the problem immediately arises of why the rate of evolution has been so much slower. It is possible that relatively infrequent sexual reproduction has been a continous feature of the Xanthophyta. This would have limited the opportunity for genetic recombination and the appearance of new forms. Biochemical features associated with the different pigmentation may also have restrained variation and evolutionary success.

HAPTOPHYTA

Habitat	Mostly marine.
Pigments	Chlorophylls a, c; β-carotene; fucoxanthin.
Food reserves	Fats, chrysolaminarin.
Cell wall components	Cellulose, hemicellulose, calcium carbonate.
Reproduction	Asexual, but life cycles little known.
Growth forms	Unicellular.
Flagella	Typically two, mostly smooth, usually with a third (haptonema) lying between them.

The Haptophyta (Prymnesiophyta) are distinguished by the haptonema, a flagellumlike organ of variable length. In some forms, it is longer than the flagella. Although often coiled up like a proboscis (Fig. 4.6), it may, in some instances, serve to attach the cells to the substratum. In transverse section, the haptonema shows three concentric membranes enclosing a number of microtubules. The outer membrane is continuous with the plasmalemma, and the inner pair represents a tubular extension of the endoplasmic reticulum.

The Haptophyta are largely marine and planktonic. Some (particularly *Chrysochromulina* spp.) are responsible for toxic "blooms" causing the widespread death of fish. The remarkable coccolithophorids also belong to the Haptophyta. The walls of these organisms are closely beset with delicately sculptured plates of calcium carbonate. These are first laid down in Golgi vesicles and then excreted to the exterior. Deposits of these scales from dead organisms (coccoliths) form the principal component of calcareous rocks such as the chalk of the Cretaceous period (Fig. 4.7). It has been estimated that chalk contains up to 800×10^6 coccoliths in 1 cm^3. Coccolithophorids rich in fats may also have contributed to oil deposits dating from Cretaceous times.

Asexual reproduction probably predominates in the Haptophyta, motile stages often alternating with nonmotile. Similar nonmotile forms, particularly of the coccolithophorids, often have dissimilar motile stages, a feature which makes a knowledge of the life cycle essential for classification.

In some instances, there is evidence of a sexual cycle, a coccolith-bearing diploid stage alternating with a small haploid filamentous plant. This may reproduce itself by zoospores or produce gametes.

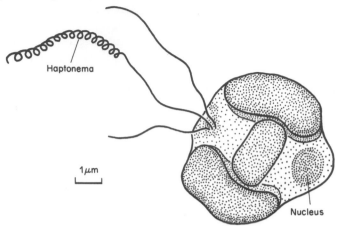

Figure 4.6. *Chrysochromulina* **showing the haptonema between the flagella. The two parietal chloroplasts, central chrysolaminarin (leucosin) granule, and nucleus are also indicated. (After Scagel, Bandoni, Rouse, Schofield, Stein, and Taylor. 1966.** *An Evolutionary Survey of the Plant Kingdom.* **Wadsworth, Belmont.)**

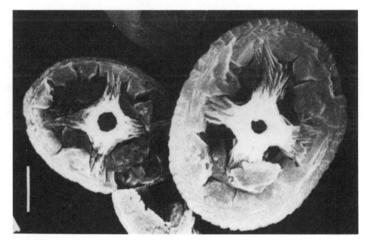

Figure 4.7. *Eiffellithus eximus,* **a coccolith from the Cretaceous. (Micrograph provided by A. J. J. Rees.) Scale bar 2 μm.**

DINOPHYTA

Habitat	Aquatic, frequently planktonic.
Pigments	Chlorophylls a, c; β-carotene; peridinin, dinoxanthin, other xanthoplylls less prominent; in a few phycobilins.
Food reserves	Starch, fat.
Cell wall components	Cellulose, hemicellulose.
Reproduction	Asexual, rarely sexual (probably isogamous).
Growth forms	Mostly unicellular, a few coccoid or filamentous.
Flagella	Two, apical or lateral, in some forms absent.

The Dinophyta (Pyrrophyta) are predominantly unicellular planktonic organisms, with walls characteristically furnished with longitudinal and transverse furrows. Although there are ultrastructural characteristics common to the whole division, two classes are usually recognized: the Desmophyceae and the Dinophyceae. The Desmophyceae are notable for having cell walls consisting of two halves, each one resembling a watch crystal. The edges are sometimes extended as elaborate borders, possibly assisting flotation. The flagella originate at the anterior end of the cell.

In the Dinophyceae, which include the dinoflagellates, the cell walls are often reinforced with hexagonal polysaccharide plates (Fig. 4.8), forming the *theca*. The arrangement of the flagella in the dinoflagellates is unique. The attachment is lateral. One flagellum, usually smooth, is directed towards the posterior and its undulations push the cell forward. The other flagellum, usually Flimmer, lies in the transverse groove. This flagellum may serve to stabilize the cell and improve the effectiveness of the extended flagellum. When the transverse flagellum is inactive, the cell rotates in one direction, and when it is active, in the converse.

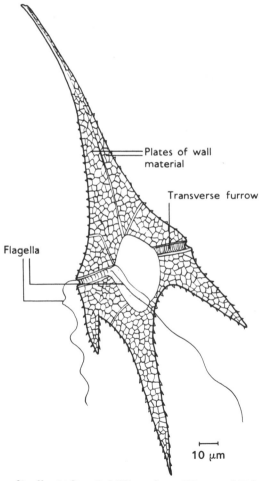

Plates of wall material

Transverse furrow

Flagella

10 μm

Figure 4.8. *Ceratium hirundinella.* **(After Schilling, from West and Fritsch. 1927.** *A Treatise on the British Freshwater Algae.* **Cambridge University Press.)**

The cells of the Dinophyta contain one or more chloroplasts, often dark brown in color as a consequence of a large proportion of the pigment peridinin. The chloroplast envelope consists of three membranes, a feature found elsewhere only in the Euglenophyta. Also as in the Euglenophyta, the thylakoids are characteristically in stacks of three. A stigma is often present in the motile species and in zoospores. Many species lack pigmented plastids and live heterotrophically, while others are fully pigmented and form symbiotic associations with invertebrates (e.g., *Amphidium klebsii*, which is also free-living, in the flatworm *Amphiscolops*). Some parasitic dinoflagellates are amoeboid. The nuclei of the dinoflagellates are peculiar in containing chromosomes which remain spiralized throughout the nuclear cycle and which, like those of prokaryotes, are deficient in histone proteins. Another prokaryote feature of some dinoflagellates is that DNA synthesis is continuous during population growth and shows no well-defined pauses in relation to nuclear division. Many dinoflagellates emit **trichocysts** when irritated. These are narrow proteinaceous needles or coils, partly crystalline, which first appear in Golgi bodies and emerge to the exterior through fine pores. This protozoan feature is encountered sporadically among the unicellular flagellate algae.

Many dinoflagellates are involved in blooms, such as the "red tides" of the Gulf of Mexico. Millions of cells may be present in a single liter of surface water. Some of these blooms are toxic, causing widespread death of fish and invertebrates. They may also be accumulated by shellfish and affect human beings.

A few dinoflagellates, principally of warmer seas, exhibit bioluminescence. *Noctiluca* is a well-known example. It is naked and wholly heterotrophic, feeding on blooms of other dinoflagellates. Some dinoflagellates emit flashes of light when cultures are shaken.

Although motile, unicellular forms predominate in the Dinophyta, a few planktonic representatives are coccoid, and in the genera *Dinothrix* (known only from marine aquaria) and *Dinoclonium* the cells, which lack flagella, are aggregated in short sparsely branched filaments.

In the motile species, asexual reproduction is only by simple vegetative division. In armored forms, the theca may either be shed and then reformed around the daughter cells, or split into two and then restored after division. Zoospores are produced by the coccoid

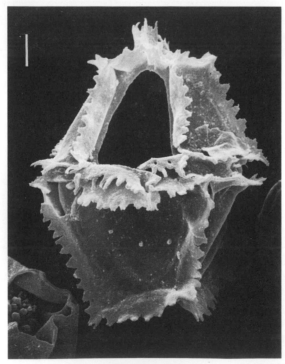

Figure 4.9. *Gonyaulacysta helicoidea*, **a dinoflagellate cyst from the Upper Jurassic. (Micrograph provided by J. Davy.) Scale bar 5 μm.**

and filamentous forms. Many dinoflagellates produce thick-walled cysts in unfavorable conditions. Although many sink and are lost, others remain sufficiently near the surface to germinate when conditions improve. These provide the "seed" for ensuing blooms. The cysts are distinctive and are well known as fossils from the Triassic onwards (Fig. 4.9).

Sexual reproduction, both isogamous and anisogamous, is known in the dinoflagellates. The zygotes either germinate immediately or form resting cysts. In most instances, meiosis occurs on germination; *Noctiluca* is unusual in being diploid.

The Dinophyta are clearly a highly specialized assemblage. The three membranes bounding the chloroplast suggest an endosymbiotic origin (as in the Euglenophyta and Cryptophyta). In one instance (i.e., *Peridinium*), it appears that the original endosymbiont may have been a chrysophyte alga.

BACILLARIOPHYTA

Habitat	Aquatic and terrestrial.
Pigments	Chlorophylls a, c; β- (and possibly ε-) carotene; diatoxanthin, diadinoxanthin, fucoxanthin
Food reserves	Fat, chrysolaminarin.
Cell wall components	Hemicellulose, silica.
Reproduction	Asexual and sexual (ansiogamous and oogamous).
Growth forms	Unicellular, colonial.
Flagella	One Flimmer (in some forms lacking the central pair of fibrils); anterior. (Present only in the male gametes of some members.)

The Bacillariophyta, commonly known as the diatoms, are a large division of microscopic algae with intricately sculptured, siliceous walls. The bilateral or radial symmetry of the cells and the regularity of the delicate markings on their walls (Fig 4.10) make the diatoms very beautiful microscopic objects, rivaling even the desmids.

The diatoms are frequent in fresh water and marine phytoplankton and are therefore of economic importance in the management of fisheries. Some species are benthic and live upon rocks or sand, or as epiphytes. The siliceous walls resist dissolution and decay after the death of the organism and accumulate as fossils on beds of lakes and seas (Fig. 4.10). Since the composition of the diatom flora is dependent upon pH of the water, floristic analysis of lake deposits can give an indication of environmental trends over a long period of time. Huge deposits of these "diatomaceous earths" (Kieselguhr) are known from the Tertiary era, and some are mined for use as abrasives, filters, and the refractory linings of furnaces.

Much study has been given to the cell wall of diatoms, and it is principally upon its characteristics that the classification rests. Although the several thousand species of diatoms occur in many different shapes, the walls of all consist basically of two parts which overlap like the halves of a Petri dish (Fig 4.11). Consequently, the appearance of the cells from the side (girdle view) is different from that from above (valve view). Two principal orders, the **Centrales** (centric diatoms) and **Pennales** (pennate diatoms), are distinguished by the difference in valve view. In the Centrales, the valve is circular, triangular, or polygonal, and in the Pennales elongate. The decoration of the valve of the centric diatoms follows a radial pattern, while in the pennate diatoms the lines of ornamentation lie parallel and transverse to the plane of symmetry of the valve (Fig. 4.12b). The intersection of the plane of symmetry and the surface of the valve is usually marked by a longitudinal fissure (raphe) at which there may be contact between protoplast and medium. Elsewhere, the sculpturing of the wall depends upon inequalities in the thickness of the outer siliceous layer. In some species, pores may be present in the thinner regions.

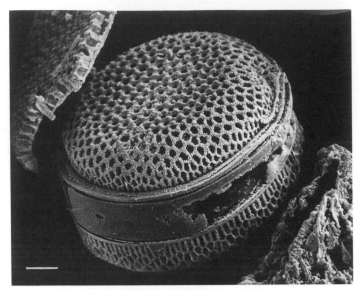

Figure 4.10. A fossil species of *Coscinodiscus* **from a diatomaceous earth. (Micrograph provided by J. Davy.) Scale bar 10 μm.**

Coscinodiscus, a genus containing both living species widely distributed in the oceans and fossil species (Fig. 4.11), is representative of the free-floating centric diatoms. *Melosira*, found in both freshwater and marine plankton, is colonial. The cylindrical cells remain attached valve to valve, forming many-celled filaments (Fig. 4.12a). In *Chaetoceras*, also filamentous, the valves have spine-like outgrowths which project from the sides of the filament and possibly assist flotation. The common free-floating pennate diatom of fresh water is *Pinnularia* (Fig. 4.12b). A similar marine organism is *Pleurosigma*, whose walls are so delicately and regulary sculptured that the valves are used as specimens to test the resolution of light microscopes. Colonial forms are *Asterionella* (Fig. 4.13), frequent in Windermere, in which the cells are grouped in starlike aggregates, and the marine *Licmophora*, where the cells form treelike colonies. Many pennate diatoms are able to perform small jerky movements, often returning irregularly to their starting point. Although the mechanism of this movement is not entirely clear, it is believed to result from jets of mucilage which issue from the raphe.

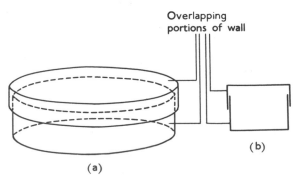

Figure 4.11. Diagrammatic representation of the arrangement of the two halves of the wall in a diatom. (a) Three-dimensional view. (b) Transverse section.

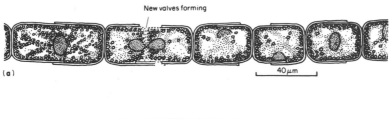

New valves forming

(a)

40μm

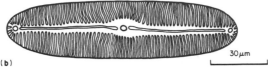

(b)

30μm

Figure 4.12. (a) *Melosira.* **(b)** *Pinnularia.* **([b] After Esser. 1976.** *Kryptogamen.* **Springer, Berlin.)**

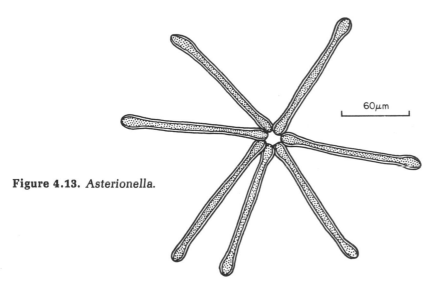

60μm

Figure 4.13. *Asterionella.*

Within the cell, the cytoplasm forms a thin layer containing one or more chloroplasts, usually brownish in color because of the accessory pigments, surrounding a large central vacuole. In the centric diatoms, the nucleus is also held in this peripheral layer, but, in the pennate diatoms, the nucleus is suspended in the center of the cell by a band of cytoplasm which traverses the vacuole. A few diatoms lack pigmentation and are heterotrophic (e.g., *Nitzschia*).

Simple asexual reproduction of diatoms takes place by cell division. This involves the separation of the two halves of the wall, one half going to each daughter cell. Each daughter then immediately secretes another half wall, thus completing its envelope. Since the mature walls are quite inflexible, and each new half fits into a pre-existing one in the way the bottom half of a Petri dish fits into the top half, the size of the cells inevitably decreases with successive vegetative divisions. Eventually, a regeneration process occurs which is normally sexual. Conditions of depleted nutrition, however, may lead to the formation of resting spores. After two divisions, the second of which is unequal, the larger cell becomes the spore. The wall surrounding the spore is of the normal kind, but thicker.

Sexual reproduction in the centric diatoms is oogamous and meiosis occurs in the production of gametes. The antheridial cell produces four spermatozoids, each of which has a single Flimmer flagellum. An unusual feature of the flagellum is the absence of the two central tubules from the axoneme. In the cell giving rise to the oogonium, one daughter nucleus of the first meiotic division degenerates, and also one of the second. The remaining nucleus, together with the whole of the cytoplasm, becomes the egg. After fertilization, which may be within or without the oogonium, the zygote surrounds itself with an extensible polysaccharide wall. It then grows to as much as four times its original size, forming an auxospore. Normal siliceous walls are then laid down, and a renewed sequence of vegetative divisions with progressively diminishing cell size begins.

Reproduction of the pennate diatoms is isogamous. Two vegetative cells become held together in mucilage. The two nuclei divide meiotically, but only two products survive as gametes in each instance. The valves then open and the gametes (which lack flagella) fuse. As in the centric diatoms, the zygotes grow into auxospores before the beginning of vegetative divisions. Sometimes two gametes of the same parent may fuse (autogamy), and formation of auxospores without sexual fusion (apogamy) has also been observed.

Although the Bacillariophyta share basic biochemical and metabolic features with the Xanthophyta and Chrysophyta (particularly with some chrysomonads), they are clearly highly specialized. If a common ancestor existed, the Bacillariophyta probably diverged at an early stage. On the other hand, there are no well-established records of Paleozoic diatoms, and they do not appear in quantity until the later Mesozoic and Tertiary eras. The earliest forms are centric, suggesting that the flagellate male gamete is a primitive feature.

CRYPTOPHYTA

Habitat	Aquatic.
Pigments	Chlorophylls a, c; α-(and rarely ε-) carotene; alloxanthin, other xanthophylls less prominent; biliproteins (phycoerythrin, phycocyanin).
Food reserves	Starch.
Cell wall components	Naked, but a pellicle may be present consisting of proteinaceous plates.
Reproduction	Asexual, sexual doubtful.
Growth forms	Flagellate, palmelloid stages occur.
Flagella	Two, anterior, Flimmer, slightly unequal.

This division includes a few species, referred to collectively as cryptomonads, occasional in freshwater and marine phytoplankton. They are unicellular and biflagellate; in some forms, one of the flagella is directed towards the posterior. The flagella are inserted at one side of a depression ("gullet") lined with refractive granules (Fig. 4.14). These are the site of "ejectosomes" which function in a manner analogous to that of the trichocysts of the dinoflagellates. The color of the cryptomonads is variable depending upon the proportion of accessory pigments in the chloroplasts. These are commonly cup-shaped and bounded not only with the usual two membranes, but also with a sheet of endoplasmic reticulum ("chloroplast endoplasmic reticulum") continuous with the outer membrane of the nuclear envelope. The thylakoids are widely spaced, but less so than those of rhodophyte plastids. The biliproteins are not in the form of discrete phycobilisomes. A body termed a nucleomorph lies between the chloroplast and the surrounding endoplasmic reticulum. It is bounded by two membranes and contains DNA, and may represent the nucleus of an ancestral endosymbiont. Starch, although closely associated with the pyrenoid, lies outside the chloroplast but within the chloroplast endoplasmic reticulum (Fig. 4.14).

Although clearly properly placed in the "$a + c$" algae, the wider relationships of the Cryptophyta are obscure. The only well-established means of reproduction is by fission.

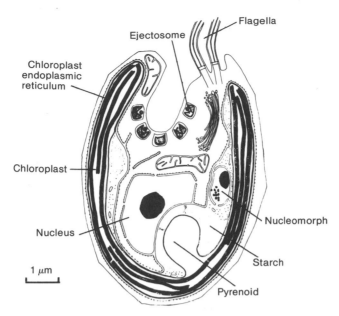

Figure 4.14. *Cryptomonas.* **Longitudinal section based on electron micrographs, partly diagrammatic. (After Gibbs, from Bold and Wynne. 1985.** *Introduction to the Algae.* **Prentice Hall, Englewood Cliffs. NJ.)**

PHAEOPHYTA

Habitat	Predominently marine.
Pigments	Chlorophylls *a*, *c*; β-carotene; violaxanthin, fucoxanthin.
Food reserves	Polyols (e.g., mannitol), laminarin.
Cell wall components	Cellulose, hemicellulose, sulfated polysaccharides.
Reproduction	Asexual and sexual (oogamous).
Growth forms	Filamentous, parenchymatous.
Flagella	Two unequal, one smooth and one Flimmer (sometimes spiny); anterior or lateral.

Of the many genera of the Phaeophyta (brown algae), only very few are freshwater, the remainder being seaweeds whose macroscopic flattened thalli are familiar inhabitants of the intertidal regions just beneath the low-tide mark, some being found solely in midocean.

The vegetative organization of the Phaeophyta surpasses that of any of the algae so far considered. The simplest thallus, consisting of branching filaments, is heterotrichous, and thus resembles the most complex found in the Chlorophyta. This morphologically advanced state is also reflected in the manner of reproduction. In the Phaeophyta, oogamy is the general rule, and the alternation of phases in the life cycle is developed to the point where gametophyte and sporophyte begin to diverge morphologically. In the epiphytic *Desmarestia antarctica* the gametophyte is an endophyte growing in the medulla of the host (*Curdiea,* Rhodophyta).

The number of chloroplasts in a cell varies from one to many, but is usually constant within a genus. As in the Cryptophyta, a chloroplast endoplasmic reticulum is present. The thylakoids are commonly in groups of three. Pyrenoids, where present, are often conspicuous, but are absent from the plastids of many advanced genera.

The Phaeophyta are regarded as a single class (Phaeophyceae) divided into at least twelve orders, but representatives of only the five more common ones will be considered here. They will, nevertheless, fully illustrate the probable evolutionary trends in the development of the morphology and reproductive cycles of the brown algae.

The **Ectocarpales** are represented by *Ectocarpus* (Fig. 4.15), mostly small heterotrichous plants resembling in habit the green alga *Cladophora,* are common along the Atlantic coast of North America and in the colder seas of the Northern Hemisphere. Around ports and docks some forms have become tolerant of metal pollution, particularly of copper. In the erect filaments, the division of the cells, which contain a few branching, ribbonlike chloroplasts, is limited to well-defined intercalary regions. This method of growth, frequent in the Ectocarpales, is termed "trichothallic." The common *Pylaiella* is very similar to *Ectocarpus* but tidal action often causes the axes to roll together, forming a characteristic cablelike bundle.

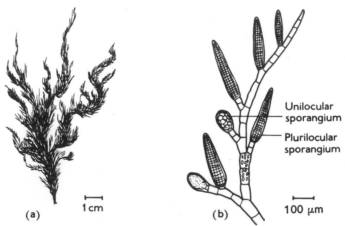

Figure 4.15. *Ectocarpus confervoides.* **(a) Habit. (b) Sporangia. (After Newton. 1937. A Hand-book of the British Seaweeds. British Museum Publications.)**

Other members of the order are more complex. *Ascocyclus,* for example, has an elaborate prostrate system, and in this respect resembles the green alga *Coleochaete.* The gelatinous cushionlike thallus of *Leathesia,* although appearing parenchymatous, is formed by the adhesion of filaments. A truly parenchymatous condition is reached in *Punctaria* (Fig. 4.16). Here cell division takes place in various directions and results in a leaflike thallus closely resembling *Ulva.*

The life cycle of *Ectocarpus* is well known. In favorable environmental conditions the two phases, except in respect of the reproductive organs, are morphologically identical, but in colder regions the haploid plant may develop little or even not at all. Reproduction of the diploid plant takes place solely by zoospores, but these are produced in two kinds of sporangia, termed respectively "plurilocular" and "unilocular" (Fig. 4.15). The plurilocular sporangia arise from cells which undergo repeated transverse and subsequently longitudinal division, giving rise to the spindle-shaped groups of small, more or less cubical compartments, each of which at maturity contains a single biflagellate zoospore. The plurilocular sporangia are commonly either lateral or terminate short lateral branches. Dehiscence is by an apical pore. Dissolution of the partitions allows the zoospores to escape. After a motile phase, the zoospores settle and yield diploid plants identical with their parent.

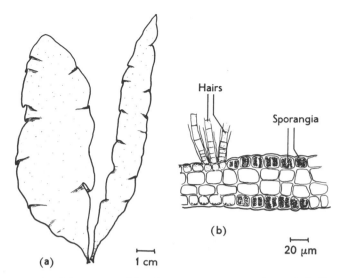

Figure 4.16. *Punctaria latifolia.* **(a) Habit. (b) Transverse section of thallus with sporangia. (After Newton. 1937.** *A Handbook of the British Seaweeds.* **British Museum Publications.)**

The cytology of the developing unilocular sporangium is more complex. The nucleus of the initial cell is conspicuously large, and its first divisions are probably meiotic. Although the cytoplasm becomes multinucleate, no walls are formed, and it is only at maturity that the protoplast becomes divided into uninucleate portions. Each portion differentiates into a biflagellate zoospore, typically bean-shaped. All are eventually released at an apical pore. Although these zoospores behave as those produced in plurilocular sporangia, they yield only haploid plants. The haploid plants bear only plurilocular sporangia. These produce motile cells which behave either as zoospores, reproducing the haploid phase asexually, or as gametes. The zygote germinates without meiosis or production of zoospores, and grows directly into a diploid plant.

Aberrations of this cycle are frequent. In cold conditions the products of the unilocular sporangia may function directly as gametes. The gametophytic plant is then eliminated from the cycle. Unmated gametes have occasionally been observed to give rise to plants bearing unilocular sporangia. This is an example of apogamy, the generation of a plant with the characteristics of the sporophyte from an unfertilized gamete.

Although most species of *Ectocarpus* are isogamous, two sexes are distinguished by the behavior of the gametes. Those regarded as female are the first to settle, and fertilization follows. The male and female gametes are borne on different plants. Anisogamy has been detected in the more complex Ectocarpales. A pheromone has been demonstrated in *Ectocarpus* and has been identified chemically as an unsaturated hydrocarbon (ectocarpene).

In some epiphytic species of *Ectocarpus* and *Pylaiella,* the gametophytes and sporophytes, although similar, grow on different plants. The sporophytes of *Pylaiella littoralis,* for example, grow on *Fucus* and the gametophytes on *Ascophyllum.*

Representative of the small order **Sphacelariales,** close to the Ectocarpales, is *Sphacelaria,* a heterotrichous alga ranging from polar to tropical seas. The upright filaments terminate in an apical cell conspicuous for its large size and brown pigmentation. Investigations have shown that photoperiod and temperature affect the form of the life cycle. Above 20°C (68°F) there is a tendency for the reproduction of both the gametophyte and sporophyte to take place solely by multicellular propagules.

Heterotrichous growth is also characteristic of the small order **Cutleriales.** The principal genus, *Cutleria,* is represented by a number of species in warmer seas of the Northern Hemisphere. There is a well-developed life cycle (Fig. 4.17), but only one component of the heterotrichous system is fully represented in each phase, so the cycle is markedly heteromorphic.

The sporophyte, which displays only the prostrate component, grows as an incrustation on rocks and was originally regarded as a distinct genus, *Aglaozonia.* The unilocular sporangia, in which, as usual, meiosis occurs, are superficial and yield haploid biflagellate zoospores. On settling, these become attached to the substratum by a suckerlike holdfast.

The gametophytes, which arise from the attached zoospores, consist principally of an erect filamentous system. The filaments run parallel and adhere to each other for much of their length, forming a pseudoparenchymatous thallus. For a short distance at the apex, however, the filaments are free, and here there is distinct trichothallic growth. The gametophytic phase of the Mediterranean *Cutleria multifida* is dioecious (Fig. 4.17). The male and female gametes are both biflagellate and motile, but the male are minute compared with the female.

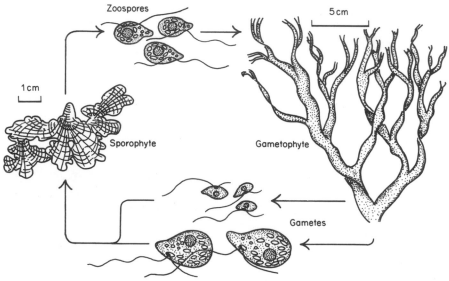

Figure 4.17. *Cutleria multifida.* **Life cycle, showing both anisogamy and the striking morphological difference between gametophyte and sporophyte. The gametophyte, although only one is shown, is dioecious. There is segregation for sex in the formation of zoospores. Gametes and zoospores diagrammatic showing relative sizes. (After Esser. 1976.** *Kryptogamen.* **Springer, Berlin.)**

The **Laminariales,** known collectively as kelps, include *Macrocystis* and *Nereocystis,* the largest known algae, together with a number of species harvested commercially as sources of the mucopolysaccharide algin. This yields alginic acid, widely used as an emulsifying agent in the food and paint industries, and in the processing of rubber. Specimens of *Macrocystis,* off the coast of southern Australia, grow by as much as 1 m (39.4 in.) a day, probably the fastest growth rate in the Plant Kingdom.

The Laminariales are usually found below low-water mark, but a few are regularly exposed at low tide. An example of the latter is the striking *Postelsia* of the Pacific coast of North America, a species with an arboreal habit, suggestive of a miniature coconut palm.

The vegetative structure of the kelps. In *Laminaria,* which may be taken as representative of the order, the large thallus is differentiated into holdfast, stipe, and blade (Fig. 4.18). Abrasion by tides and turbulence continually wears away the end of the blade, but the loss is made good by continued growth from a meristematic region at the base. A trans-

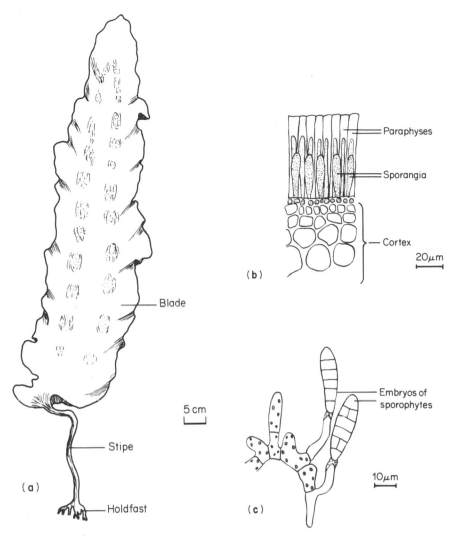

Figure 4.18. *Laminaria.* **(a)** *L. saccharina.* **Habit. (b)** *L. cloustoni.* **Transverse section of fertile region of lamina. (After Newton. 1937.** *A Handbook of the British Seaweeds.* **British Museum Publications.) (c)** *L. digitata.* **Portion of female gametophyte. (After Sauvageau, from Fritsch. 1945.** *The Structure and Reproduction of the Algae, 2.* **Cambridge University Press.)**

verse section of the tough, pliable stipe reveals three anatomically distinct zones. Chloroplasts are confined to an outer zone (**meristoderm**), covered by a layer of protective mucilage. Within the meristoderm, in which some cell division persists indefinitely, is a zone of paler, elongated cells forming a well-defined cortex. Towards the center of the stipe, the longitudinal walls become increasingly gelatinous and, at the center itself, is a mucilaginous matrix (**medulla**) containing intertwined and branching filaments. The innermost part of the cortex and the medulla also contain columns of elongated cells broadened at each end and hence referred to as "trumpet hyphae." The transverse walls of these cells are often perforated by groups of pits and are frequently callosed. These are features of the sieve plates of higher plant phloem. There is now evidence that the trumpet hyphae of the Laminariales have a phloemlike function and take part in the basipetal transport of assimilates.

Reproductive areas, referred to as **sori,** develop on the blades at certain times of the year. The sori consist of many unilocular sporangia interspersed with thick, sterile, protective hairs (paraphyses) (Fig. 4.18b). Meiosis occurs in the development of the sporangia, which eventually yield haploid zoospores. These, in turn, develop into haploid gametophytic plants, much smaller than, and totally different in morphology from, the highly organized sporophytes (Fig. 4.18c). The gametophytes are dioecious and, since it has been shown in a number of instances that spores giving rise to male and female gametophytes are produced in equal numbers in a sporangium, it appears that sex determination is genotypic.

Although the gametophytes show a tendency towards heterotrichous growth, any cell seems capable of yielding a gametangium. Oogamy is fully developed, the oogonium producing a single egg which escapes at maturity through a pore at the apex of the cell. The egg, in at least one species bearing a pair of caducous flagella, does not become free, but remains seated in a cup formed by the thickened margins of the pore. The male plant produces a number of terminal antheridia, from each of which is liberated a single spermatozoid with two lateral flagella. A substance has been isolated from female gametophytes which stimulates the release of spermatozoids and subsequently acts as a pheromone. Following fertilization, the zygote secretes an external membrane and develops into a new sporophyte without any resting period. The young embryo may remain attached to the female gametophyte for a short period, but it is doubtful whether this represents anything more than purely physical adhesion.

There is no specialized asexual reproduction of either the haploid or diploid generations.

The **Fucales,** the various species of which are known as "wracks," are probably the most familiar of all seaweeds, particularly in the British Isles. The intertidal (littoral) regions of rocky shores often show a number of distinct horizontal bands, each consisting of an almost pure stand of a member of the Fucales. *Fucus* (Fig. 4.19), *Pelvetia,* and *Ascophyllum* are genera frequent in these habitats. Desiccation of the thalli during expo-

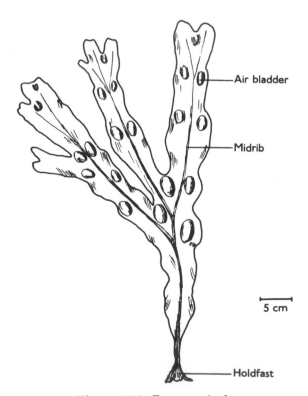

Figure 4.19. *Fucus vesiculosus.*

sure to air is prevented by the secretion of mucilage, and photosynthesis probably continues during low tide. A few Fucales grown in deeper water, and some (e.g., *Sargassum*) are free-floating. Although principally confined to the warmer regions of the great oceans, a Japanese species of *Sargassum* has become naturalized along the English Channel. Air bladders in which the proportions of oxygen and nitrogen often differ from those of normal air are frequently present in the thalli of the Fucales.

The thallus of *Fucus* is a much smaller structure than that of *Laminaria,* rarely exceeding 50 cm (20 in.) in length, but the differentiation into holdfast and blade is still evident. The flattened thallus is dichotomously branched and each branch grows from an apical meristem. The cells in the apical regions produce polyphenols. These are released on to the surface by discharge of vacuoles (exocytosis). The polyphenols are probably bacteriocidal and account for the freedom of the growing regions from adhering microorganisms. A distinct midrib usually lies at the center of each segment, branching in register with the thallus. Small cavities containing tufts of hairs (**cryptostomata**) are scattered in the marginal wings. The wings in the older parts are gradually worn away by the action of the sea until, in the region of the holdfast, only the midrib remains, giving the impression of a short stipe (Fig. 4.19). Small, densely branched ecotypes of *Fucus* often form swards in salt marshes subject to periodic inundation.

Both monoecious and dioecious species of *Fucus* are known. In both, the reproductive structures form at the apices and these in turn become swollen with mucilage.

Microscopic examination reveals flask-shaped invaginations (conceptacles) in this swollen region, some of which contain female gametangia and others male, both interspersed with paraphyses (Figs. 4.20 and 4.21). Meiosis occurs during gametogenesis. The gametes are liberated at high tide. Following discharge of the gametes the swollen apices are often shed.

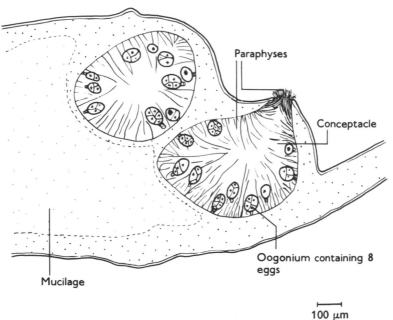

Figure 4.20. *Fucus vesiculosus.* **Transverse section of thallus with female conceptacles.**

As in the Laminariales, oogamy is fully developed, but in *Fucus,* the eight eggs produced in an oogonium become quite free and drift passively in the sea. The spermatozoids, each of which has two lateral flagella, are attracted to the eggs. The pheromone, named fucoserratene, is a hydrocarbon. Although hybrids have been described, fertilization in experimental conditions is species-specific and appears to

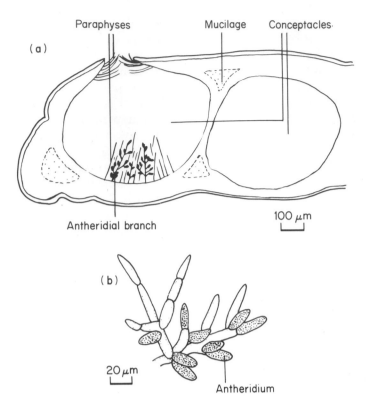

Figure 4.21. *Fucus vesiculosus.* **(a) Transverse section of thallus with male conceptacles. (b) Portion of antheridial branch. ([b] After Thuret, from Fritsch. 1945.** *The Structure and Reproduction of the Algae, 2.* **Cambridge University Press.)**

depend on a glycoprotein recognition system on the surfaces of the egg and spermatozoid. Eggs treated with fucose-binding lectins, for example, are fertilized far less readily, presumably because of masking of the binding sites. Investigations with monoclonal antibodies to sperm surface antigens show that these binding sites are species-specific.

Following fertilization the zygote continues to drift and secretes a mucilaginous envelope. It eventually settles, becomes anchored by the mucilage, and germinates. The first division of the zygote establishes the polarity of all subsequent growth, since one daughter cell gives rise to the rhizoid (and ultimately the holdfast) and the other to the blade. It has been found experimentally that, if the zygote is allowed to germinate in an environment that is not uniform, such as in a gradient of light, temperature or hydrogen ion concentration, the dividing wall always forms transversely to the direction of the gradient. The subsequent behavior of the daughter cells depends upon the nature of the gradient. With a gradient of temperature, for example, the cell on the warmer side yields the rhizoid. In nature, of the two cells formed in the first division of the zygote, the rhizoid usually develops from that in greater contact with the substratum. It seems likely that the orientation of this first division under natural conditions is determined by a combination of the environmental gradients that have been found effective in influencing the polarity in laboratory cultures.

Once polarity is established, a minute electric current passes through the embryo parallel to the axis, the rhizoid pole becoming increasingly negative. This current naturally influences the distribution of the ions and charged molecules within the embryo, and may promote regional differentiation. In *Pelvetia* (closely allied to *Fucus*) calcium ions move into the tip of the growing rhizoid.

Two different interpretations have been made of sexual reproduction in *Fucus*. One brings it into line with that of the Laminariales, interpreting the gametangia as homologous with unilocular sporangia, the gametophyte being reduced to nothing more than a gamete. The other draws an analogy with sexual reproduction in animals such as *Homo sapiens,* where there is no question of an independent haploid phase.

There is no specialized asexual reproduction in the Fucales, but fragments of thalli may regenerate in favorable conditions to yield independent plants. The free-floating species of *Sargassum* and the salt marsh forms of *Fucus* reproduce solely in this manner.

The thallus of *Dictyota* (**Dictyotales**), a widely distributed genus, is flattened and dichotomously branched, each branch growing from a conspicuous apical cell (Fig. 4.22). The thallus is only three cells thick, lacking a midrib. The two outer layers of cells are assimilatory, and the central, consisting of large cells, may act as a storage region.

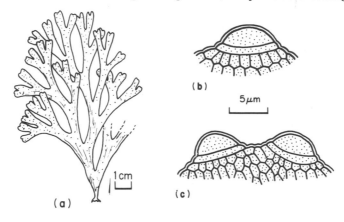

Figure 4.22. *Dictyota.* **(a) Habit. (b) Apical cell. (c) Incipient dichotomy of the thallus.**

The reproduction of *Dictyota* is notable in involving an isomorphic alternation of generations. Unilocular sporangia arise scattered or in groups over the surface of the diploid plant, and each yields, as a consequence of meiosis, four nonflagellate tetraspores (Fig. 4.23), a feature which distinguishes the Dictyotales from all other brown algae. During meiosis in *Dictyota* there is a 2 : 2 segregation for sex, and consequently male and female gametophytes are present in approximately equal numbers. In other genera the gametophytic phase is commonly monoecious.

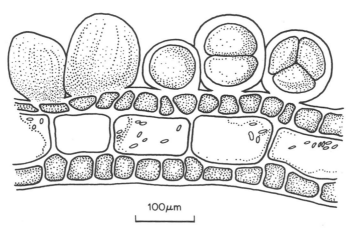

100μm

Figure 4.23. *Dictyota.* **Sporophyte producing tetrasporangia.**

The oogonia and antheridia are produced in groups (sori) on the surface of the thallus. Each oogonium produces a single egg which, like that of *Fucus,* drifts passively in the water when released. The antheridia are plurilocular, and the sorus is surrounded by a ring of sterile cells. The spermatozoids (antherozoids) have a single lateral flagellum (Flimmer), but electron microscopy has shown that a second is present as a rudiment enclosed in the cytoplasm, recalling the situation in some uniflagellate Chrysophyta. Spermatozoids are attracted to the egg by a pheromone, now known to be a complex hydrocarbon similar to ectocarpene. The zygote germinates very soon after fertilization.

There is no specialized asexual reproduction of either phase of the life cycle.

The relationships of the Phaeophyta. The Phaeophyta are a circumscribed division of the algae and little can be said of their wider relationship since no forms are known simpler than the Ectocarpales. They presumably arose from some primitive flagellate ancestor and proceeded to exploit a particular kind of pigmentation and metabolism that proved especially satisfactory in marine environments. There is no evidence that the Phaeophyta have ever contained forms becoming adapted, as some of the Chlorophyta, to terrestrial life. The few varieties of the Fucales (e.g., *Fucus vesiculosus* var. *muscoides*), all of limited reproductive capacity, which occur in saltmarshes mixed with halophytic flowering plants, appear to be instances of specialization without any far-reaching significance. It is noteworthy that photosynthesis in *Laminaria,* which often dominates a canopy, becomes saturated only at high irradiances. This is a characteristic of "sun plants" of land vegetation and is perhaps the consequence of an analogous specialization of the photosynthetic system.

EVOLUTIONARY TRENDS WITHIN THE ALGAE

From the foregoing survey of the principal features and interrelationships of the algae, we can now proceed to a consideration of the evolutionary and morphological trends displayed by the algal kind of organization as a whole.

Aquatic Habitat and Evolutionary Change

A point of general significance in relation to the evolution of the algae arises from their predominantly aquatic habitat. One of the main factors influencing evolution within the Plant Kingdom has undoubtedly been environmental change. Nevertheless, aquatic plants, particularly those that are marine, are to some extent protected from such change, at least in a catastrophic form. The volume of the sea, in particular, is so vast that changes in such features as salinity and temperature must of necessity be gradual. Algae, therefore, have exploited to an extent greater than that of any other component of the world's vegetation, an environment which demands only comparatively slow adaptation to changing conditions. This possibly accounts for the persistence of numerous states of algal organization intermediate between the simplest unicellular and the complex multicellular, and of simple isomorphic life cycles (as in *Ulva* and *Ectocarpus*) without spores or zygotes suited to withstand unfavorable periods.

Closer inspection of individual groups reveals, of course, that continuity of structure and pattern is, to some extent, illusory. The Charales among the Chlorophyta and the Florideophycidae among the Rhodophyta provide examples of the numerous groups of living algae which lack close relatives, and which consequently provide few indications of the paths along which they may have evolved.

The Antiquity of the Algae

Geological evidence undoubtedly points to algae being an extremely ancient form of life. Calcareous nodules (stromatolites) structurally similar to those produced by some living blue-green algae (e.g., *Lithomyxa*), are known from beds of pre-Cambrian age, possibly over 2500 million years old. Filamentous remains, containing organic matter and suggestive of blue-green or even green algae, have been found in pre-Cambrian rocks not less than 2000 million years old. Possible coccoid forms of blue-green algae have been described from even older rocks. The record of algal life begins about 3000 million years ago.

Naturally fossilization has favored those algae with calcarious skeletons. Remains of the calcified Siphonales, for example, are found in the mid-Proterozoic (Table 1.1), and the record continues (with evidence of increasing complexity) into the Mesozoic. Similarly, the calcareous Rhodophyta appeared in the Cambrian, and were well represented by the Ordovician. The fossil record is thus to some extent biased. Nevertheless, current techniques are able to extract identifiable remains of coccoid and filamentous eukaryotic algae from rocks almost as ancient as those bearing calcified forms.

The fossil record, although admittedly still very fragmentary, fully supports the view that prokaryotes preceded eukaryotes (perhaps by as much as 1000 million years), and that simple unicellular and filamentous algae preceded the pseudoparenchymatous and parenchymatous forms encountered today in the three major divisions: Rhodophyta, Chlorophyta, and Phaeophyta.

Evolution of the Vegetative Thallus

If the increase in morphological complexity from flagellate unicell, through coccoid, filamentous, and pseudoparenchymatous states to large parenchymatous forms such as *Macrocystis* is an evolutionary progression, the problem remains of what has caused and directed this progression. The cause presumably lies in the mutability characteristic of all life, and the direction is no doubt a consequence of natural selection. The nature of selection in an aquatic environment with its uniformity in space and time is, however, little understood. This is particularly true of the marine environment and of planktonic algae. Consequently discussion of algal evolution involves considerable conjecture.

Nevertheless, reasoned speculation is not to be discouraged, and algae undoubtedly show many features of evolutionary significance which merit consideration. Among the flagellate forms, for example, it is striking that the development of motile colonies and aggregate organisms has not proceeded beyond *Volvox*. Presumably, with a diameter exceeding about 1 mm (0.04 in.) the *Volvox* system would become physically unstable, and the coordination of the thallus impossible. If the earliest forms were indeed flagellate, the evolution of a sedentary form from a motile would result in the energy that would otherwise be expended in swimming becoming available for growth and reproduction. The enhanced reproductive capacity would result in proportionately greater numbers and provide the opportunity for the establishment of a line of sedentary organisms.

The tendency for daughter cells to remain united appears to be a basic one and would account for the occurrence of some kind of colonial or multicellular forms in all the divisions of the algae. In the major divisions, we must assume that the advantages of association, possibly again residing in metabolic and reproductive efficiency, led to the elaboration of multicellular thalli in which there gradually appeared divisions of labor among the cells. In a multicellular form such as *Ulva*, all the cells, except possibly the basal anchor, divide and liberate reproductive bodies simultaneously. There is no somatic tissue, and the individual is destroyed in reproduction. A form in which the parent persists through several reproductive phases clearly has an advantage in a situation (as might arise with prolonged turbulence) where conditions become temporarily intolerable for the reproductive bodies, but remain tolerable for the mature plant.

The evolution of complex thalli in which reproduction was confined to special areas or branch systems would thus be favored. The further opportunity would then arise for various parts of the somatic tissue to become specialized and assist either in the support or protection of the reproductive structures, as in the Floridophycidae and many other algae, or in the exploitation of particular habitats, as with the air bladders of many Fucales. Nevertheless, the extent to which the algal thallus can diversify is clearly limited by its anatomical simplicity. This and the uniformity of the aquatic environment can be held responsible for the similar forms which have evolved in the different divisions (cf. *Ulva, Punctaria,* and *Porphyra*). Parallelisms in reproductive regions are also notable (e.g., the trichogyne of the oogonia in the Coleochaetales and Florideophycidae).

Evolution of Sexual Reproduction

With regard to sexual reproduction, it seems beyond doubt that oogamy has evolved from isogamy. With forms inhabiting moving water, isogamous reproduction must be extremely wasteful, and there are evident advantages if one gamete remains relatively stationary, especially if it secretes chemotactic pheromones causing the male gametes to accumulate around it. Moreover, a zygote which begins life with a copious food reserve has a better chance of survival than one with little. Increasing size, however, severely limits mobility, so again advantages can be envisaged in a situation in which one gamete, the male, remains small and motile, and the other, the female, loses motility and specializes in the laying down of food reserves.

A nonmotile zygote may, of course, be disadvantageous if it settles in a situation unfavorable for the plant. This is compensated for in those algae, such as *Coleochaete,* in which the zygote germinates to produce zoospores. Another development, possibly limiting the wastage of zygotes, is shown in the prostrate species of *Coleochaete* where the zygote germinates while still embedded in the gametophyte, foreshadowing a feature of the archegoniate plants.

Life Cycles of Algae

Sexual reproduction inevitably involves a cyclic alternation between a haploid and a diploid condition. The simplest life cycle found among the algae is that in which the diploid condition, generated by the fusion of morphologically identical gametes, is represented only by the zygote (Fig. 4.24a). Meiosis occurs on germination of the zygote, thereby initiating a new haploid (gametophytic) phase. A cycle of this kind, termed "haplontic," is frequently encountered in the simpler algae, and is typical of the filamentous Chlorophyta. Here, however, it may have been retained and developed as an adaptation facilitating survival in unfavorable conditions, since only rarely does meiosis immediately follow syngamy, and the intervening diploid phase is often spent as a thick-walled, resting zygote.

Closely related to the haplontic cycle, and possibly evolved from it by a delaying of meiosis, is that in which the zygote generates a multicellular, diploid (sporophytic) phase. This eventually produces reproductive bodies, almost always zoospores, by a process involving meiosis. The haploid condition is thus restored and the cycle recommences (Fig. 4.24b). Where, as in *Ulva,* the haploid and diploid plants are morphologically similar, the alternation of phases is isomorphic. Where the phases are morphologically different, as, to take an extreme example, in *Cutleria,* the cycle is heteromorphic. The evidence available does not warrant any general conclusion about whether isomorphic life cycles have evolved from hetermorphic, or the converse. Life cycles, in which both haploid and diploid phases are present as multicellular individuals, are termed "diplohaplontic."

Both haplontic and diplohaplontic cycles show, in addition to isogamy, various states of anisogamy and oogamy. The complexities of the diplohaplontic cycles of the Florideophycidae of the Rhodophyta are unique among the algae. The early appearance of the coralline rhodophytes in the fossil record suggests, nevertheless, that these complexities had been acquired by Paleozoic times.

A third kind of life cycle, similar to that of most animals, occurs in the diatoms, Fucales, and in several other isolated instances throughout the algae. The diploid condition predominates, and the haploid is represented only by the gametes, meiosis occurring during gametogenesis (Fig. 4.24c). Again, the evidence does not allow any general conclusion about how this kind of cycle, termed "diplontic," originated. As with the haplontic, it may in certain instances have selective value. In *Fucus,* for example, as compared with *Laminaria,* it is perhaps an advantage to have the gametophytic phase, possibly vulnerable to the vicissitudes of inter-tidal life, reduced to the unilocular sporangia and gametes.

It should be noted that, throughout the algae, variations of the regular cycle have been observed both *in vivo* and *in vitro.* Gametes, for example, may occasionally develop apogamously into sporophytelike plants without fertilization. The converse, the development of gametophytelike plants from zygotes (apospory) is known in experimental conditions (*Oedogonium*), but not apparently in nature.

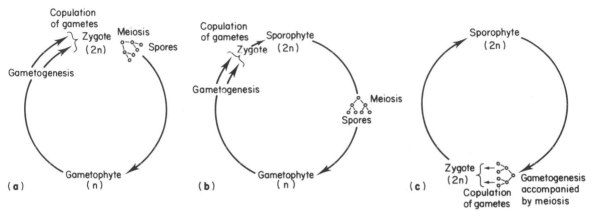

Figure 4.24. **The three kinds of life cycles found in the algae. (a) Haplontic. (b) Diplohaplontic. (c) Diplontic.**

The Importance of the Algae in the Evolution of Plants

Thus in the algae, the simplest of phototrophic organisms, a number of progressions, which can be regarded as representing channels of evolution, can be traced from unicellular to parenchymatous organization, from isogamy to oogamy, and from simple to elaborate life cycles. These all represent steps of fundamental importance in the evolution of plant life, and this fully justifies giving the algae considerable attention in any review of the Plant Kingdom. At its upper limit, the structural and reproductive complexity of an alga approaches that of a land plant. *Laminaria,* for example, possesses a thallus with not only morphological but also anatomical differentiation, that in the stipe being accompanied by a phloemlike function. Additionally, the life cycle is strikingly heteromorphic, and the sporophyte is attached in the early stages of its development to the gametophyte. Considered solely in terms of the level of organization, the transition from an advanced alga such as *Laminaria* to an archegoniate or even angiospermous land plant is small when compared with the evolution of that alga from a unicellular flagellate.

__ 5 __
THE SUBKINGDOM BRYOPHYTA
(Mosses and Liverworts)

GENERAL FEATURES OF THE BRYOPHYTES

The mosses and liverworts, although vegetatively dissimilar, are classified together as the Bryophyta. Because of their distinctive features they are treated as comprising a subkingdom of the Plant Kingdom, on a rank equal to that of the Algae and Tracheophyta (Table 1.2). Three classes are recognized, namely, Hepaticae, Anthocerotae, and Musci. They comprise a single division.

BRYOPHYTA

Habitat	Mainly terrestrial.
Plastid pigments	Chlorophylls a, b; β-carotene; xanthophyll (lutein).
Food reserves	Starch, to a lesser extent fats.
Cell wall components	Cellulose, hemicelluloses.
Reproduction	Heteromorphic life cycle, the gametophytic phase normally the more conspicuous, and the sporophytic determinate and partly dependent upon it. Sex organs with a jacket of sterile cells. Zooidogamous, spermatozoids with two whiplash flagella. Embryogeny exoscopic. Sporophyte producing nonmotile, cutinized spores, in some species with heavily thickened and sculptured walls, usually all of one size (homospory). Vegetative propagation of the gametophyte by fragmentation or specialized gemmae.
Growth forms of gametophyte	Thallus flattened, with some internal differentiation, or consisting of a main axis with leafy appendages.

Although the simplest terrestrial plants, the bryophytes in some parts of the world form a conspicuous component of the vegetation. They are, for example, prominent among the epiphytes of mist forests of tropical mountains. Some species form dense communities submerged in Antarctic lakes. Vast bogs in the Northern Hemisphere has been built up largely by the growth of the moss *Sphagnum*. The dead stems and leaves accumulating below the growing surface become consolidated to peat, often several meters in depth. In some places, peat is an important fuel and, in granulated form, is widely used in horticulture as a source of humus.

The largest bryophyte, *Dawsonia,* is a tufted moss of swampy places in southeastern Asia and Australasia. Individual stems of this genus may reach or even exceed a meter in length, but dimensions of this order are quite atypical of bryophytes. Most are lowly plants and many are inconspicuous and not easily seen without a hand lens. The cellular differentiation within the larger bryophytes is greater than in the algae, but lacks the complexity found in vascular plants. Among the mosses, it reaches its maximum in *Polytrichum,* and among the liverworts in *Symphogyna.* In both genera, groups of elongated cells occur in the central region of the stem. In form they approach the tracheids of higher plants, but detailed patterns of thickening are absent.

Although a cuticle has been demonstrated in some bryophytes, in general they are little able to resist desiccation and are consequently found principally in damp and humid localities. The exceptions, such as species of the liverwort *Metzgeria* and of the moss *Orthotrichum,* common on tree trunks, are able, by virtue of colloids present in the cell sap, to retain sufficient water to maintain life even in prolonged dryness.

The morphology of the bryophytes is on the whole more complex than that of the algae. In the mosses, for example, the mature thallus regularly takes the form of a leaf-bearing stem. On the other hand, the immature gametophyte (**protonema**) of many mosses closely resembles a heterotrichous green alga and may, as, for example, in the moss *Pogonatum aloides,* persist for many months. Like the algae, the bryophytes produce no roots, although both the mature and immature forms of the thallus bear rhizoids. In a few forms, such as the thallose liverwort *Hymenophytum* and the moss *Polytrichum,* the aerial parts of the plant are continuous with subterranean creeping axes, superficially resembling the filiform rhizomes of the smaller filmy ferns. Some leafy liverworts also produce fine underground axes. These may descend to depths of 1 m (39.4 in.) or more in a peaty substratum and provide a means of vegetative propagation if the site is disturbed. So far as is known, only one bryophyte (*Cryptothallus mirabilis,* a thallose liverwort) is subterranean in habit and heterotrophic. The thallus lacks chlorophyll and provides the only instance known of the occurrence of leucoplasts in the somatic cells of a liverwort.

Of the two phases in the bryophyte life cycle, the haploid gametophyte is usually persistent and the sporophyte (**sporogonium**) of limited life span. The sporophyte consists of little more than a capsule, a stalk (**seta**), and a basal foot inserted into the gametophyte. The developing sporophyte is wholly or partly parasitic upon the gametophyte, drawing its sustenance by way of the foot. The apposed walls of the gametophytic and sporophytic cells at the insertion of the foot are frequently labyrinthine, either on both sides of the boundary or only on one side. In *Polytrichum,* for example, the labyrinthine walls are confined to the sporophytic face. The passage of materials into the sporophyte is entirely through the apoplast, since in all archegoniates the boundary between the two phases lacks plasmodesmata. In *Polytrichum* (one of the few instances where the function of the so-called "transfer cells" has been investigated) the cells of the sporophytic face develop a transmembrane potential of the order of −200 mV. This promotes the absorption of amino acids leaked into the apoplast from the gametophyte. In other respects, the labyrinthine walls may act as a filter restricting the transfer of informational molecules. This would ensure the continued separation of the distinctive developments and functions of the two phases. The formation of the labyrinthine walls possibly arises from an interaction of the gametophytic and sporophytic systems (analogous to an antigen-antibody reaction), resulting in a perturbation of the cell wall metabolism and the consequent architecture.

Sexual reproduction of the bryophytes is zooidogamous, and all depend upon free water for fertilization. The mobility of the male gametes is very limited, restricting the spread of genetic variation within a population. Dispersal depends upon the spores, which are light and readily wind-borne. They contain chloroplasts and are surrounded by a two-layered wall: an inner **intine** and an outer **exine.** Sporopollenin, often deposited in a pattern characteristic of the species, is a feature of the exine. This complex polymer protects the spores from rapid dehydration. Although much of the sporopollenin is produced by the spore itself, some may come from degenerating cells adjacent to the sporocytes and be added to the exine in the final stages of differentiation. These degenerating cells do not, however, form a distinct tissue with a clear function as in most tracheophytes.

Many bryophytes produce capsules only rarely or (so far as known) not at all. These species evidently rely largely or exclusively on asexual means of reproduction.

The features used in the classification of the bryophytes are, (1) the nature of the thallus and, where present, of the leaves; (2) the extent of the development of the juvenile phase of the gametophyte; and (3) the presence or absence of an opening mechanism in the capsule.

Hepaticae (Liverworts)

Despite the diversity of the liverworts, there is little doubt that they form a natural group. The protonemal phase of the gametophyte is usually ill-defined and the mature thallus almost always shows recognizable dorsiventrality. A characteristic feature of many liverworts is the presence of oil bodies, often of complex morphology, in the cytoplasm. These possibly render the tissues unpalatable to grazing insects. Liverworts also produce bacteriocidal substances, some of which may have commercial value.

The antheridia of liverworts break open irregularly instead of by a distinct cap cell. The capsule of the sporophyte matures before the elongation of its stalk, the converse of the situation in the mosses.

Of the seven orders of Hepaticae, the common **Marchantiales, Jungermanniales,** and **Metzgeriales** will be considered in some detail, and the small orders **Sphaerocarpales** and **Calobryales** mentioned on account of special features which claim attention.

The thallus in the Marchantiales. The Marchantiales are exclusively thallose. Although some species are simple in appearance and structure, internal organization more complex than that found in any other thallose liverwort is also encountered in this order.

The thallus of *Marchantia* itself (Fig. 5.1), frequent on damp soil and areas of burnt ground, is dichotomously branched, with a thickened central rib, and the surface divided into hexagonal areas visible with the naked eye. On examination with a hand lens, a pore can be observed at the center of each hexagonal area, which, in transverse section, is seen to consist of an air chamber containing photosynthetic tissue (Fig. 5.2). The pore, like the

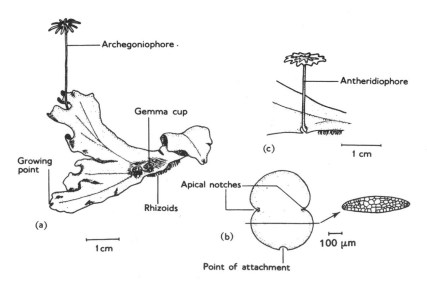

Figure 5.1. *Marchantia polymorpha.* **(a) Habit of female plant. (b) Structure of gemma. (c) Antheridiophore.**

stoma of a higher plant, probably allows aeration of the thallus with the minimum dehydration, but is incapable of significant change in its aperture. Below the chlorophyllous tissue is a compact body of cells largely lacking chloroplasts. The lower side of the thallus bears up to eight rows of scales and unicellular rhizoids, the walls of some of which bear peglike invaginations ("peg rhizoids"). The growth of the thallus is sensitive to photoperiod and ceases in long days. This is accompanied by an accumulation of lunularic acid, an endogenous growth regulator possibly found in liverworts generally.

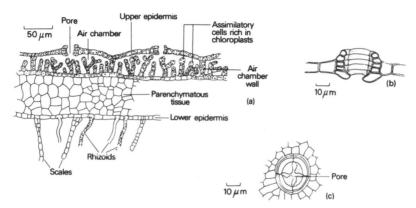

Figure 5.2. *Marchantia polymorpha.* **(a) Transverse section of thallus. (b) Transverse section of pore. (c) Surface view of pore.**

Among other members of the Marchantiales showing a chambered thallus are *Conocephalum*, the thallus of which yields a characteristic fragrance when crushed, and *Preissia*. In *Conocephalum* the pore is much simpler than in *Marchantia*, but in *Preissia* the pore is not only complex but its aperture also diminishes with falling humidity. This feature may help to limit loss of water from the chamber and account for *Preissia* being able to tolerate drier habitats than *Marchantia*. The midrib of the thallus of *Preissia* also contains conspicuous elongated fibrous cells.

Riccia represents the simplest kind of structure found in the order. The lower part of the thallus is again a compact colorless tissue, but the upper part consists of columns of chlorophyllous cells, separated by narrow air channels. The upper cells of the columns are colorless and fit closely together, leaving no distinguishable pores. In *Riccia fluitans* (Fig. 5.9) the narrow, dichotomously branched, floating thallus is divided almost entirely into air chambers separated by partitions one cell thick.

Reproduction in the Marchantiales. In *Marchantia*, which is dioecious, sexual reproduction is induced by increasing day length. The male and female gametes are produced on upright, umbrella-shaped structures termed **antheridiophores** and **archegoniophores** respectively (and **gametangiophores** collectively) (Fig. 5.1a and c). Both structures develop from one-half of a dichotomy and are therefore homologous with a bifurcation of the thallus. Their morphological nature is clearly demonstrated by the rhizoids which grow down grooves in the stalks (Figs. 5.3 and 5.4) and by the characteristic photosynthetic chambers which develop in the caps of the mature gametangiophores. The female organs (archegonia) arise in radial rows on the upper surface of the cap. During the maturation of the archegonia, the cap grows more above than below, with the result that the archegonia become transferred to the lower surface. In the mature archegoniophore (Fig. 5.5) each row of archegonia is separated from its neighbors by a curtainlike out-growth, termed a **perichaetium**. In addition, sterile processes emerge radially from the upper surface of the cap between the rows of archegonia, giving the whole its familiar stellate appearance (Fig. 5.3).

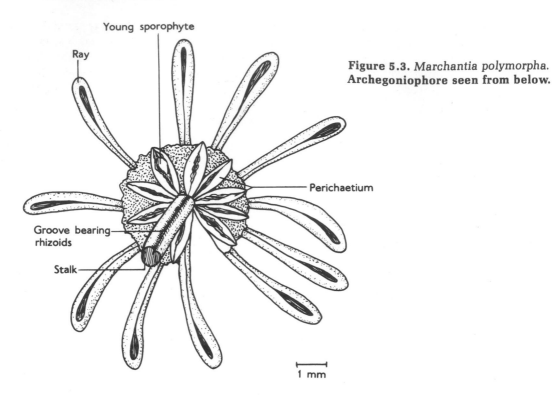

Ray

Young sporophyte

Perichaetium

Groove bearing rhizoids

Stalk

1 mm

Figure 5.3. *Marchantia polymorpha.*
Archegoniophore seen from below.

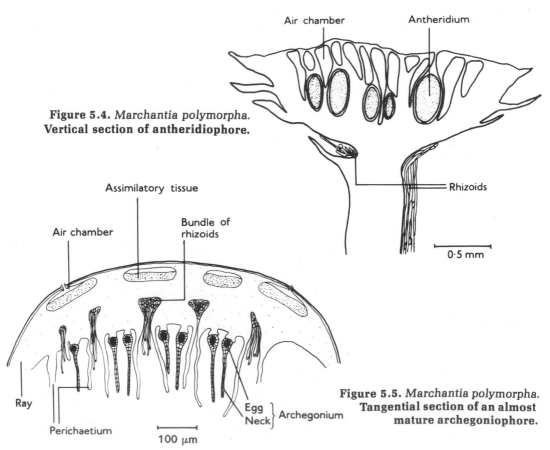

Air chamber

Antheridium

Rhizoids

0·5 mm

Figure 5.4. *Marchantia polymorpha.*
Vertical section of antheridiophore.

Assimilatory tissue

Bundle of rhizoids

Air chamber

Ray

Perichaetium

Egg
Neck } Archegonium

100 µm

Figure 5.5. *Marchantia polymorpha.*
**Tangential section of an almost
mature archegoniophore.**

The archegonium, as always in the bryophytes, is formed in its upper parts by a simple column of cells and has a strikingly long neck (Fig. 5.6). The egg lies at the dilated base of the ventral canal and, when mature, appears to be suspended in fluid. It is surmounted by a ventral canal cell and a number of neck canal cells. These degenerate at maturity, and their products, when hydrated, give rise to a mucilage through which the spermatozoids swim to reach the egg.

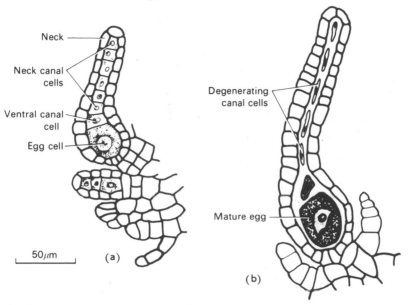

Figure 5.6. *Marchantia polymorpha.* **(a) Archegonium before breakdown of neck canal cells. (b) Mature archegonium.**

The antheridiophore lacks the complexity of the archegoniophore, being merely an elevated cap (Fig. 5.4), with the antheridia on the upper surface. Although superficial in origin, the mature antheridia are sunk in pits, each opening to the exterior by a narrow pore. Each antheridium is borne on a short stalk and bounded by a single layer of sterile jacket cells. When mature, it contains a mass of small cubical cells (**spermatocytes** or **antherocytes**) in each of which differentiates a biflagellate spermatozoid. This is in the form of a single gyre of a helix and consists of a head piece, an elongated nucleus, and a cytoplasmic tail (Fig. 5.7). The head piece includes the two posteriorly directed flagella, one inserted behind the other, and a large mitochondrion. Close to this is a lamellate body, the multilayered structure. A ribbon of microtubules, lying at an angle of 45° to the plates of the multilayered structure, extends from the surface of this body and follows a helical path along the outside of the nucleus. The tail contains a large plastid, parts of which extend as a flap over the end of the nucleus, and a mitochondrion lying in a depression in the plastid. The differentiation of these highly specialized gametes from cells which are initially more or less isodiametric presents many problems of gene activation and cell mechanics. Particularly interesting is the state of the chromatin. Although fully condensed in the mature gamete, intermediate stages show coarse fibrils becoming oriented parallel to the longitudinal axis of the nucleus.

The mature antheridia open in moist conditions and the spermatocytes, in contact with a film of water, are rapidly dispersed, possibly a consequence of the high surface tension of the cell membrane. After a short time, the spermatocytes themselves break open and release the spermatozoids. For fertilization to be possible the male and female plants must be growing together. It seems likely that the gametangia become mature and fertilization occurs before elongation of the gametangiophores.

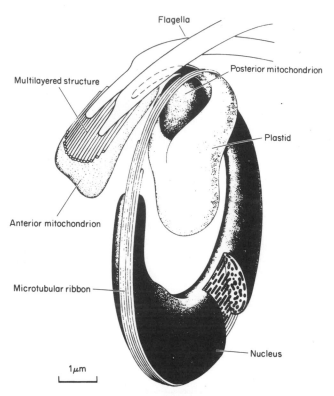

Flagella

Posterior mitochondrion

Multilayered structure

Plastid

Anterior mitochondrion

Microtubular ribbon

Nucleus

1μm

Figure 5.7. *Marchantia polymorpha.* **Diagram showing the disposition of the organelles in an almost fully differentiated spermatozoid. The nucleus is cut open to show the condensing chromatin. (After Z. B. Carothers. 1975.** *Biological Journal of the Linnean Society* **7, Supplement 1.) Scale approximate.**

Although detailed observations are few, germination of the zygote of *Marchantia* probably begins within 48 hours of fertilization. The first division is by a horizontal wall, transverse to the longitudinal axis of the archegonium. Since it is from the outer cell that the apex of the sporophyte arises, embryogenesis is said to be exoscopic. The products of the inner cell form the foot, by which the sporophyte remains anchored in the gametophyte. Continued growth and differentiation, which are dependent upon nutrients drawn from the gametophyte, lead to an embryonic sporophyte consisting of three distinct regions. At the summit is the immature capsule containing the sporogenous cells, below this is a short seta, and at the base the foot (Fig. 5.8).

At this stage, the young sporophyte is not only enclosed by the proliferated jacket cells of the archegonium, which form a *calyptra*, but is also surrounded by a further tubular outgrowth of the gametophyte called a **pseudoperianth** (or perigynium). Division of the sporogenous tissue (**archesporium**) inside the capsule remains mitotic until eventually spore mother cells (sporocytes) are formed. In these, the nucleus undergoes meiosis. The protoplast, without initial furrowing, is divided into four spores, each with a haploid nucleus. The spores of the tetrads separate in the capsule, become rounded in outline, and develop walls. Not all the cells inside the capsule become sporocytes; some (referred to as **elaters** because of their subsequent behavior) elongate and lay down spiral thickenings.

Elongation of the cells of the seta eventually causes the calyptra to rupture, and, once exposed to air, the single layer of cells surrounding the capsule soon bursts, so revealing the mass of yellow haploid spores. The loosening of this mass and the dispersal of the spores are now assisted by the contortions of the elaters. These contortions are

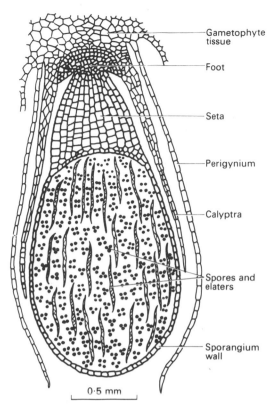

Figure 5.8. *Marchantia polymorpha.* **Longitudinal section of sporophyte rupturing the calyptra. Note the parallel alignment of the elaters. (After Parihar. 1967.** *Bryophyta.* **Allahabad.)**

caused by the spiral bands in their walls, presumably consisting of cellulose microfibrils, altering their curvature and pitch as they dry. In response to the strains generated in this way, the cell as a whole makes jerky twisting movements.

The spores germinate rapidly on a damp surface, giving rise to short, algalike filaments of cells. Division of the apical cell then ceases to be confined to one plane and subsequent growth leads to the mature form of the gametophyte.

In most capsules of *Marchantia,* a number of dyads can usually be found in addition to tetrads. The spores of the dyads are larger and contain an unreduced chromosome number. They yield diploid gametophytes. In *Marchantia polymorpha,* these are either purely male or purely female, but in some tropical species bisexual diploid forms are known.

No other genus has gametangiophores as elaborate as those of *Marchantia.* In *Conocephalum,* for example, the archegoniophore is a simple cap without emergent rays, but its surface has complex pores in contrast to the simple ones elsewhere. In *Riccia,* gametangiophores are entirely absent, both archegonia and antheridia merely lying at maturity in pits in the dorsal surface of the thallus. The sporophyte generation is again dependent on the gametophyte for nutrition, but at maturity it consists of only a sac of spores, with no seta or foot (Fig. 5.9). By the time the spores are mature, no living diploid tissue remains, and dispersal must await the decay of the gametophyte.

Asexual reproduction of the Marchantiales often follows from bifurcation of the thallus being accompanied by progressive decay of the older posterior region. In this way, an area becomes quite rapidly colonized by many, seemingly, individual plants. Additional to this, *Marchantia* has notably elegant means of asexual reproduction.

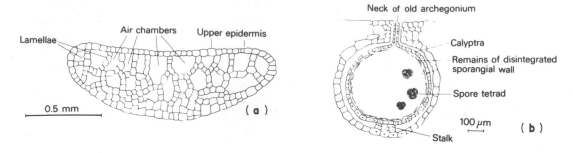

Figure 5.9. *Riccia fluitans.* **(a) Transverse section of thallus. (b) Thallus containing mature sporophyte.**

Multicellular bodies, called gemmae, develop inside cuplike growths on the upper surface of the thallus (Fig. 5.1a). Each gemma is slightly biconvex, with two diametrically opposed, marginal notches, each containing a small meristem (Fig. 5.1b). When mature, the gemmae become detached from the short stalk on which they are borne and are readily dispersed. Experiments have shown that the newly detached gemmae have no innate dorsiventral symmetry. This becomes fixed at germination by gradients of light, temperature, and other factors in the immediate environment. Each meristem grows out to form a new thallus and, finally, two individuals result from decay of the central portion.

Culture experiments have shown that short days promote the production of gemmae. The germination of the gemmae while in the cup is inhibited by growth-regulating substances diffusing basipetally from the apical meristem of the parent thallus.

Evolution of the Marchantiales. *Marchantia* seems to represent the highest level of organization achieved by a wholly thallose gametophyte. Are we therefore to regard the simple *Riccia* as a primitive Marchantialean plant, and *Marchantia* as an advanced form? Although this would appear plausible, some striking breeding experiments with *Marchantia* point in the other direction. A number of mutants were raised from species of *Marchantia* in culture and hybridized in various ways, with the result that a whole series of forms was obtained which reproduced features found in other genera of the order. The thallus of the var. *dumortieroides,* for example, lacks air chambers and resembles that of *Dumortiera,* a genus which, except for this feature, is close to *Marchantia.* Similarly the var. *riccioides* resembles *Riccia* in its narrow branching and the immersion of sex organs in the prostrate thallus. This reservoir of variation in *Marchantia* suggests that its evolutionary antecedents may have yielded the other genera of the order by a process of simplification. *Riccia* would then be regarded as a reduced form.

On the other hand, perhaps both *Marchantia* and *Riccia* should be regarded as evolved forms. The archegoniophores of *Marchantia,* which elevate the setaless capsule and facilitate the wide dispersal of the relatively thin-walled spores in air currents, can reasonably be regarded as an advantageous development. In *Riccia,* however, the spores are thick-walled and long-lived. Despite its apparent rudimentary sporophyte, *Riccia* is probably no less well adapted than *Marchantia,* but to a different, Mediterranean environment. Clearly identifiable remains of both *Marchantia*- and *Riccia*-like Marchantiales have been found in Upper Triassic coals of Sweden.

Possibly allied to the Marchantiales is the largest known thallose liverwort, *Monoclea,* of New Zealand. There are no specialized gametangiophores, but the manner of growth is similar to that of *Marchantia,* and there is no furrowing of the sporocyte protoplast preliminary to spore-formation. The thallus may reach a length of 25 cm (10 in.), and its branches a width of 5 cm (2 in.).

Morphology of the Jungermanniales and Metzgeriales. The Jungermanniales and Metzergiales, the largest orders among the hepatics, contain both thallose and leafy forms. Most achieve only a small size in temperate regions, but leafy forms reaching several centimeters in length are common in the humid tropics, where they are frequently epiphytic. The thallose genera are typified by the common *Pellia*, which grows dichotomously as *Marchantia*, but differs in outward appearance. A poorly defined midrib of elongated cells extends to each apical region. Examination under the microscope shows that the thallus is indeed simple, having none of the specialized photosynthetic tissue of the Marchantiales. Some genera have much more distinct midribs (*Pallavicinia*), while others, in which the thallus is regularly dissected (*Fossombronia*), begin to resemble the leafy forms. Morphological differentiation in the thallose forms reaches its peak in some of the tropical representatives. In *Hymenophytum*, for example, the filiform underground rhizome gives rise, in a sympodial fashion, to a sequence of erect aerial thalli. These, to which photosynthesis is confined, and on which the sex organs are borne, may reach a height of 2 cm (0.75 in.).

The leafy liverworts also show a wide range of vegetative morphology, but here principally in the form of the leaves. These may be simple, more or less circular, plates of cells, as in the common *Odontoschisma sphagni*, or more often twice or several times lobed. Sometimes the leaves achieve great delicacy. In *Blepharostoma*, for example, the lobes consist only of a single file of cells. In *Frullania*, the leaf has two lobes, the lower of which is shaped like a minute helmet, and possibly serves as a water sac.

In most leafy liverworts, the stem is inclined or prostrate, and its symmetry clearly dorsiventral. Although the leaves are usually in three ranks, only those on the dorsal side are fully developed. Those of the third, ventral row (termed **amphigastria** or underleaves) remain small and are often shed a short distance behind the apex.

Classification of the leafy liverworts is based largely on the features of the leaves, including the orientation of their insertions on the stem. When the anterior margins of the leaves lie regularly beneath the posterior of those in front, the arrangement is said to be *succubous* (Fig. 5.10b) and when the converse *incubous*. The leaves of liverworts regularly lack nerves of the kind seen in mosses, but the lower lobes of the bilobed leaves of *Diplophyllum* possess a conspicuous central row of elongated cells. Conduction of water along the stems of leafy liverworts is probably largely by surface capillarity. Experiments have shown that such transport is more rapid with succubous arrangements of leaves than with incubous.

Reproduction of the Jungermanniales and Metzgeriales. Sexual reproduction is essentially similar to that described for the Marchantiales, except that specialized gametangiophores are never produced. The antheridia, superficial in origin, usually occur singly, lying either in cavities in the upper surface of the thallus (e.g., *Pellia*, Fig. 5.11), or, in leafy forms, in the axils of leaves of special branches of limited growth (Fig. 5.10a). The spermatozoids are basically similar to those of *Marchantia*, with differences in detail of the multilayered structure and microtubular ribbon. In at least one species of *Pellia* the spermatozoids are relatively large, the nucleus extending for several gyres. Also in *Pellia* the microtubular ribbon, as seen in transverse section, is not closely applied to the nucleus across its whole width, but is inclined at an angle of about 45° to its surface. The archegonia are usually grouped and are produced either laterally, as in *Pellia* (Fig. 5.12), or at the tip of the main shoot, as in most leafy liverworts. When archegonia are apical, they, and ultimately, the sporophyte, terminate the growth of the main shoot, so that vegetative growth is continued by a lateral, resulting in sympodial branching. Both monoecious and dioecious forms occur, sometimes in the same genus. The common *Pellia epiphylla*, for example, is monoecious, but *P. fabbroniana*, frequent in calcareous districts, is dioecious.

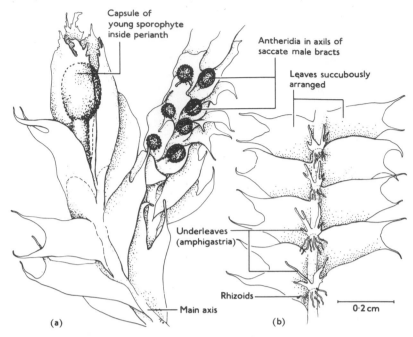

Figure 5.10. *Lophocolea cuspidata*. **(a) Fertile shoot seen from above. (b) Ventral surface showing amphigastria and the succubous arrangement of the leaves.**

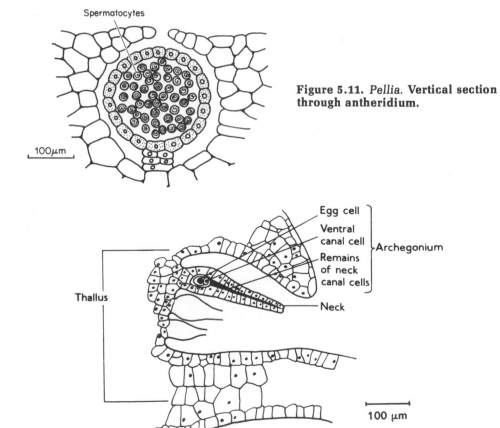

Figure 5.11. *Pellia*. **Vertical section through antheridium.**

Figure 5.12. *Pellia*. **Vertical section of thallus showing archegonia.**

The sporophyte of the Jungermanniales and Metzgeriales (Fig. 5.13) has a higher proportion of sterile tissue than that of the Marchantiales. The sporophytes are green when young and capable of appreciable photosynthesis. Minerals are probably largely transmitted from the gametophyte, but some may be absorbed directly. The capsule develops while still enclosed in the calyptra and the ultimate extension of the seta, which may reach 1 cm (0.4 in.) or more, is extremely rapid (rates of 1 mm (0.04 in.) per hour have been recorded in *Pellia*). This extension is brought about solely by cell elongation and is accompanied by the disappearance of starch from the cells. Extension is markedly diminished if the activity of the Golgi bodies in the cells is inhibited, indicating that it is dependent upon a supply of hemicelluloses reaching the walls. The extending setae display phototropism, but the curvature is confined to the regions illuminated. There is no transmission of the stimulus, a conspicuous feature of phototropism in coleoptiles and seedlings. This difference is presumably related to the absence in the sporophyte of *Pellia* of any localized site of growth.

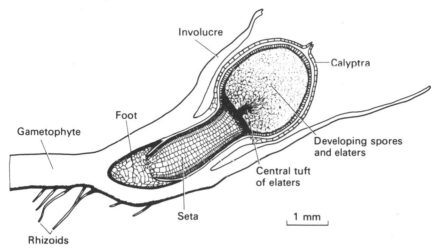

Figure 5.13. *Pellia*. **Longitudinal section of maturing sporogonium.**

Meiosis in the sporocytes of the Jungermanniales and Metzgeriales is preceded by a lobing of the protoplast, the furrows indicating the planes which will separate the four spores of the tetrad. Elaters are commonly present among the spores. The capsule usually opens by four valves which, as a consequence of differential thickenings in the wall, become sharply reflexed. The manner in which the elaters assist the distribution of the spores varies. In *Pellia* the elaters, some of which remain as a brush attached to the top of the seta (Fig. 5.13), are similar to those of *Marchantia*. In some genera, however, the elaters are "explosive." In *Cephalozia bicuspidata,* for example, the elaters are loosely attached to the ends of the valves of the capsule. After the capsule opens they begin to dry and in consequence to twist. Suddenly they violently untwist, hurling both the elater and its adhering spores into the air. The sudden expansion of the elater is believed to be caused by the shrinking column of fluid in the drying cell being put under such tension that it eventually spontaneously vaporizes, thus increasing its volume many times.

The spores of *Pellia* begin to develop before being shed (Fig. 5.14), but this does not appear to be a common feature. Some species of the Jungermanniales and Metzgeriales are rarely fertile. Capsules of *Plagiochila tridenticulata* are wholly unknown. This may be a consequence of the two sexes having acquired different distributions. The North American populations appear to be female and the European male.

Asexual reproduction of many species takes place by regeneration from fragments of mature plants. Multicellular gemmae are not uncommon and are sometimes con-

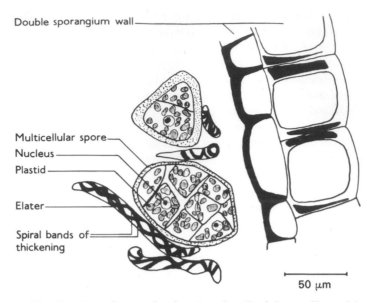

Double sporangium wall

Multicellular spore

Nucleus

Plastid

Elater

Spiral bands of thickening

50 μm

Figure 5.14. *Pellia*. **Portion of capsule showing detail of the wall, multicellular spores, and elaters.**

spicuous, as, for example, the clusters of reddish two-celled gemmae on the margins of the upper leaves of *Sphenolobus exsectiformis*, a plant frequent on rotting wood. In *Blasia pusilla*, a thallose form, multicellular gemmae are produced in remarkable flask-shaped receptacles on the dorsal side of the thallus.

Remains very suggestive of members of the Jungermanniales and Metzgeriales are known from Devonian and Carboniferous rocks.

The Sphaerocarpales. The best-known member of the Sphaerocarpales is *Sphaerocarpus*. The thallus resembles that of the simpler Marchantiales, such as *Riccia*, but the margin is frilly and suggestive of leaves. The antheridia and archegonia occur in small clusters, each cluster enclosed in a distinctive involucral sheath. As in *Riccia*, the sporophyte is little more than a capsule. The spores can be stored indefinitely in de-aerated distilled water. In these conditions they do not germinate, but viability is preserved.

Sphaerocarpos, which is regularly dioecious, has figured prominently in genetic investigations. It was found, for example, that the wavelength of ultraviolet light most effective in producing mutations in nuclei of immature spermatozoids was that most strongly absorbed by nucleic acids, providing the first evidence for the special role of these acids in heredity. *Sphaerocarpos* was also the first plant in which sex chromosomes were demonstrated. The female gametophyte possesses a large X chromosome and the male a small Y, in each instance accompanied by seven autosomes. The usefulness of the plant for Mendelian studies derives from the tendency of the spores in many species to adhere in their original tetrads. The products of each meiosis can then be separated and cultured individually, allowing a direct demonstration of genetic recombination ("tetrad analysis"). The spores are also of interest in relation to the inheritance of the pattern of thickening of the wall, a feature used to discriminate species. The spores in hybrid capsules show the thickening characteristic of the female parent, suggesting that the factors determining the pattern are maternally inherited through the cytoplasm of the egg cell.

Although *Sphaerocarpos* is normally found on damp earth, *Riella* is submerged, some species being able to withstand brackish conditions. The thallus takes the form of an erect filiform stem, rarely exceeding 2 cm (0.75 in.) in height, bearing along one side a ruffled "wing" only one cell thick. The rhizoids are confined to the base. The fossil

Naiadita from the Upper Triassic is strikingly similar to Riella (except that there is no well developed involucre around the sex organs), and it probably grew in similar situations. Its reference to the Sphaerocarpales seems well established.

The Calobryales. The small order Calobryales, of which Haplomitrium is representative, is highly distinctive and the two genera it contains have a similar growth form. The upright stems, which are radially symmetrical and bear three ranks of leaves, rise from a creeping rhizomelike axis lacking rhizoids. The archegonia are effectively terminal. There are no involucral leaves protecting the young sporophyte, but the calyptra is particularly conspicuous. Some bryologists have considered the sexual reproductive structures of the Calobryales to be the most primitive among the bryophytes as a whole. Possibly allied to the Calobryales is Takakia, a curious liverwort of Japan and the Pacific Northwest with an upright stem bearing linear appendages ("phyllids"), usually in groups of two or three. The archegonia are terminal. The ultrastructure of the base of the sporophyte resembles that of the moss Andreaea.

Anthocerotae (Hornworts)

This class contains the single order, **Anthocerotales**. Although formerly included with the Hepaticae, the Anthocerotales are now usually placed in a separate class, mainly on account of their unique sporophyte. Anthoceros is representative of its class.

Morphology of the Anthocerotales. The gametophyte of Anthoceros recalls Pellia in external morphology except that there is neither regular dichotomous growth nor a midrib (Fig. 5.15a). The thallus is undifferentiated, apart from internal cavities which contain mucilage and occasionally the blue-green alga Nostoc, a genus known to fix atmospheric nitrogen. In most species, a single chloroplast, containing a complex pyrenoid, occurs in each cell, a situation unknown elsewhere in the Bryophyta or in higher plants (with the exception of some species of Selaginella), but common in the algae. This has led to the suggestion that the Anthocerotae are closer to an algal ancestry than other Bryophyta.

Some species of Anthoceros (such as those of the Mediterranean region) regularly form tubers, enabling them to tide over a dry season unfavorable for growth.

Reproduction in the Anthocerotales. In Anthoceros, as in Marchantia, the formation of the sex organs is dependent upon photoperiod. Here, however, gametogenesis in most species is initiated by diminishing day-length, so that fertilization occurs during winter.

The antheridia arise from a cell beneath the surface, one to several antheridia (depending upon species) coming to lie in a closed chamber (Fig. 5.15b). The roof remains intact until the antheridia are mature. The spermatozoids resemble those of the liverworts generally, but have some distinguishing features of ultrastructural detail.

Archegonia arise superficially, but the wall of the archegonium is continuous with the thallus, the neck opening at the surface (Fig. 5.15c). This resembles the situation in some vascular archegoniates (see, for example, Fig. 6.12). Development of the diploid zygote leads to a slender cylindrical sporophyte (Fig. 5.16) with a relatively small proportion of fertile tissue. The sporophyte remains inserted into the gametophyte by a conspicuous lobed foot, and the basal part is surrounded by an upgrowth of the thallus, the *involucre*.

The development of the archesporium begins as a domelike layer within the summit of the sporophyte. The sporophyte continues to grow for several weeks or longer from a meristem close to its base. During this growth the archesporium differentiates basipetally as a hollow cylinder, the center of which is occupied by a sterile columella. Multicellular elaters (which lack distinct spiral thickenings and are often referred to as pseudoelaters) differentiate among the sporogenous cells. When the spores at the top of the capsule are ripe, the capsule dehisces basipetally (Fig. 5.15a) along two longitudinal slits, the opening

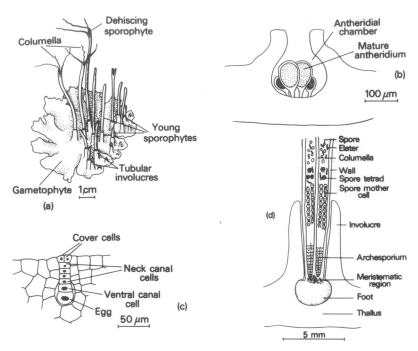

Figure 5.15. *Anthoceros laevis.* **(a) Female thallus with sporophyte. (b) Section of antheridial chamber. (c) Section of nearly mature archegonium. (d) Diagrammatic longitudinal section of a sporophyte showing the different regions.**

Figure 5.16. *Anthoceros laevis.* **Female plants bearing young sporophytes. Scale bar 5 cm.**

beginning near the tip. As the upper part of the capsule dries, the valves separate completely above and begin to twist longitudinally. The consequent contortions of the valves expose the spores and elaters adhering to the central column, and dispersal begins. The separation of the valves continues downwards as the spores mature, and meanwhile the basal meristem generates new sporophytic tissue at about the same rate. Consequently a single sporophyte continues to yield spores over a considerable period. The spores have conspicuously thickened and sculptured walls.

Anthoceros and its allies further differ from other liverworts, but resemble the mosses, in possessing photosynthetic tissue in the outer layers of the extending sporophyte. Stomata are also present, as in the capsules of some mosses. The sporophyte is thus not entirely dependent on the gametophyte for nutrition, but, since the sporophyte will still mature even if it is covered with a tinfoil cap, it seems likely that it is able to draw a substantial proportion of its essential metabolites from the parent gametophyte.

The evolutionary position of the Anthocerotales. The Anthocerotales are remarkable among the liverworts in recalling the features of both the algae (the presence of the pyrenoid in the chloroplast) and the mosses and higher plants (the presence of stomata and the continued growth of the sporophyte). They are consequently thought by some to stand close to the line of evolution leading from the algae to terrestrial vegetation. The intermediate position of the Anthocerotales extends even to their ultrastructure. The electron microscope confirms that the chloroplasts of *Anthoceros* resembles those of the algae, but the chloroplasts of *Megaceros,* which are several in each cell, often lack pyrenoids and possess well-defined grana. They thus resemble those of land plants generally.

Another feature of the Anthocerotales which has excited much attention is the axial form of the sporophyte. Since the sporophyte in some forms is long-lived and may even persist for a time after the death of the parent gametophyte, it may indicate how simple axial plants, such as those of the Silurian and early Devonian, have evolved. Alternatively the *Anthoceros* condition may be derived, the sporophyte having become reduced and almost deprived of its independence. The presence of stomata in the wall of the capsule has been held to support this view.

Although evidence relating to either possibility is lacking, it seems quite plausible that a growth form similar to that shown by the living Anthocerotales did play some part in the evolution of land plants from algal ancestors. Unfortunately, the only fossils so far known attributable to the Anthocerotales are spores from the Tertiary.

Musci (Mosses)

The mosses are a class much greater in number and more widely distributed than the liverworts, occurring in almost every habitat supporting life. Apart from being the dominant vegetation in acid bogs and alpine and Arctic regions, they are a familiar feature of woodlands and hedgerows. Some species even survive the polluted atmosphere of urban areas, often forming dark green cushions between paving stones and in other damp crevices.

The features which distinguish the mosses from the liverworts are found in both gametophyte and sporophyte. The protonemal stage of the gametophyte is often conspicuous, the mature form of the gametophyte is always leafy, and the rhizoids are multicellular. The sporophyte grows from an apical cell, the capsule, in many instances, has a complex opening mechanism which affects the distribution of the spores, and sterile elaters are never present.

The three orders—**Sphagnales, Andreaeales,** and **Bryales**—differ principally in the nature of the protonema and the structure of the capsule. There are also differences in the ultrastructural detail of the spermatozoids (which resemble those of other bryophytes) which may have taxonomic significance.

Morphology of the Sphagnales. The Sphagnales, a very distinctive order, are represented by a single genus, *Sphagnum*, confined to acid waterlogged habitats. The more or less continuous spongy cover of peat bogs consists very largely of a range of *Sphagnum* species.

The adult gametophyte comprises an upright main axis from which whorls of branches arise at regular intervals (Fig. 5.17). The leaves, which are closely inserted, have a peculiar structure which is diagnostic of the genus, and also, in its finer details, of the many species (Fig. 5.17c and d). When first formed, the leaves are made up of many diamond-shaped cells. These then cut off narrow daughter cells, but on two sides only. The daughter cells develop chloroplasts, but the mother cell remains colorless. It eventually dies, but before doing so pores are formed in its upper and lower walls and spiral bands of thickening are often laid down around the cell. The dead cells are able to take up water and act as reservoirs. This unique leaf structure accounts for the outstanding bog-building properties of *Sphagnum*. Additionally, the cell walls of *Sphagnum* contain phenolic substances which act as ion-exchange resins. The growing plants extract metallic cations from inflowing waters and release hydrogen ions, so maintaining the acidity of the bog.

Well-preserved leaves similar to those of *Sphagnum* have been found in Permian deposits in the U.S.S.R., indicating beyond doubt the great antiquity of this kind of construction.

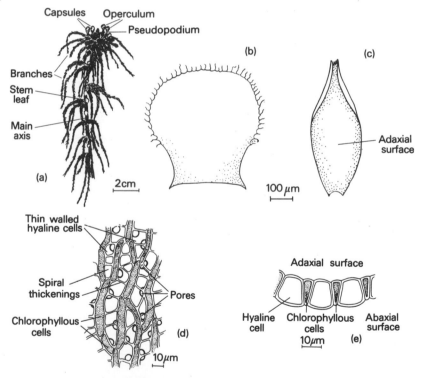

Figure 5.17. *Sphagnum fimbriatum.* **(a) Upper portion of shoot with sporophytes. (b) Stem leaf. (c) Branch Leaf. (d) Arrangement of cells in leaf, adaxial surface. (e) Section of leaf.**

Reproduction in the Sphagnales. The reproduction of *Sphagnum* is probably principally vegetative, the decay of the older parts eventually causing branches to separate and thus to become new individuals. Disks of *Sphagnum* peat from depths as great as 30 cm (12 in.) will produce innovative growths when exposed to light and a saturated atmosphere. These probably arise from the outer cortex of the buried stems. Some cells evidently remain viable and capable of division, possibly for as long as 60 years.

Mature plants in favorable situations produce sex organs freely. Both monecious and dioecious species occur. The antheridia, each of which begins its development from a single apical cell, lie in the axils of leaves towards the tips of small upper branches (Fig. 5.18). These antheridial branches are often strongly pigmented and clustered in a conspicuous comal tuft. The female inflorescence consists of a budlike aggregate of archegonia and bracts borne laterally near the summit of the main stem.

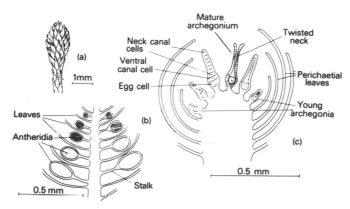

Figure 5.18. *Sphagnum* **sp. (a) Antheridial branch. (b) Longitudinal section of antheridial branch. (c) Longitudinal section of archegonial branch.**

After fertilization, the zygote yields a sporophyte (Fig. 5.17a) consisting principally of a capsule, containing a dome-shaped archesporium, and a foot. The seta remains inconspicuous, and the function of elevating the capsule is taken on by the base of the female inflorescence which, as the capsule matures, grows up as a leafless axis, or *pseudopodium* (Fig. 5.19). Release of the spores is brought about by air pressure which builds up in the lower half of the capsule as it dries. Eventually, this pressure is sufficient to dislodge the clearly differentiated lid (operculum) with explosive force, and the spores are effectively dispersed.

Developing *Sphagnum* capsules are occasionally parasitized by a fungus (*Bryophytomyces*). Although the mature capsule may appear perfect, the small spores of the fungus replace those of the host. This gave rise to reports of heterospory in *Sphagnum* before the situation was understood.

Sphagnum spores germinate to form a filament, but this is rapidly replaced by a small thallose protonema. This, in turn, gives rise to a bud which develops into the familiar leafy gametophyte, the protonema meanwhile becoming moribund and disappearing. Protonemata also appear on disks of old peat maintained in humid conditions. Many appear to arise from spores within the compacted peat which have remained viable. Others are secondary in origin from viable cells in the old stems.

Morphology and reproduction of the Andreaeales. The Andreaeales are another order containing only a single genus, which is distinguished by its peculiar capsule.

The leafy gametophyte of *Andreaea* (Fig. 5.20) rarely exceeds 1 cm (0.39 in.) in height. It is usually found growing on rock, chiefly in cold, exposed and relatively dry regions. The leaves are olive-brown in color, composed of rounded cells, and in most species showing no distinct midrib.

Sex organs are formed apically. The sporophyte resembles that of *Sphagnum* in having a domed archesporium (Fig. 5.20c) and in being borne on a pseudopodium at maturity. Dehiscence of the capsule takes place by four longitudinal slits which do not meet at the tip (Fig. 5.20b). The hygroscopic properties of the wall cause the slits to close in damp conditions and to open again in dry (Fig. 5.20a).

The protonema of *Andreaea* is similar to that of *Sphagnum.*

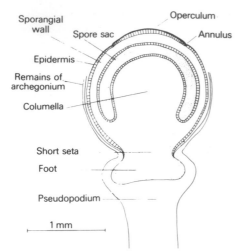

Figure 5.19. *Sphagnum sp.* Longitudinal section of nearly mature sporophyte.

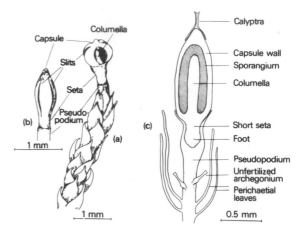

Figure 5.20. *Andreaea nivalis.* (a) Habit of fertile plant showing dehisced capsule in dry condition. (b) Dehisced capsule in wet condition. (c) Longitudinal section of mature sporophyte.

The Bryales. The 600 or so genera of the Bryales form a well-defined order. Although there is a common basic morphology and life cycle within the order, the variation in size, detailed structure, and habitat preferences is considerable. Many mosses are confined to permanently damp situations in woodlands and by springs, but others, for example, *Tortula ruraliformis,* are able to survive periods of drought in sand dunes and other arid habitats. The cells of these species appear to have acquired the capacity to continue metabolism at a reduced rate while partially dehydrated. At the other extreme are a few subaquatic species, such as *Fontinalis antipyretica.*

Some mosses are remarkably tolerant of heavy metals and in some areas serve as "indicator" plants. The worldwide genus *Mielichhoferia,* for example, contains a number of species characteristic of acidic copper-bearing soils and rocks.

The most familiar form of the moss plant is the adult gametophyte (Figs. 5.21 and 5.26). This consists of a main axis growing from an apical cell and bearing leaves which, although usually spirally inserted, may in some forms come to lie in one plane. This gives the shoot a complanate appearance (e.g., *Neckera,* common on banks and rocks). In a few species (e.g., *Fissidens*) the leaves are equitant and arranged in two ranks. The leaves of most mosses consist of a single sheet of cells, although the central region may be thickened and contain a well-defined midrib (often referred to as a "nerve"), sometimes excurrent in a hyaline point. The most complex leaf is found in *Polytrichum* and its allies. Here a number of parallel longitudinal lamellae grow up from the upper surface (Fig. 5.22), and the chloroplasts occur principally in these cells. The shape of the leaf and the nature and development of the midrib and of the cells at the margin of the leaf are important features in the taxonomy of the mosses.

Anatomically, moss gametophytes offer little that is outstanding among land plants, the most complex differentiation being found in the stem of *Polytrichum* (Fig. 5.23). Not only is there an approach here to the development of tracheids (but no evidence of lignification), but also a clear radially symmetrical zonation in structure, resembling that of the axes of some of the smaller ferns. A core of thin-walled cells (**hydroids**), which are dead and empty at maturity, lies at the center. These have inclined end walls and are able to conduct water and solutes. In the rhizome of *Polytrichum* (but less frequently in the upright stem) the hydroids are mixed with **stereids**, living cells usually with thickened

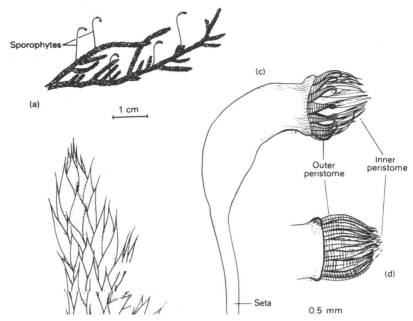

Figure 5.21. *Hypnum cupressiforme.* **(a) Fertile shoot system (b) A portion of the shoot showing the closely inserted leaves. (c) Capsule, showing peristome in dry state. (d) Peristome in wet condition.**

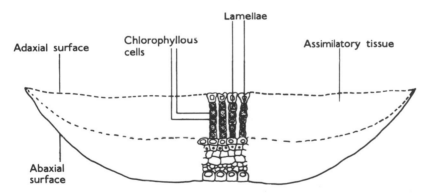

Figure 5.22. *Polytrichum commune.* **Transverse section of leaf showing the assimilatory lamellae.**

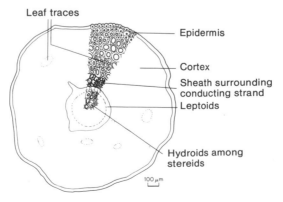

Figure 5.23. *Polytrichum commune.* **Transverse section of stem. Stereids are more conspicuous in rhizomes than in aerial stems.**

lateral walls. Surrounding the hydroids is a zone of elongated elements tending to be swollen at their ends. These are the *leptoids*. Their end walls are inclined, callosed, and perforated by plasmodesmata. Although the nuclei are often degenerate, some protoplasmic contents persist. These include plastids with rudimentary thylakoids and plastoglobuli, mitochondria, and often conspicuous endoplasmic reticulum. Leptoids are believed to have a phloemlike function. There is a close structural similarity to the sieve cells of gymnosperms.

Although leptoids are found only in *Polytrichum* and its allies, a core of hydroids is found in many moss stems. Experiments have shown that, where present, the hydroids are the principal site of the internal conduction of water. In most mosses, however, there is also substantial apoplastic conduction, or even capillary conduction external to the stem facilitated by appressed leaf bases or matted rhizoids. In some instances (e.g., *Thuidium tamariscinum*) the stem is clothed with a tomentum of green-branched filamentous outgrowths (*paraphyllia*) which acts like a wick.

Development of the gametophyte of the Bryales. In all Bryales the protonemal phase is conspicuous. In *Funaria* and in some other mosses, two distinct phases of protonemal development have been recognized. In the first (*chloronema*), which resembles a branching filamentous alga, the dividing walls are transverse and the cells bright green. This gives way to a strongly heterotrichous phase. The upright filaments resemble the chloronema, but the prostrate (*caulonema*) have oblique septa and are yellowish in color. The buds yielding the mature form of the gametophyte (Fig. 5.29) arise solely on the caulonema. A culture growing from a single spore tends to be radially symmetrical. The buds arise in a series of concentric rings, indicative of a form of developmental periodicity within the culture.

Reproduction in the Bryales. In sexual reproduction, the Bryales show every possible arrangement of the archegonia and antheridia. Both monoecious and dioecious species occur. Among the monoecious species, the gametangia may be either mixed together in a budlike inflorescence or separate. Whatever the arrangement, the antheridia and archegonia are often numerous and interspersed with sterile hairs or paraphyses (Figs. 5.24 and 5.25), the whole cluster of sex organs commonly being surrounded by a

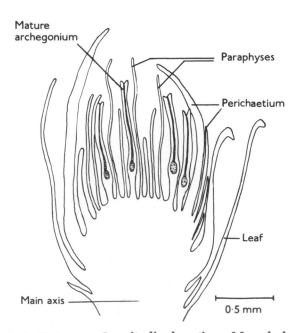

Figure 5.24. *Mnium* sp. Longitudinal section of female head.

whorl of closely appressed leaves (perichaetium). The archegonia usually have long necks, each consisting of several tiers of cells, and the central canal may contain as many as ten cells. The antheridia are stalked, and one or more cells at the apex usually form a distinct lid at maturity, opening as if on a hinge while the mass of spermatocytes is discharged.

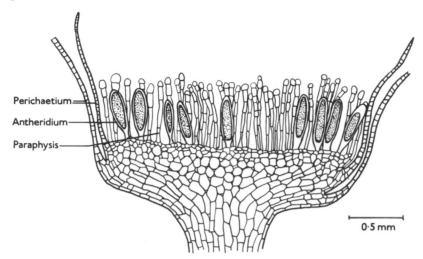

Figure 5.25. *Mnium hornum.* **Longitudinal section of male head.**

There is a striking correlation in the Bryales between the position on the plant where the sex organs are produced and the growth habit. Where the reproductive organs terminate the main axis, and growth is consequently sympodial, the main axis is almost invariably upright. These are the acrocarpous mosses. In the remainder, where the sex organs are produced laterally (the pleurocarpous mosses), the main axis is usually creeping (Fig. 5.21a). With only a few exceptions, the tufted mosses are acrocarpous. Photoperiod and temperature affect the onset of the sexual phase in many mosses, and these factors are probably responsible for the annual reproductive cycle seen in many temperate species.

The development of the sporophyte begins immediately after fertilization, and the regions of the embryo yielding the foot, seta, and capsule are soon distinguishable. The venter of the archegonium is also stimulated into growth by germination of the zygote, and a caplike calyptra is formed over the young sporophyte. Extension of the seta tears the calyptra away from the main body of the gametophyte, but it continues to ensheathe the developing capsule and exert formative effects on the upper part of the sporophyte. Premature removal results in anomalous growth and a deformed capsule. The influences emanating from the calyptra seem to be partly physical and partly chemical. Compared with that of a typical liverwort, the moss sporophyte develops slowly, and the extension of the seta begins well before the capsule is mature.

The capsule itself (Figs. 5.21, 5.26b, 5.27, and 5.28) is a complex organ, but its differentiation follows a regular radial pattern, and two concentric regions of tissue can be recognized which follow distinct developmental paths. An inner region, termed the **endothecium**, gives rise to the archesporium, which in the Bryales is never domed but is always a cylinder, often with a central sterile columella. Outside the endothecium is the **amphithecium** which, in most Bryales, gives rise to a ring of remarkable toothlike structures, the peristome. This remains as a fringe around the mouth of the opened capsule (Figs. 5.21 and 5.26). The peristome is developed from three layers of cells. Thickenings are laid down on both sides of the tangential walls bounding the middle layer. Only these thickenings remain at maturity. They then fall into a number of columns, each of which

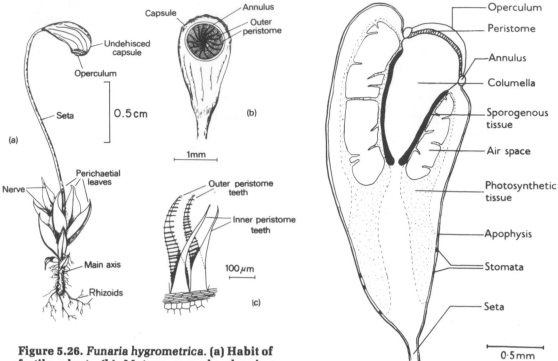

Figure 5.26. *Funaria hygrometrica.* (a) Habit of fertile plant. (b) Mature capsule showing intact peristome. (c) Portion of peristome viewed from the inside.

Figure 5.27. *Funaria hygrometrica.* Median longitudinal section of immature capsule.

Figure 5.28. *Bryum pallens.* Fruiting condition. Scale bar 1 cm.

yields a peristome tooth. Because of the ordered sequence of mitoses in the differentiation of the amphithecium, and the regular spacing of the cells giving rise to the peristome, the number of teeth is constant in any given species and is always a power of two. Although the peristome is fully developed in a moss such as *Funaria*, in other species it may be imperfect or rudimentary, or even absent.

Chlorophyllous tissue occurs in the immature capsule, particularly in the basal region (apophysis), where stomata are also found in the epidermis. The sporophyte is thus to some extent autotrophic. In some species (e.g., *Funaria hygrometrica*), air spaces occur between the archesporium and the wall of the capsule. This conspicuous aeration has perhaps been developed in relation to the respiratory demands of the developing archesporium. The seta frequently contains a strand of hydroids ascending from the foot into the base of the capsule. Leptoids are never present in the seta.

Meiosis within the spore mother cells is foreshadowed by the appearance of furrows indicating the site of the future dividing walls, accompanied by the siting of bundles of microtubules in anticipation of the imminent nuclear divisions. The columella usually begins to break down at this stage and ultimately the cavity of the capsule is occupied solely by spores. When these are mature, the operculum detaches itself, exposing the peristome (Fig. 5.26b and c), which now begins to play an important role in the dispersal of the spores. The polysaccharide material forming the peristome teeth is hygroscopic and, since the macromolecular orientations of the thickenings in the two columns of cells giving rise to a tooth differ, tensions are generated in the tooth with changes in its hydration. These are released by sharp twisting and bending movements. The peristome thus forms a very effective scattering mechanism, activated by changes in atmospheric humidity. The exact nature of the movements of the teeth varies with the species. In some Bryales, for example, *Funaria hygrometrica,* the peristome is incurved when wet and recurved when dry, but in others the movements are less regular. In *Polytrichum* the capsule remains closed by a diaphragm after shedding of the operculum. The spores escape through pores around its edge (Fig. 5.31).

The number of spores in a capsule varies from a few thousand in species with small capsules to a million or more in those with large capsules. Although in dioecious mosses the male plants are often smaller than the female, differences in spore size correlated with sex are rare. The best-known example is provided by *Macromitrium* of the Southern Hemisphere. In some species of this genus, the spores fall distinctly into two size classes, the smaller giving rise to diminutive male plants. The four spores in each tetrad are initially similar, but subsequently two become larger. Evidently, there is segregation of sex at meiosis.

The spores of the Bryales have thin walls and germinate rapidly on a damp surface. The duration of the protonemal phase varies widely. In some Bryales it is comparatively brief, but in others prolonged. In *Pogonatum aloides,* common on shaded acidic banks, the protonema persists as a dark green felt from which mature plants of limited life arise over a long period. A tropical example is the Southeast Asian *Ephemeropsis* which forms an extensive protonema on leaves. The mature plants arising from it are minute, consisting only of a few bracts surrounding sex organs. In species in which the protonema is of limited duration the protonemal state is often continued by small amounts of secondary protonema issuing from the bases of the mature plants. The protonema of *Tetraphis* is unusual in producing small leaflike outgrowths.

Vegetative propagation undoubtedly plays a large part in the asexual reproduction of the Bryales. Almost any part of the gametophyte—leaf, stem, or even rhizoid—is capable of regeneration, usually yielding initially a secondary protonema. Ultimately, buds are formed, as on the primary protonema, giving rise to new individuals (Fig. 5.29). Many species produce gemmae of characteristic shape. These serve as an ordered means of asexual reproduction. Their production may be general, or confined to certain regions

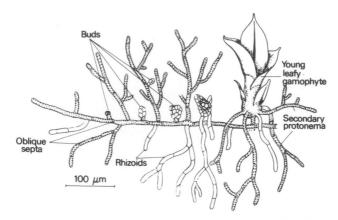

Figure 5.29. *Funaria hygrometrica.* **Development of buds on protonema.**

of the plant, or even to specialized structures (Fig. 5.30).

   Campylopus atrovirens is an example of a species which, although widely distributed, is hardly known in the sporophytic condition. Its dispersal must be almost entirely by asexual means.

   The Bryales as experimental material. The ease with which many mosses can be grown in pure culture on artificial media makes them very suitable for the experimental investigation of morphogenesis. Mutants of *Physcomitrella,* for example, are known in which the protonema will not produce buds unless cytokinin is added to the medium. The existence of mutants of this kind facilitates the investigation of the genetical and physiological factors bringing about the change in form of growth. The results of experiments *in vitro* have nevertheless to be interpreted with caution. When *Funaria,* for example, is grown in pure culture on medium containing activated charcoal the protonema produces buds as readily as on soil, but in the absence of charcoal they are delayed. It seems likely that the charcoal absorbs substances secreted by the protonema into the medium which would otherwise accumulate and have morphogenetic effects.

   The Bryales were the first archegoniate plants in which it was discovered that portions of the sporophyte placed on a mineral-agar medium would give rise directly to gametophytic outgrowths without reduction of chromosome number (apospory). Although the diploid gametophytes are initially larger in all their parts than the haploid (the so-called *gigas* condition), this difference in size may be spontaneously lost and the haploid and diploid become indistinguishable. The diploid gametopohytes are reproductively perfect and yield tetraploid sporophytes. There is evidence that autodiploid gametophytes exist in nature, and they may have arisen in this way or by the chance production of dyads of unreduced spores. In the Australian moss *Hypopterygium,* for example, several different intraspecific polyploids were found in a small population on a single log.

   The transition from gametophyte to sporophyte can also be induced directly (apogamy), but probably only with gametophytes which are at least diploid in constitution. *Phascum cuspidatum,* for example, will begin to produce sporogonia at the tips of leaves and elsewhere if the medium on which it is growing is allowed to dry out. Rehydration results in the resumption of gametophytic growth. Some Bryales will produce sporogonia from secondary protonema if the medium is enriched with sugar. These results show that in many Bryales the expression of the genes controlling the kind of growth can be readily modified by nutrition and environmental stress.

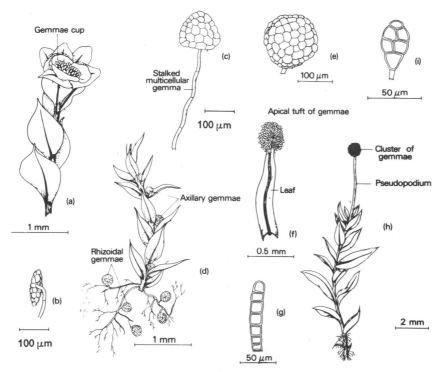

Figure 5.30. Asexual reproduction in mosses. (a–c) *Tetraphis pellucida;* **(a) habit of gem-miferous plant; (b,c) mature gemma, front and side views. (d,e)** *Bryum rubens;* **(d) habit; (e) gemma. (f,g)** *Ulota phyllantha;* **(f) young leaf with apical tuft of gemmae; (g) gemma. (h,i)** *Aulacomnium androgynum;* **(h) habit; (i) gemma.**

RELATIONSHIPS OF THE BRYOPHYTA

Origin of the Bryophytes

The significant similarities, particularly in the protonemal phase, between the bryophytes and the algae provide compelling evidence that the mosses and liverworts had their origin in some algal form. Further, it is clear that the Bryophyta share many more basic features, such as the nature of the photosynthetic pigments, cell wall components, and food reserves, with the Chlorophyta than with any other algae. Among the Chlorophyta are many examples of the heterotrichous habit and of oogamy, and in the Charophyceae of developments possible foreshadowing the archegonium. The Charophyceae also share with the bryophytes the possession of glycolate oxidase. The flagella of the spermatozoids are also of the same kind, and the multilayered structure underlying the flagellar bases recalls similar structures in the motile stages of the Charophyceae. In the Charophyseae, however, the two centrioles which become flagella bases arise at right angles to each other, whereas in biflagellate land plants they are initially coaxial. Another significant difference is that, whereas the microtubular ribbon lies transverse to the plates of the multilayered structure in the Charophyceae, it is at an angle of 45° in the bryophytes. This results in the ribbon following a curved path in the bryophyte spermatozoid, contrasting with the situation in *Coleochaete.* It is also clear that the Bryophyta are considerably more highly organized than any Chlorophyta. This is shown by their terrestrial habit, differentiated thallus, regular phasic alternation of gametophyte and sporophyte, and the production of aerial spores. The manner in which the reproductive organs are enclosed in a wall of sterile cells also contrasts with the naked gametangia and sporangia customary in most Chlorophyta.

Although the Charophyceae indicate how archegonia and antheridia may have evolved, little information is available about the origin of the bryophytes themselves. A possible form transitional between algae and bryophytes is *Protosalvinia* of the Upper Devonian. This was a thalloid plant which probably grew on lake shores. The thallus, a few centimeters in length, bifurcated and produced spores in tetrads in cuplike regions on its dorsal surface. Chemical analysis of the remains (which are locally abundant) reveals traces of lignin- and cutinlike substances, and sporopollenin in association with the spores. *Protosalvinia* thus has indications of land plant features, but is generally considered closer to the algae.

In the absence of well-defined remains we can only speculate how a bryophyte might have evolved from algal ancestors. A thallose liverwort like *Pellia*, for example, might have been derived from a heterotrichous form in which the prostrate component had become parenchymatous (a tendency already evident in some species of *Coleochaete*) and adopted dichotomous growth, the aerial component meanwhile being lost. Heteromorphic life cycles in the transmigrant forms would have been no novelty since they are well represented in the algae, but examples of nonmotile spores, such as those produced by *Dictyota*, are few. The evolution of nonmotile spores with a well-developed exine, an essential for terrestrial life, thus probably accompanied the transmigration, and those forms which remained as algae have never developed this feature.

The fossil record gives little information about the earliest bryophytes. The possible representatives from Devonian rocks are usually so ill-preserved that their exact status is uncertain. The most promising candidate is perhaps *Sporogonites* from the Lower Devonian, a form in which a number of capsules with stalks a few centimeters in height seem to be arising from a creeping thallus. Unfortunately, no detail can be discerned and the bryophytic nature of the fossil must remain conjectural. So far, the fossil record is wholly silent on how the ancestral organism gave rise to the sporophytes characteristic of bryophytes.

The morphological and reproductive differences between the mosses and liverworts appear to extend as far back as the Carboniferous, since the general classification of the fossil bryophytes from these ancient rocks is readily apparent. Examples of liverworts have already been mentioned. Among the mosses, *Muscites* from the Upper Carboniferous can be accepted as a member of the Bryales. Several convincing fossil mosses are known from the Permian. These findings strengthen the view that mosses and liverworts have been independent evolutionary lines from a very early period, and it is even possible that they had independent origins from transitional archegoniate forms.

Apart from their archegoniate reproduction, bryophytes have no obvious relationships with even the simplest vascular plants, living or fossil. A number of fossils from the Rhynie Chert may have belonged to plants intermediate between Bryophyta and Tracheophyta, but the evidence is not conclusive. The Bryophyta appear never to have been a major component of the earth's vegetation, although they may have been conspicuous in the Carboniferous forests. They have probably remained isolated from the main line of evolution of land plants, changing only slowly, and exploiting a relatively circumscribed ecological niche. The failure to reach the morphological complexity characteristic of the remainder of the land flora may have been a consequence of the restriction of independence to the gametophytic phase in the life cycle. A haploid organism has no possibility of carrying a reservoir of variability in the form of recessive genes, capable of being advantageously expressed in future chance recombinations. This argument clearly has to be used with caution in view of the notable number of diploid mosses now known. Such doubling of the gametophytic chromosome numbers probably however, came too late to disrupt the well-established morphology of the cycle. Despite the factors limiting the evolutionary advance of the gametophytic phase, a number of anatomical and morphological trends have been strikingly similar in bryophytes and vascular plants. Bryophytes, for example, have acquired both leafiness and, in the stem of *Polytrichum*, a rudimentary vascular strand, the elements of which resemble structurally and functionally xylem and phloem. Gametophytic and sporophytic organisms have evidently responded in a similar manner to common environmental factors. Nevertheless, the achievement of the bryophytes has in comparison been modest.

Evolutionary Relationships Within the Bryophytes

The evolutionary relationships within the mosses and liverworts themselves are also obscure. The liverworts, for example, show a whole series of forms from the creeping thallus (e.g., *Pellia*), to thallose with two rows of ventral scales (e.g., *Blasia*), thallose in which the margin of the thallus is so deeply crenulate that the lobes resemble leaves (e.g., *Fossombronia*), and ultimately leafy forms with upright stems and radial symmetry (e.g., *Haplomitrium* of the Calobryales). There has been much argument about whether this series represented a phylogenetic advance or whether the first liverworts resembled *Haplomitrium,* the other forms being derived. The view that the radially symmetrical leafy form is primitive is strengthened by the surprisingly complex features present in many of the thallose forms (e.g., in *Marchantia* and *Anthoceros*). Also a thallose form such as *Pellia* is so outstandingly simple that it stands under the suspicion of being reduced and specialized.

Similar arguments apply to the mosses, although here the relative uniformity of the group makes comparative morphology even less informative. Few would regard those species whose capsules possess only rudimentary peristomes, or even lack them altogether, as anything other than reduced. The moss capsule with its clearly differentiated operculum and peristome must therefore have been an early and distinguishing fea-

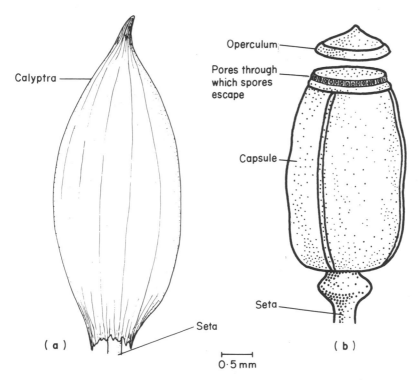

Calyptra

Operculum

Pores through which spores escape

Capsule

Seta

Seta

(a)   (b)

0·5 mm

Figure 5.31. *Polytrichum juniperinum.* **Capsule. (a) Before removal of calyptra. (b) Operculum removed showing "pepper-pot" mechanism.**

ture of the class. We have no direct evidence of how the peristome evolved, but it is possible that the "pepper-pot" mechanism present at the mouth of the capsule of *Polytrichum* (Fig. 5.31) and the bristlelike peristome found in a few other genera, indicate steps in a developmental pathway that culminated in the typical peristome of the Bryales.

Although we must regard the basic morphological features of the bryophytes as having arisen very early in the evolution of land plants, evolution of a more superficial nature has no doubt continued in the subkingdom, its course probably being influenced by evolution of vegetation as a whole. Among the mosses, widespread polyploidy probably facilitated this more limited adaptative variation. We can envisage that the rise of the angiospermous forests in the Cretaceous (Chapter 9) provided many new surfaces well suited for colonization. A renewed burst of evolution, consisting principally of diversification of already established morphological forms, may have occurred at this time leading to the substantial epiphytic element in the existing bryophyte flora.

THE SUBKINGDOM TRACHEOPHYTA
Part 1

GENERAL FEATURES OF THE TRACHEOPHYTES

All vascular plants (i.e., those possessing the lignified conducting tissue, xylem) are placed in a single subkingdom, the Tracheophyta. Discussion of the many classes which comprise the Tracheophyta and their interrelationships will occupy the remaining chapters of this book.

The general characteristics of the Tracheophyta may be defined as follows:

TRACHEOPHYTA

Habitat	Predominantly terrestrial or epiphytic.
Plastid pigments	Chlorophylls *a*, *b*; carotenoids (principally β-carotene); xanthophylls (usually principally lutein).
Food reserves	Starch; to a lesser extent fats, inulin and other polysaccharides. Proteins.
Cell wall components	Cellulose, hemicelluloses, lignin.
Reproduction	Heteromorphic life cycle; sporophyte the conspicuous phase, its growth usually indeterminate. Sex organs with or without a jacket of sterile cells. Male gametes in some flagellate. Embryogeny rarely exoscopic, embryo in many enclosed in a seed. Spores rarely green, usually with well-defined wall (exine) impregnated with sporopollenin. Often of two sizes, produced in different sporangia, the larger (megaspores) female and smaller (microspores) male (heterospory). Specialized vegetative reproduction of the sporophyte infrequent.
Growth forms	Predominantly axial.

The Tracheophyta fall into a number of divisions (Table 6.1), some of which are represented solely by fossils. Those living tracheophytes which reproduce by spores (as opposed to seeds) are often referred to collectively as the Pteridophyta. The Tracheophyta as a whole have a rich fossil record.

Most of the living tracheophytes are seed bearing, and most of the seed-bearing tracheophytes are, in turn, angiosperms (flowering plants), in which the female gametophyte is characteristically the embryo sac. The angiosperms number about 200,000 species, and in many plant communities they are the principal component of the vegetation. They are nevertheless the most recent Tracheophyta to have been evolved, and their dominance is comparatively recent.

The remainder of Tracheophyta, with the exception of a few gymnosperms, are archegoniate and their fossil record extends back in some instances well into the

Paleozoic. Taken as a whole, the archegoniate Tracheophyta show a progressive ability to exploit terrestrial habitats. The steps which led to the development of heterospory from homospory, and the elimination of the necessity of free fluid for fertilization, are indicated convincingly in the living and fossil archegoniates. The progression from simple heterospory to elaborate seeds is also evident in fossil forms. These reproductive changes were accompanied by increasing protection of the plant body from unfavorable climatic conditions, thus widely extending the range of environments available to plant life. The emergence and diversification of the angiosperms involved the exploitation of many features which originated in archegoniate evolution.

Table 6.1. Classification of subkingdom Tracheophyta.

Division	Class/Subclass	Order
Rhyniophyta	Rhyniopsida	Rhyniales
Zosterophyllophyta	Zosterophyllopsida	Zosterophyllales
		Asteroxylales
Trimerophyta	Trimerophytopsida	
Psilopsidophyta	Psilopsida	Psilotales
Lycophyta	Lycopsida	Lycopodiales
		Selaginellales
		Isoetales
		Drepanophycales
		Protolepidodendrales
		Lepidodendrales
Sphenophyta	Spenopsida	Equisetales
		Calamitales
		Sphenophyllales
		Pseudoborniales
Filicophyta	Filicopsida	Ophioglossales
		Marattiales
		Filicales
		Hydropteridales
		Cladoxylales
		Zygopteridales
		Coenopteridales
Progymnospermophyta		Archaeopteridales
		Aneurophytales
		Protopityales
Gymnospermophyta	Pteridospermopsida	Lyginopteridales
		Medullosales
		Cycadales
		Bennettitales
		Caytoniales
		Gnetales
	Cordaitopsida	Cordaitales
		Ginkgoales
	Coniferopsida	Voltziales
		Coniferales
		Taxales
Angiospermophyta	Angiospermopsida	Numerous orders of
	Monocotyledonae	flowering plants
	Dicotyledonae	

Although the simpler and, on the basis of fossil evidence, more primitive vascular plants are both homosporous and archegoniate, they are, nevertheless very different from the mosses and liverworts. Contrary to the situation in the Bryophyta, in the archegoniate Tracheophyta it is the sporophyte which is the conspicuous phase, being longer lived and possessing considerably greater anatomical complexity than the gametophyte. There are no living plants clearly intermediate between bryophytes and tracheophytes. Intermediates may have existed in the Devonian, but the evidence is unconvincing. Further, the common ancestors of the diverse, living homosporous tracheophytes were probably remote and perhaps present only at the time when plants were beginning to colonize the land. If such ancestors possessed little or no lignified tissue they may have largely escaped preservation as macrofossils.

A feature common to all living homosporous Tracheophyta is the presence in the sporangium of one or more concentric layers of cells surrounding the sporocytes. These cells break down progressively during meiosis. The materials so liberated are utilized in the final differentiation of the spores, including the ultimate stages in the laying down of the exine. This specialized tissue is referred to as the **tapetum**. In the tracheophytes which are both heterosporous and seed bearing, the tapetum is well represented in the microsporangia, less so in the megasporangia.

SPORE-BEARING TRACHEOPHYTES KNOWN ONLY AS FOSSILS

RHYNIOPHYTA

Although trilete spores are known from the late Ordovician and early Silurian, undisputed vascular plants are not found until almost half-way through the Silurian. One of the earliest of these spore-bearing plants is *Cooksonia*. Several species are now known ranging from the mid-Silurian to the Lower Devonian. It was evidently widespread, occurring in a number of localities in North America and Europe. The plants was dichotomously branched (Fig. 6.1) and probably formed swards, perhaps in swampy places, a few centimeters in height. The axes, which were bare of any appendages, terminated in discoidal or globose sporangia with little evidence of predetermined dehiscence. So far as is known *Cooksonia* was homosporous. Vegetative axes have been found with a simple strand of tracheids, but the presence of stomata has been confirmed only in forms from the earliest Devonian. There is as yet no evidence of extensive aerating systems in *Cooksonia* which might support the suggestion that carbon dioxide was taken up through the underground organs (as in a few living plants).

Cooksonia is placed in the division Rhyniophyta, together with a later axial plant, *Rhynia,* from the Rhynie Chert, a siliceous rock of Lower Devonian age in Scotland. *Rhynia gwynne-vaughanii* had branching aerial stems arising from a horizontal rhizome attached to the substratum by fine rhizoids. The ascending axes, with both dichotomous and lateral branching, reached a height of about 20 cm (8 in.). Some of the axes terminated in fusiform sporangia. The sporangium was apparently shed after release of the spores, the fertile axis then being overtopped by the subjacent lateral so giving the branching a monopodial character. Both the rhizome and aerial shoots had a central strand of tracheids, the protoxylem lying at the center and the tracheids surrounded by cells resembling the leptoids of mosses. *Aglaophyton* (Fig. 6.2) is another Rhynie plant, similar to *Rhynia* (and formerly known as *R. major*), but the branching of the aerial axes was wholly dichotomous. The central strand consisted of uniformly thickened cells. *Horneophyton* is also of the Rhynie assemblage. This plant had dichotomizing aerial axes reaching a height of about 20 cm (8 in.). They arose from a sequence of cormlike structures, each richly furnished with unicellular rhizoids. The axes terminated in sporangia in which the sporogenous tissue surrounded a central columella, recalling the situation in the sporogonium of *Anthoceros*. Unlike *Anthoceros*, the sporangia of *Horneophyton* often dichotomized, occasionally as many as three times.

The Rhynie plants (which were beautifully preserved in the siliceous matrix) had all the characteristics of land plants. They possessed a well-defined cuticle, stomata, and air spaces in the cortex. The spores, as evidenced by the triradiate scar, were produced in

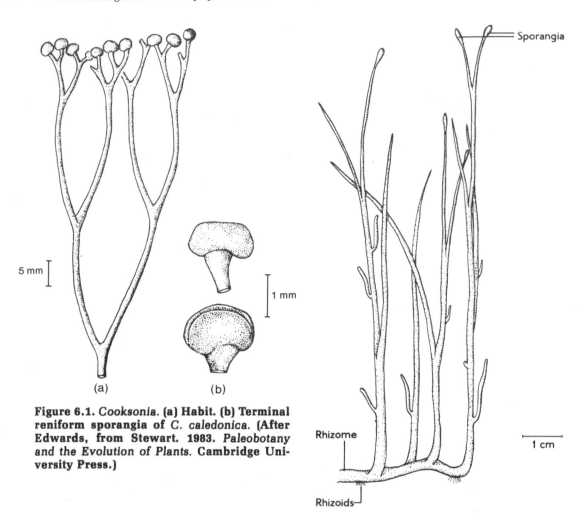

Figure 6.1. *Cooksonia.* (a) Habit. (b) Terminal reniform sporangia of *C. caledonica.* (After Edwards, from Stewart. 1983. *Paleobotany and the Evolution of Plants.* Cambridge University Press.)

Figure 6.2. *Aglaophyton (Rhynia) major.* (After Delevoryas. 1962. *Morphology and Evolution of Fossil Plants.* Holt, Rinehart and Winston, New York; from a reconstruction in the Chicago Natural History Museum.)

tetrahedral tetrads (Fig. 6.8) and were protected by a resistant wall, presumably containing sporopollenin. It appears very likely that they were examples of early archegoniate plants, and, from the presence of tetrads in the sporangia, that they represented the diploid sporophytic phase of a life cycle. The existence of a gametophytic phase has long been suspected, but examples have proved extremely elusive. The most convincing candidate is *Lyonophyton,* also found in the Rhynie Chert. This appears to have consisted of ascending axes, similar to those of *Rhynia,* but terminating above in a bowl-shaped structure. On the inner surface of the bowl were rounded emergences with dark contents. These are believed to have been antheridia. Other sections show short tubular extensions arising from the central region of the disc. These may have been the necks of archegonia. If *Lyonophyton* was indeed a gametophyte, and of an early tracheophyte and not a bryophyte, it would provide strong evidence for the homologous origin of the two phases of the life cycle. The alternative antithetic theory holds that the establishment of a distinct sporophytic phase was coincident with the colonization of the land. Although at first a simple structure, it became increasingly elaborate as the land flora evolved. However, the plants of the Lower Devonian, despite their great age, give no indications of the sporophyte being a comparatively recent development.

ZOSTEROPHYLLOPHYTA

The early vascular plants also included forms in which the sporangia were borne laterally. They are placed in a separate division, the Zosterophyllophyta. One representative, *Zosterophyllum* itself, is the earliest plant known possessing tracheids, a cuticle and stomata. It is known from the lowermost Devonian in several localities in Europe. It appears to have had a tufted habit, with the branches diverging at wide angles ("H-shaped" branching), the whole reaching a height of about 15 cm (6 in.). The xylem was exarch, and often elliptical in section. In *Zosterophyllum* the sporangia were clustered towards the ends of the aerial branches and dehisced along the distal convex surface. Also placed in the Zosterophyllophyta is *Sawdonia* (Fig. 6.3), found at a number of levels in the early Devonian. This had a creeping rhizome from which aerial branches, showing a monopodial tendency, rose to a height of about 30 cm (1 ft.). The aerial axes were clad with tapering spines, and the ultimate branches were circinately coiled at the tips. Stomata were present in the epidermis, and within the axes a solid core of tracheids with reticulate thickening. The sporangia were borne laterally in loose spikes towards the ends of the branches. *Deheubarthia*, widespread in the Lower Devonian of South Wales, is similar to *Sawdonia*.

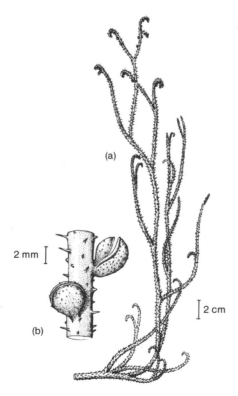

Figure 6.3. *Sawdonia.* **(a) Habit. (b) Sporangia showing transverse dehiscence. ([a] After Ananiev and Stepanov. [b] After Gensel. Both from Stewart. 1983.** *Paleobotany and the Evolution of Plants.* **Cambridge University Press.)**

A Rhynie plant, *Asteroxylon,* has also been referred to the Zosterophyllophyta. The branching was monopodial with dichotomizing lateral branches, the whole probably reaching a height of approximately 50 cm (19 in.). The aerial axes bore spinelike enations about 5 mm (0.2 in.) long, lacking a vascular strand but furnished with stomata. The sporangia were reniform and terminated distinct stalks between the enations towards the tips of the branches. The stem in section shows a stellate stele (actinostele) with mesarch xylem. The cortex displayed at least three zones, the middle consisting of plates of cells separated by large air spaces. Vascular strands traversed the cortex. They entered the stalks of the sporangia but stopped short of the enations. The anatomy suggests that *Asteroxylon* was a marsh plant.

TRIMEROPHYTA

The Trimerophyta, the third group of ancient plants, were not unlike the Rhyniophyta, but are distinguished by a general tendency to a monopodial habit and by laterals which branched freely, either dichotomously or even trichotomously. The stems were either naked or sometimes furnished with small spines. The xylem was in the form of a protostele with the protoxylem towards the center and the metaxylem tracheids often with scalariform thickening. The genus *Psilophyton* (Fig. 6.4) from the Lower Devonian is, as currently emended, representative of the trimerophytes. In some forms the monopodial habit was well established. In *Pertica,* for example, from later Lower Devonian beds in North America there was a distinct main axis with the lateral branches inserted in four ranks. The sporangia of the trimerophytes were terminal, usually fusiform, and often in tassels. In *Pertica,* 50 or more occurred crowded together at the end of a spray of branchlets. So far as is known all trimerophytes were homosporous.

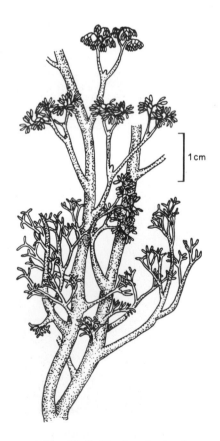

1 cm

Figure 6.4. *Psilophyton dawsonii.* **(After H. P. Banks, S. Leclercq, and F. M. Hueber. 1975.** *Palaeontographica Americana* **8.)**

SPORE-BEARING TRACHEOPHYTES WITH LIVING REPRESENTATIVES: THE FERN ALLIES

The plants falling into divisions Psilopsidophyta, Lycophyta, and Sphenophyta are commonly referred to as the "fern allies."

PSILOPSIDOPHYTA

The division Psilopsidophyta is represented solely by a few living forms. They constitute a single class, the Psilopsida, with the following characteristics.

Sporophyte consisting of more or less dichotomously branching axes, often with small leaflike appendages. Roots absent, the subterranean axes bearing rhizoids. Vascular tissue consisting of tracheids and ill-defined phloem. Sporangia terminal on short branches, homosporous. Gametophyte subterranean, sometimes with vascular tissue, resembling portions of the sporophyte rhizome. Spermatozoids multiflagellate. Embryogeny exoscopic.

Only two genera of living psilopsids are known, *Psilotum* (Fig. 6.5a) and *Tmesipteris* (Fig. 6.5b). The former is pantropical and not uncommon, but the latter, although locally abundant, is confined to Australasia and Polynesia.

The sporophyte of *Psilotum*, which may be either terrestrial or epiphytic, consists of upright (or, in one epiphytic species, pendulous), dichotomously branching axes arising

Figure 6.5. (a) *Psilotum nudum.* **Fertile region. The trilocular sporangia are subtended by small forked bracts. Scale bar 0.5 cm. (b)** *Tmesipteris tannensis.* **Fertile shoot. The bilocular sporangia are attached at the forks of conspicuous bifid bracts. Scale 0.5 cm.**

from a horizontal system of similarly branching rhizomes. The rhizomes bear rhizoids and contain an endophytic fungus, probably in symbiotic association (**mycorrhiza**). Small scales, which are at first green but soon become scarious, occur at irregular intervals on the stem. They possess no vascular strand, although in one species a strand does approach the base of the scale but stops below the insertion. *Psilotum* possesses stomata, but here they are confined to the epidermis of the stem.

The stem of *Psilotum* contains a simple stele (protostele), frequently enclosing in the upper regions a central parenchymatous medulla. The xylem, consisting solely of tracheids, is often stellate in transverse section, the arms narrowing and standing opposite poorly defined ribs at the exterior of the stem (Fig. 6.6). The xylem is exarch in the aerial stems, but tends to be mesarch in the rhizome. Phloem surrounds the xylem, but, apart from the lateral sieve areas, the sieve cells are little different from elongated parenchyma cells. An endodermis, separated from the phloem by a narrow zone of pericyclic parenchyma, marks the boundary of the stele. *Psilotum* is the simplest plant in which an endodermis is found. Its cells, forming a usually single-layered sheath around the stele, are characterized by the **Casparian strip,** a band of fatty material embedded in the radial walls. This provides a hydrophobic barrier to the lateral movement of solutes in the apoplastic space, possibly giving the living protoplasts of the endodermal cells greater control of the movement of substances, including mineral ions, in and out of the stele. The cells of the inner cortex, close to the endodermis, often contain accumulations of phlobaphene, a condensation product of tannin.

The branches of both the aerial and terrestrial systems of the sporophyte grow indefinitely from single or small groups of apical cells. There is no evidence that the dichotomy of the axes follows median longitudinal division of a single apical cell, as in certain algae (e.g., the brown alga, *Dictyota dichotoma*).

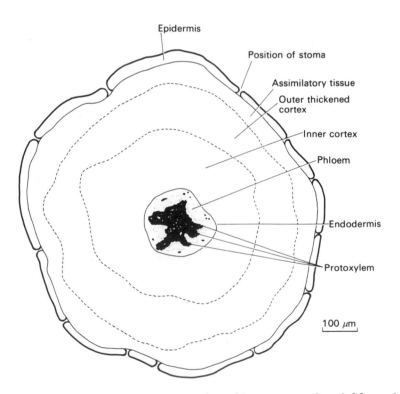

Figure 6.6. *Psilotum nudum.* **Transverse section of lower part of aerial branch.**

Tmesipteris is frequently an epiphyte with trailing stems (Fig. 6.5b). The general morphology is similar to that of *Psilotum,* but branching is much less frequent. The appendages are larger and more leaflike, remaining green and possessing stomata, and also frequently a vascular strand. The insertion of the appendages is however, peculiar, being longitudinal instead of transverse, so that they appear more as flangelike outgrowths of the axis than as normal foliage leaves.

Reproduction of the Psilopsidophyta. Spores are produced in the upper region of the sporophyte of *Psilotum* (Fig. 6.5a). The spore-bearing organs, which are distant from each other, are three-lobed, and each is subtended by a bifid appendage. The lobes correspond to three internal chambers (Fig. 6.7), separated by septa, and each filled with spores. It is still not clear whether this spore-bearing organ is to be interpreted as a trilocular sporangium or as a synangium formed by the fusion of three sporangia. Three primordia are, however, visible early in the ontogeny of the organ, perhaps indicating its synangial nature. Since a distinct vascular strand extends into the base of the synangium, it is usually regarded as terminating a lateral axis rather than as arising in association with a sporophyll. This view is strengthened by the existence of a cultivated form, "Bunryu-zan," in which the synangia terminate vertical axes.

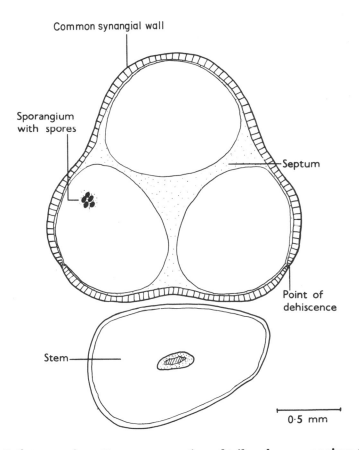

Figure 6.7. *Psilotum nudum.* Transverse section of trilocular sporangium (synangium).

Each of the three groups of archesporial cells in the synangium arises from several cells and is thus **eusporangiate** in origin. Part of the peripheral archesporium, although not regularly layered, functions as a tapetum. The remainder yields sporogenous cells. Meiosis leads to tetrads of spores, each bilaterally symmetrical with a single linear scar (**monolete** spores). Such spores can be considered as formed by the bisection of two conjoined hemispheres by a plane perpendicular to the equator (Fig. 6.8). The linear scar thus represents the line of common contact between the four segments. The exine is lightly reticulate with no very distinct ornamentation. The massive wall of the sporangium, some five cells thick at maturity, dehisces at three sites symmetrically placed with respect to the loculi.

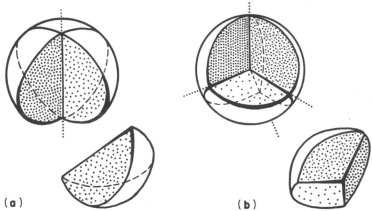

Figure 6.8. Diagrammatic representation of manner of formation of (a) monolete and (b) trilete spores.

The gametophyte of *Psilotum* is a subterranean axial structure (Fig. 6.9), dichotomously branching, and resembling short lengths of the sporophytic rhizome. The similarity extends to the anatomy, the finer axes being wholly parenchymatous, and the broader containing a central vascular strand. The peripheral cells, like those of the rhizome, house an endophytic fungus. Both the gametophyte and the sporophytic rhizome produce globular multicellular gemmae, a means of vegetative propagation.

Antheridia and archegonia arise from superficial cells in the region of the growing points of the gametophyte (Fig. 6.9). The antheridia, depending upon their size, liberate up to 250 spermatozoids. The archegonium has four tiers of neck cells, but at maturity all but the lower one or two tiers degenerate. Fertilization, which depends upon the presence of a film of water, is brought about by spirally coiled, multiflagellate spermatozoids.

The first division of the zygote is in a plane transverse to the longitudinal axis of the archegonium, as in the bryophytes. The outer (or **epibasal**) cell yields the apex of the embryo, and the inner (or **hypobasal**) the foot. The embryogeny is thus exoscopic.

In *Tmesipteris,* spore-bearing organs occur in the upper parts of some of the shoots, each subtended by a bifid appendage (Figs. 6.5b and 6.10). These organs are regarded as synangia terminating very short lateral branches, but in *Tmesipteris* each consists of only two fused sporangia.

The reproduction of *Tmesipteris* is very similar to that of *Psilotum*. The foot of the young embryo of *Tmesipteris* is lobed, and the whole bears a striking resemblance to the young sporophyte of *Anthoceros*. It is doubtful, however, whether this bears the phylogenetic significance that some have claimed.

Origin of the Psilopsidophyta. The anatomical and reproductive features of the Psilopsidophyta recall the Rhyniophyta and Trimerophyta of the Paleozoic. There is, however, no continuity in the fossil record, and the origin of the psilopsidophytes remains conjectural. Their general high chromosome number (52 to 210 in the gametophytic phase) may indicate a complex polyploid series developed with little evolutionary change over geological time, but evidence of this kind is inconclusive. A remarkable similarity, both in sporophyte and gametophyte, between the psilopsids and certain New Caledonian ferns suggests that the affinities of the division may lie with the Filicales (Chapter 7) rather than with the Rhyniophyta or Trimerophyta.

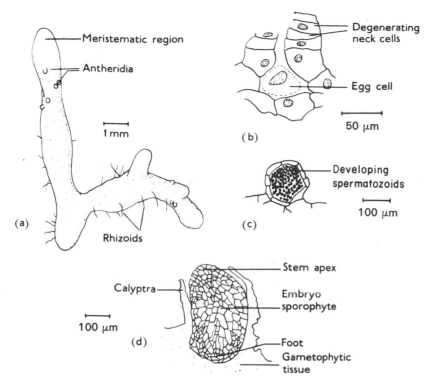

Figure 6.9. *Psilotum nudum*. **(a) Gametophyte. (b) Mature archegonium. (c) Antheridium. (d) Young sporophyte. (All after D. W. Bierhorst. 1953.** *American Journal of Botany* **40; and D. W. Bierhorst. 1954.** *American Journal of Botany* **41.)**

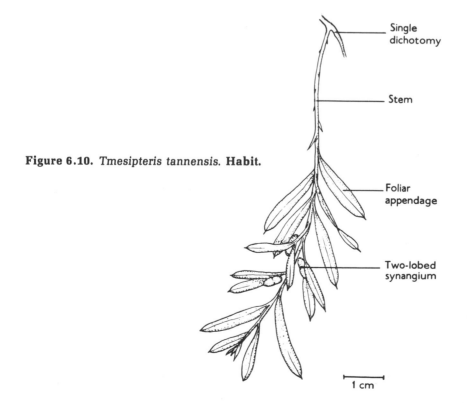

Figure 6.10. *Tmesipteris tannensis*. **Habit.**

LYCOPHYTA

The division Lycophyta contains both living and fossil plants. It represents one of the most complete records of plant evolution currently known, extending from the Lower Devonian to the present day. The characteristics of the division can be summarized as follows:

Sporophyte consisting of more or less dichotomously branching axes, but differentiated into root and shoot. Shoot bearing microphylls, each containing a single vein, but leaf trace leaving no gap in the stele. Vascular tissue consisting of tracheids and phloem. Sporangia in or near axils of microphylls, homo- or heterosporous. Gametophyte (in living forms) terrestrial or subterranean. Spermatozoids flagellate. Embryogeny endoscopic.

The Lycophyta fall into a single class, the Lycopsida. Six orders are recognized: three containing the living species and their undoubted fossil relatives, and three wholly fossil. The living lycophytes, although more numerous than the psilopsidophytes, are again a minor component of contemporary vegetation. The fossil record, however, shows that the lycophytes were abundant in the Carboniferous period, many then being represented by substantial trees. Only herbaceous forms have persisted until the present day.

The Lycopodiales. Members of the order Lycopodiales, which include the genera *Lycopodium* (club mosses) and the Australasian *Phylloglossum,* are distinguished by their eligulate leaves and homospory.

Lycopodium (Fig. 6.11) with about 20 species is distributed throughout the world from Arctic to tropical regions, with a related range in the growth form of the sporophyte. The colder climates favor species with short, erect stems, or creeping stems giving rise to short upright side branches, while those in the tropics have much laxer growth, are sometimes stoloniferous, and are often epiphytes. Taxonomists are now inclined to assign species of *Lycopodium* to a number of smaller genera. For present purposes however the concept of *Lycopodium* will be retained, and alternative nomenclature will be indicated where appropriate.

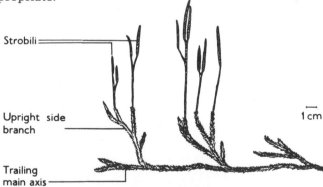

Figure 6.11. *Lycopodium clavatum.* **Habit.**

The stem is surrounded by microphylls, a form of leaf which is typically small, simple in outline, and with a single median vascular strand. Some species of *Lycopodium,* especially the epiphytic, are heterophyllous, the lateral rows of leaves being expanded in the plane of the shoot system and the upper and lower rows appressed and smaller. The fertile leaves (sporophylls) may be similar to the sterile, as in *L. (Huperzia) selago,* and the fertile regions not clearly set off from the sterile along the axis. More usually, however, the sporophylls differ from the sterile leaves in size, shape, and the extent of the chlorophyllous tissue, and are often grouped together, as in *L. clavatum,* in distinct cones (strobili) of determinate growth.

Roots arise endogenously, emerging from the underside of the stem in the prostrate species, or from near the base of the stem in upright species. In some of the latter, initiation of the roots occurs near the shoot apex, but, instead of emerging there, the roots grow down inside the cortex and break out only when they reach the level of the soil.

The stem, which grows from a group of initial cells (Fig. 6.12), contains a central stele. Although basically a protostele, the detailed anatomy shows considerable variation with species. In its simplest form, the stele consists of a core of tracheids, more or less stellate in section, with phloem lying between the arms. In other species, the xylem and phloem form parallel bands (Fig. 6.13) (*plectostele*) or may be intermingled, anastomosing strands of sieve cells being scattered among the tracheids. In every instance, differentiation of the xylem begins at the exterior and then proceeds centripetally, leaving no undifferentiated tissue at the center. The xylem is thus uniformly exarch.

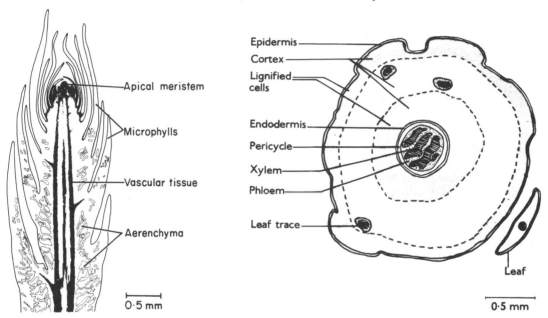

Figure 6.12. *Lycopodium (Diphasiastrum) alpinum.* **Longitudinal section of apical region of vegetative shoot.**

Figure 6.13. *Lycopodium clavatum.* **Transverse section of stem.**

The sieve cells of the phloem are elongated with steeply inclined end walls. The cell contents are partly degenerate, but associated parenchyma cells have well structured protoplasts. The sieve plates occur on both the lateral and tapering end walls. The regular presence of callose has not been confirmed. The vascular tissue, entirely primary, is usually surrounded by a narrow zone of parenchyma, and this in turn by an endodermis, the cells of which have a distinct Casparian strip.

The leaves, the chief site of photosynthesis, are structurally simple. There are abundant stomata on both surfaces, and internally numerous intercellular spaces. There is, however, no clearly differentiated mesophyll.

Reproduction of the Lycopodiales. The sporangia, borne singly in the axil of a sporophyll or close to its insertion (Fig. 6.14), are eusporangiate in origin. Continued cell division within the primordium leads to a central mass of spore mother cells surrounded by a wall several cells thick. The inner layers of the wall function as a tapetum, breaking down to provide materials which contribute to the development and maturation of the spores. Following meiosis, the four young spores arrange themselves in a tetrahedral fashion so that each spore has three triangular faces at its proximal pole where it is in contact with its fellows (Fig. 6.8). As the spores separate, these areas become less thickened than elsewhere, and the edges between them conspicuous, often appearing as a prominent triradiate scar in the mature spore (**trilete** spores). The sporangia dehisce transversely along a line of thin-walled cells (stomium) and the minute spores (each about 50 μm in diameter) are distributed by wind.

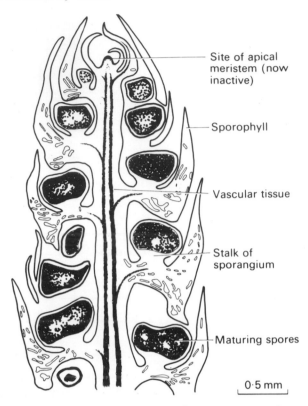

Figure 6.14. *Lycopodium clavatum.* **Longitudinal section of upper region of young strobilus.**

In some species of *Lycopodium,* the exine of the spore has quite elaborate reticulate ornamentation, while in others it is almost smooth. The rough spores are not easily wetted and their germination may be delayed for several years, probably until weathering and attrition have rendered the coat permeable. The process may be simulated in the laboratory by immersing such spores in concentrated sulfuric acid. Following this treatment, the spores germinate freely in pure culture. In natural conditions, initially unwettable spores may be washed deep into the soil before germination occurs, and this is reflected in the nature of the gametophyte. For example, in *L. clavatum,* the spores of which have a pronounced reticulate relief, the gametophyte is a subterranean saucer-shaped structure, growing saprophytically and persisting for several seasons. The sex organs are produced on a cushion in the central region. *Lycopodium volubile* has a similar spore and gametophyte (Fig. 6.15). In the tropical *L. (Lycopodiella) cernuum,* where the spores are smooth, germination and development occur rapidly, producing lobed, cup-shaped gametophytes which possess chlorophyll and last for little more than a single season. In *L. (Huperzia) selago,* in which the spores are somewhat intermediate in the development of the wall, there is a corresponding ambivalence in the habitat of the gametophyte. In all instances, the gametophyte of this species is a small carrot-shaped body growing saprophytically and producing sex organs on the upper cushion. It may, however, be either buried or at the surface, and, if the latter, the upper part becomes pale green. There is thus a general relationship between the nature of the spore wall and the form and duration of the gametophytic plant.

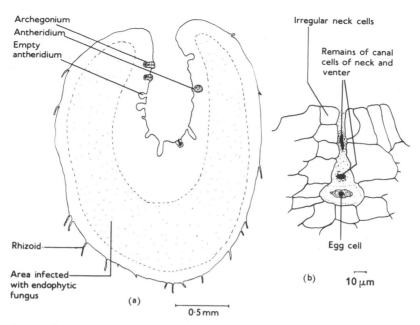

Figure 6.15. *Lycopodium volubile.* **(a) Vertical section of subterranean gametophyte. (b) Mature archegonium.**

The gametophytes of many species, including all those gametophytes which are subterranean, are invested with a mycorrhizal fungus. Provided the medium is appropriate, such gametophytes can nevertheless be grown in axenic culture.

The sex organs of *Lycopodium* are similar to those of *Psilotum*, except that in many species the archegonia, though immersed, have more conspicuous necks, and both the male and female gametangia are enclosed to a greater extent by vegetative tissue (Fig. 6.15). The spermatozoids are biflagellate, about 8 μm long, and spindle-shaped. The ultrastructure is basically similar to that of bryophyte spermatozoids, but the nucleus is less elongated and not coiled. Fertilization is probably facilitated by chemotactic attraction of spermatozoids to archegonia. The zygote divides by a wall transverse to the axis of the archegonium. The outer cell, termed the **suspensor**, divides no further, but the inner remains meristematic and gives rise to two regions of cells. The central region becomes the foot, while the inner differentiates into the root, first leaf, and stem apex of the embryo proper. This development results in the embryo being directed inwards, and the embryogeny is consequently said to be endoscopic.

At first the apex of the embryo points downwards, but expansion of the foot region pushes it to one side and the young sporophyte finally breaks out of the surface of the gametophyte. Before it becomes fully established, the young plant depends upon food materials absorbed from the gametophyte, probably through the foot. In some species of *Lycopodium* (e.g., in *L. [Lycopodiella] cernuum*) the differentiation of the embryo is delayed and it emerges as a parenchymatous protuberance, termed a protocorm. This eventually gives rise to one or more growing points, each of which yields a normal plant. Apogamy has been observed in *L. cernuum* in culture, but is not known in nature. *Lycopodium selago* commonly produces budlike gemmae in the axils of the upper leaves.

Phylloglossum, confined to Australasia, is the only other living genus of the Lycopodiales. Its single species, *P. drummondii*, consists of an upright sporophyte, reaching 5 cm (2 in.) or less, with a basal whorl of leaves and a pedunculate strobilus (Fig. 6.16). During growth, a lateral axis arises near the base of the plant and extends down into the soil, its tip eventually becoming transformed into a tuber. This tuber forms an organ of perennation, persisting through the dry period (when the remainder of the plant perishes) and giving rise to the following year's growth. The gametophyte of *Phylloglossum* is similar to that of *Lycopodium (Lycopodiella) cernuum,* and there is an analogy between the protocorm stage in the development of the sporophyte of this species and the perennating tuber of *Phylloglossum.* Although strikingly different from *Lycopodium* in habit, *Phylloglossum* is clearly not distant in the basic features of anatomy and reproduction. It is no doubt a form that has become specialized in relation to a particular kind of habitat.

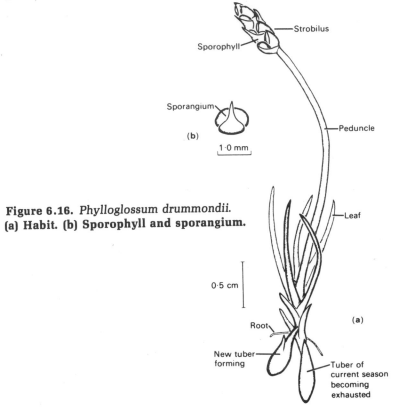

Figure 6.16. *Phylloglossum drummondii.*
(a) Habit. (b) Sporophyll and sporangium.

The Selaginellales. The habit of the order Selaginellales resembles that of the Lycopodiales, but the microphylls are ligulate, the sporophylls are always aggregated into distinct strobili, and the spores are of two kinds (produced in separate sporangia), differing in size and in the sexes of the gametes they eventually produce.

There are more than 700 species of *Selaginella,* most of which are tropical, ranging from small epiphytes to large climbing plants. A few species of arid areas are capable of flexing their branches and, in a rolled-up form, resisting prolonged desiccation. The basic cell structure and enzymes are conserved in the dehydrated cells. Normal form and metabolism are rapidly resumed on moistening ("resurrection plants"). The stems are commonly much more branched than in *Lycopodium.* In scandent species, the branching of the laterals often takes place in a single plane. Leaves originate in a spiral sequence, but the arrangement may be modified during subsequent development. In *Selaginella selaginoides* the leaves are all similar (**isophylly**) and remain radially arranged. Many species, however, display anisophylly (**heterophylly**). The two lower rows of leaves are then expanded laterally, the two upper remaining smaller and appressed (Fig. 6.17a). In scandent species, the anisophylly is frequently confined to the lateral branches, enhancing the frondlike appearance of the whole spray.

In young plants (Fig. 6.17b), roots are initially produced from the base of the stem.

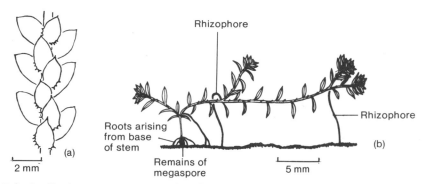

Figure 6.17. *Selaginella kraussiana.* **(a) Anisophyllous shoot viewed from above showing arrangement of upper and lower rows of leaves. (b) Young sporophyte.**

Although in a few species roots continue to be produced from a tuberous stem base, in others later roots emerge principally from leafless **rhizophores** (Fig. 6.17b). These originate from the stem at points of branching and grow down towards the soil, dichotomizing as they approach its surface. Contact apparently acts as a stimulus, causing roots to arise endogenously and penetrate the substratum.

Structurally the leaves of *Selaginella* resemble those of *Lycopodium,* but in a few species (such as the Mexican *Selaginella schaffneri*), instead of a single vein, a number of veins radiate from the insertion and branch dichotomously at their tips. In some tropical species (e.g., *S. wildenowii*) leaves shaded by other foliage develop a striking blue sheen (iridescence). This depends upon the presence of two superimposed layers of polysaccharide in the outer walls of the upper epidermis. These act as an interference filter. When the leaf is flooded, water enters the space between the two layers and the iridescence fades. In a few species of *Selaginella* (e.g., *S. wildenowii*) the cells of the upper epidermis of the leaf contain a single cup-shaped chloroplast, similar to that of *Anthoceros* but lacking a pyrenoid.

Diagnostic of *Selaginella* is the insertion of a minute tonguelike ligule into the upper side of the leaf close to the axis. This organ appears early in the life of the leaf, but remains colorless since the plastids fail to develop internal lamellae. Its cells contain prominent Golgi bodies. The whole structure is evidently specialized for the production and secretion of extracellular mucilage. Its activity, however, is of limited duration. The ligule dies and shrivels to a papery fragment as the leaf expands.

The axis of *Selaginella,* unlike that of *Lycopodium,* grows from a well-defined apical initial whose divisions are wholly anticlinal. In general, the anatomy of the axis is little different from that of *Lycopodium,* but the endodermis is of a kind unknown elsewhere in the Plant Kingdom. At maturity, it consists of elongated hyphalike cells which suspend the stele in a central cavity (Fig. 6.18). The development of the endodermis begins just beneath the apex, and its peculiar form must be a consequence of the greater expansion of the cortical region than of the vascular, so generating the intervening space. The endodermal cells remain alive and, as is usual in a well-formed endodermis, the plasmalemma forms a tight junction with the Casparian strip embedded in the radial walls.

The stele of *Selaginella* is basically a protostele lacking internal parenchyma, but it is often ribbonlike instead of cylindrical, and, especially in aerial axes, several steles may ascend the stem together (polystely). The xylem is regularly exarch. The tracheids of the metaxylem often lack end walls, and in several instances can be regarded as forming authentic vessels. The end walls of the sieve cells tend to be more or less transverse, and pores are found in both the end and lateral walls. In the rhizomes of a few species the xylem is in the form of a hollow cylinder lined on both surfaces by phloem. An endodermis also occurs both externally and internally. Such a stele, which is said to be

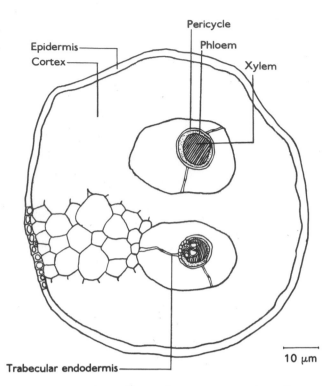

Figure 6.18. *Selaginella* **sp. Transverse section of stem.**

amphiphloic and is referred to as a *solenostele,* clearly exhibits a more complicated pattern of differentiation than a protostele, but the factors controlling differentiation of this kind are still little known.

The meristems giving rise to rhizophores are indistinguishable from those of shoots in their early stages, but they appear to be indeterminate. If a branch is removed, the adjacent primordium will grow out to form a shoot instead of a rhizophore. However, if the branch is replaced by a source of the growth-regulating substance indole-3-acetic acid (IAA), the primordium behaves normally. This was one of the first and most striking demonstrations of how growth-regulating substances, diffusing from one area to another in a plant, maintain the familiar pattern of morphogenesis. The form of growth in *Selaginella* seems to be particularly labile. The apices of roots in pure culture in the absence of auxin may begin to produce leaf primordia and generate a shoot.

Reproduction of the Selaginellales. The sporangia of *Selaginella* are borne in strobili, which terminate the main or lateral axes. The sporophylls are usually in four ranks. Each sporangium lies in the axil of a sporophyll, between the ligule and the axis (Fig. 6.19). There are two kinds of sporangia, producing mega- and microspores respectively, located in different regions of the cone. The initial development of each kind of sporangium is the same, and closely resembles that of the sporangia of *Lycopodium.* Development diverges with the formation of the spore mother cells. In the microsporangia all the mother cells undergo meiosis and form microspores, but in the megasporangia not all the sporogenous cells complete their development. Only one or a very few mother cells yield spores, and not all of these may persist. The consequence is that the number of spores reaching maturity is very small and ranges, depending upon species, from one (e.g., *Selaginella sulcata*) to about twelve. The fully differentiated megaspores are conspicuous for their size, their store of food materials built up at the expense of the resorbed abortive sporogenous tissue and spores, and the thickening and ornamentation of their walls. They are in consequence some of the most remarkable spores in the Plant Kingdom; those of *S. exaltata,* for example, may exceed 1 mm (0.04 in.) in diameter.

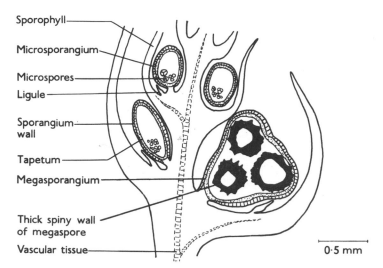

Figure 6.19. *Selaginella kraussiana.* **Longitudinal section of strobilus.**

In some species of *Selaginella* the wall of the fully ripe microsporangium everts violently and the spores are forcibly ejected. Other species lack any special form of dehiscence and the release of the microspores, like that of the megaspores generally, is largely passive. When the megaspores are shed, microspores are often found clinging to them. This early association may lessen the hazards of sexual congress.

Some nuclear divisions occur in the spores while they are still in the sporangia, but growth of the gametophytes, leading to rupture of the spore wall, does not resume until after the spores are shed. In the microspore, an unequal mitosis produces a large antheridial cell and a small cell, termed a prothallial cell, which represents the sole development of the somatic tissue of the gametophyte. The antheridial cell continues to divide and yields a normal antheridium, from which 128 or 256 biflagellate spermatozoids are eventually liberated. Each is about 25 μm long, with a narrow elongated nucleus in which longitudinal strands, probably of chromatin, become apparent during differentiation. In the megaspore, the food material comes to occupy a central position, and free nuclear division occurs at its periphery. Subsequently, cell formation begins beneath the triradiate scar, which is eventually forced open by the general swelling, exposing a cap of gametophytic tissue (Fig. 6.20). The somatic tissue of the female gametophyte is thus more extensive than that of the male, although chlorophyll is quite absent from both. There is some specific variation in the extent to which the cellular portion of the female gametophyte is delimited from the food supply below. In some species, the boundary is imprecise and cell formation gradually extends down into the lower region, but in other species a distinct diaphragm separates the upper cellular region from a largely acellular food reserve.

The female gametophyte, once exposed, protrudes as an irregular cushion bearing rhizoids at its margin and, in the central region, archegonia. The necks of the archegonia are very short, consisting of no more than two tiers of cells (Fig. 6.21). Fertilization necessarily depends upon a microspore germinating close to a megaspore, and a film of water being present when the gametangia are mature. The rhizoids around the female gametophyte may, in these conditions, serve to retain a "fertilization drop" above the archegonia in which the spermatozoids congregate.

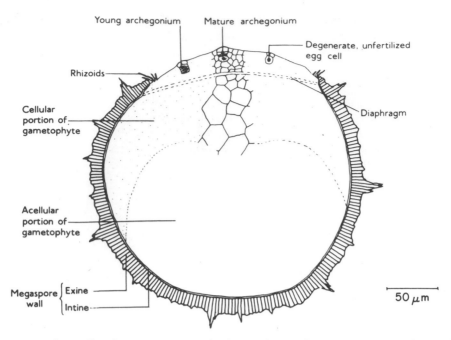

Figure 6.20. *Selaginella kraussiana*. **Vertical section of megaspore with endosporic gametophyte.**

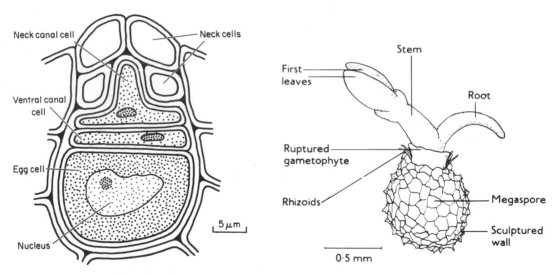

Figure 6.21. *Selaginella kraussiana*. **Mature archegonium. (From an electron micrograph.)**

Figure 6.22. *Selaginella kraussiana*. **Sporophyte emerging from gametophyte.**

The embryogeny of *Selaginella* is like that of *Lycopodium* in being endoscopic, but differs in the greater development of the suspensor, and the wide variation in detail between species. In *S. selaginoides,* for example, elongation of the suspensor pushes the developing embryo down into the food reserve. In *S. kraussiana,* however, in which the food reserve is cut off by a diaphragm, the development of the suspensor is markedly less, but a curious downward extension of the archegonial canal carries the embryo through the diaphragm into the center of the gametophyte. Apart from these features, there is also variation in the development of the foot region of the embryo and in the relative positions in which the various parts of the embryo arise. In all species the embryo eventually emerges from the upper surface of the gametophyte (Fig. 6.22).

The substantial thickening of the megaspore wall in *Selaginella* probably protects the spore for considerable periods from desiccation and decay. Once conditions are favorable for germination, development is rapid and, because of the considerable food reserve of the megaspore, independent of an external supply of nutrients. Compared with the life cycle of the homosporous *Lycopodium,* there is considerably less time spent in the gametophytic phase. Since this, in view of the delicacy of the gametophytic tissues and their dependence on the maintenance of humid conditions, is the most vulnerable phase of the life cycle, the modifications that lead to its curtailment no doubt confer a considerable selective advantage on *Selaginella.* This is perhaps reflected in its numerous species.

In some species of *Selaginella* (e.g., *S. rupestris*) young plants emerge from the female regions of strobili. This has been regarded as following from the lodging of microspores between the megasporophylls and the fertilization of an egg while the megaspore was still *in situ,* thus simulating an early step in the evolution of a seed. It is doubtful, however, whether this can be substantiated. In *S. rupestris,* at least, reproduction is apogamous, the megaspore producing an embryo directly without fertilization.

The Isoetales. The members of the order Isoetales have a remarkable, rushlike habit, quite unlike that of any other lycopsid. The short, fleshy, upright rootstocks occasionally show one or two dichotomies. Each branch bears a tuft of quill-like microphylls (Fig. 6.23). The microphylls are ligulate, and reproduction is heterosporous. All living representatives contained in the two genera *Isoetes* and *Stylites* are aquatic or plants of situations subject to periodic or seasonal inundation. The leaves of the aquatic species lack stomata.

Isoetes is widely distributed. The rootstock in all species is mostly below the level of the substratum and is rarely branched. The meristem at the upper end of the rootstock is depressed. Although it often shows a conspicuous apical cell, divisions of this cell are

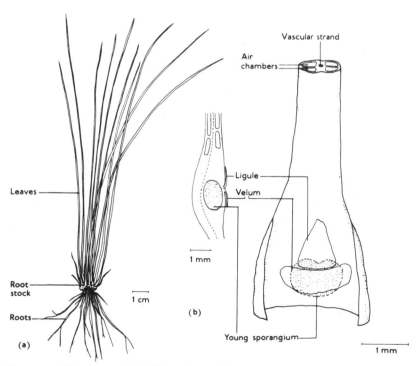

Figure 6.23. *Isoetes echinospora.* **(a) Habit. (b) Leaf base with young sporangium in face view and longitudinal section.**

both anticlinal and periclinal, not strictly anticlinal as in *Selaginella*. The leaves, which in some aquatic species may reach a length of 70 cm (27 in.), are arranged, at first distichously, but subsequently in a dense spiral around the meristem. The roots arise from the lower end of the stock where the meristem is again depressed, but here lies extended along a transverse cleft. The roots are initiated at the base of the cleft (Fig. 6.24).

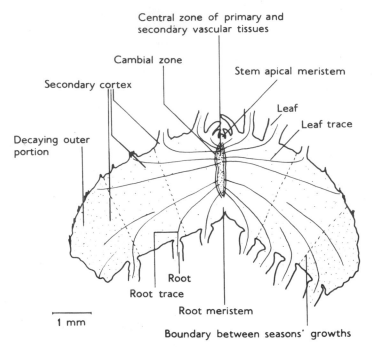

Figure 6.24. *Isoetes echinospora.* **Median vertical section of rootstock in plane perpendicular to that of the basal cleft.**

Accompanying this singular morphology is an equally remarkable manner of growth. A cambial zone consisting of more or less isodiametric cells arises around the small amount of primary vascular tissue in the stock, but it contributes more to the cortex than to the stele. The activity of this cambium, in temperate species at least, is seasonal. In step with the addition of new material within, a girdle of outer tissue, complete with its decaying leaves above and roots beneath, sloughs away. Consequently, having reached its mature diameter (which may exceed its length), the stock remains more or less the same size, the new leaves and roots being carried up on to the shoulders of their respective meristems by the expansion of the products of the anomalous cambium.

The anatomy of the stock presents a number of peculiar features. The primary xylem consists of more or less isodiametric tracheids, and at the base of the stele they are arranged in an anchorlike bifurcation lying in the same plane as the basal cleft of the stock. The tissue produced on the inside of the anomalous cambium differentiates as a mixture of tracheids, sieve cells, and parenchyma. The remainder of the tissue in the stock is parenchymatous, and no recognizable endodermis delimits the vascular tissue.

The leaves contain a single vascular strand, often very tenuous, surrounded by four air canals, interrupted at intervals by transverse septa. These canals are especially striking in the aquatic species. The broadened leaf bases lack chlorophyll and overlap widely, forming a tight comal tuft. The roots possess a single vascular strand surrounded by a cortex of two distinct zones: an outer fairly resistant to decay, and an inner of more delicate tissue with numerous air spaces.

In a North American mat-forming species of *Isoetes* the leaf arrangement remains distichous, and the rootstock extends laterally in the plane of the leaves. Adventitious buds arise at the extremities of these extensions and give rise to additional plants. Closely allied to *Isoetes* is *Stylites,* hitherto found only the high Andes where it forms dense cushions by the sides of glacial lakes. The rootstock is dichotomously branched and may reach a height of 15 cm (6 in.). The roots are confined to a single furrow which runs along one side of the stock. The leaves regularly lack stomata. Carbon dioxide is apparently taken in through the roots. It has been suggested that *Stylites* may be a relic of astomatal forms which existed in Paleozoic times when archegoniate plants were beginning to invade the land.

Reproduction of the Isoetales. Mature plants of *Isoetes* are usually abundantly fertile. The leaves first formed in a season's growth bear megasporangia, and those formed later bear microsporangia, although the sporangia frequently abort on the last formed leaves. The sporangium is initiated much as in *Selaginella* between the ligule of the sporophyll and the axis, but distinctive features emerge as development proceeds. Part of the central tissue, for example, remains sterile and differentiates as trabeculae which divide the mature sporangium into a number of compartments. Also, the ripe sporangium becomes enclosed in a thin envelope (**velum**). This originates just below the ligule and grows down over the sporangium, leaving a central pore (foramen).

In the megasporangium, part of the sporogenous tissue degenerates, resulting in many fewer sporocytes than in the microsporangium. Mature megasporangia contain 100 or more megaspores, ranging from 200 μm to almost 1 mm in diameter. The microsporangia, in which there is no loss of sporocytes, may contain up to 10^6 spores, each about 40 μm in maximum diameter. *Isoetes* and *Stylites* are peculiar in that the microspores are monolete and the megaspores are trilete. Not only are these genera unique among the living lycopsids in producing monolete spores, but they are also among the few plants in which monolete and trilete spores are produced by the same individual. In a few species of *Isoetes,* microspores and megaspores are said to occur in the same sporangium, but the viability of the spores has not been tested. Reproduction in these species may be aberrant.

The spores are liberated by the decay of the sporophylls. Their subsequent germination and development are similar to those of the micro-and megaspores of *Selaginella.* In *Isoetes,* however, the male gametophyte is wholly endosporic. The single antheridium yields only four spermatozoids, differing from those of *Lycopodium* and *Selaginella* in being multiflagellate. They are released by rupture of the microspore wall. The female gametophyte is similar to that of *Selaginella.* Cell formation usually extends down into the body of the spore, and the diaphragm seen in many species of *Selaginella* is absent. Although initially endosporic, the expanding gametophyte ruptures the megaspore at the site of the triradiate scar, but, as in *Selaginella,* fails to develop chlorophyll. The archegonia are also similar, but the necks consist of four tiers of cells in place of two.

The first division of the zygote is slightly oblique. No suspensor is formed, but the embryogeny can still be termed endoscopic since the outer cell gives rise to the foot and the remainder of the embryo comes from the products of the inner cell. Differential growth causes the embryo to turn round so that it becomes directed towards the upper surface of the gametophyte. It eventually breaks through, but the young plant remains for some time partially enclosed by a sheath of gametophytic tissue.

Apogamy has been reported in triploid species of *Isoetes.* Asexual reproduction by the production of a bud in place of a sporangium is not uncommon.

Although *Isoetes* and *Stylites* are like other lycophytes in essentials, they share a number of outstanding features unrepresented elsewhere in archegoniate plants. They appear to be the products of a line of lycophyte evolution that has been independent for a considerable period.

The early Lycophyta. The **Drepanophycales** are represented by the remarkable *Baragwanathia* from Upper Silurian or Lower Devonian rocks of Australia. The terminal dichotomizing axes of this distinctive plant reached a diameter of 1–2 cm (0.4–0.8 in.), the basal axes probably being more massive. They bore needlelike leaves, arranged in a dense spiral and reaching up to 4 cm (1.6 in.) in length (Fig. 6.25). The stem contained a stellate stele from which slender strands ascended into the leaves. Some of the shoots had fertile regions in which reniform sporangia, containing cutinized spores, lay among the leaves. Although the precise attachment of the sporangia is still unknown, the general resemblance of *Baragwanathia* to a lycopod is so striking that an affinity seems undoubted.

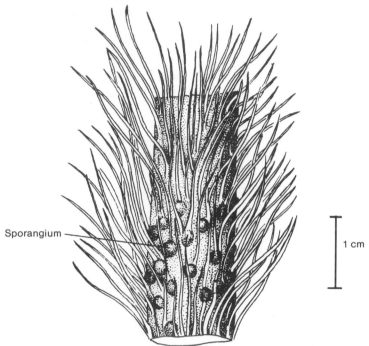

Figure 6.25. *Baragwanathia.* **Shoot bearing sporangia. (After Lang and Cookson, from Stewart. 1983.** *Paleobotany and the Evolution of Plants.* **Cambridge University Press.)**

The **Protolepidodendrales** are distinguished by their microphylls branching at the tips. In *Protolepidodendron scharyanum* from the Middle Devonian (Fig. 6.26a) the leaves forked only once, but in *Leclercqia* (Fig. 6.26b) also from the Middle Devonian, the leaf forked several times. This plant is particularly noteworthy since a ligule was clearly present. In both *Protolepidendron* and *Leclercqia* the sporangia were attached to the adaxial surface of the microphyll close to its insertion. The habit of these plants, which probably thrived in the warm, humid climate of the early Devonian, seems to have been similar to that of the worldwide *Lycopodium clavatum*. They were also probably homosporous. In the herbaceous lycophytes, living and fossil, there is a correlation between exarch xylem and the lateral position of the sporangia, contrasting with the endarch xylem and terminal sporangia of *Rhynia*.

Carboniferous rocks yield an abundance of lycophyte fossils, of both vegetative and reproductive structures, and isolated spores. Of the many forms present, some were evidently herbaceous. Those having a general resemblance to living *Lycopodium* are placed in *Lycopodites,* but others, in which heterospory has been demonstrated, are believed to have been more like *Selaginella* and are placed in *Selaginellites*. In each instance, however, the attribution is less certain than with Mesozoic forms. The most impressive Lycophyta of the Carboniferous, however, were undoubtedly the arborescent **Lepido-**

dendrales, some of which achieved a height of 30 m (98 ft.). *Lepidodendron* (Fig. 6.27), for example, consisted of a trunk, 1 m (39 in.) or more in diameter at its base, which rose as a single column until it broke up by numerous dichotomies into the dense crown of branchlets. The upper parts of the tree bore simple ligulate microphylls, up to 20 cm (8 in.) long, triangular in cross-section, and arranged in regular spirals.

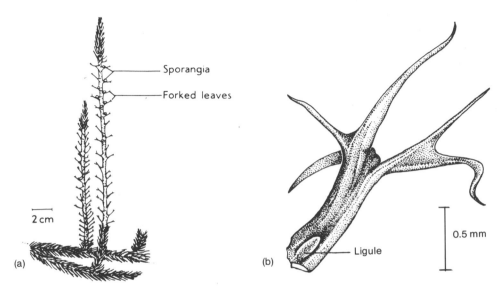

Figure 6.26. (a) *Protolepidodendron scharyanum.* **Habit. (After Kräusel and Weyland, from** Delevoryas. 1962. *Morphology and Evolution of Fossil Plants.* **Holt, Rinehart and Winston, New York.) (b)** *Leclercqia.* **Leaf showing ligule. (After Grierson and Bonamo, from Stewart. 1983.** *Paleobotany and the Evolution of Plants.* **Cambridge University Press.)**

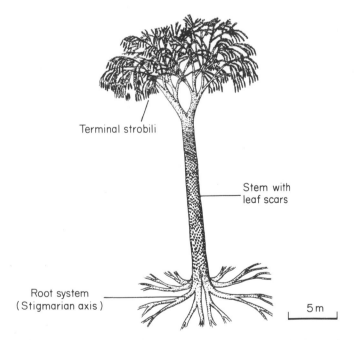

Figure 6.27. *Lepidodendron.* **Habit. (After Hirmer. 1927.** *Handbuch der Palaeobotanik.* **Oldenbourg, Munich.)**

The trunk and lower branches of *Lepidodendron* and its relatives retained a characteristic pattern of diamond-shaped leaf cushions (Fig. 6.28). In the upper part of each is a pit which indicates the site of the ligule, and at the center is a scar showing where the leaf was attached. Within the scar the emerging vascular bundle can be discerned and two small depressed areas, one on each side of the bundle. These are the remains of strands of aerenchyma (parichnos) which ran from the cortex of the stem to the leaf. The uniformity of the leaf scar suggests a distinct abscission mechanism. The trunk was anchored at ground level by four radiating arms which, since they were first found detached and not immediately recognized, were named *Stigmaria*. The Stigmarian axes dichotomized freely and the smaller bore rootlets, anatomically similar to those of *Isoetes*, in spiral sequence.

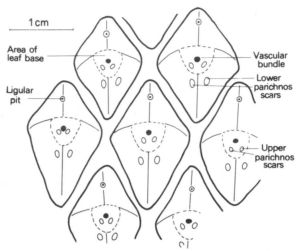

Figure 6.28. *Lepidodendron*. **Surface of cast of stem showing leaf cushions and scars. The thin-walled tissue of the parichnos strands has collapsed in the leaf cushions, producing the depressions referred to as secondary (lower) parichnos strands below the leaf scars.**

Despite the girth of *Lepidodendron*, the anatomy of the trunk was comparatively simple, and its manner of growth consequently puzzling. The primary vascular tissue is often well preserved, and even sieve cells have been seen in sufficient detail to reveal a close resemblance to those of *Lycopodium*. An exarch protostele was present at the base of the trunk, but above it became solenostele. A vascular cambium evidently arose at an early stage, but it added only a narrow zone of secondary xylem to the primary. The formation of secondary phloem is doubtful. Additional secondary activity occurred in the outer cortex, resulting in a hard sclerotic periderm which undoubtedly provided the principal mechanical support to the trunk. The inner cortex contained elongated cells, a so-called "secretory tissue," which may have been a primitive form of phloem. The central zone of cortex, usually fragmentary or missing in the fossilized material, probably consisted of thin-walled aerenchyma, continuous with that in the Stigmarian axes and leaves. The curious anatomy, particularly the small amount of secondary xylem, has led to the view that the growth of these trees did not continue indefinitely, but was determinate. It is envisaged that the plant first generated a massive apical meristem, the activity of which then produced an axis of considerable height before dichotomy began. At each dichotomy the apices became smaller, until eventually they ceased to be active.

Sigillaria was also prominent in the forests of the Upper Carboniferous. Although similar to *Lepidodendron* in its general features it was conspicuous for its long, grasslike leaves, sometimes reaching 1 m (39 in.) in length. *Chaloneria* is an Upper Carboniferous lycophyte with a strikingly different form. The main axis, about 2 m (6.5 ft.) tall, was wholly unbranched. Although about 10 cm (4 in.) in diameter, and with conspicuous secondary thickening below, the stem tapered with progressive reduction in the amount of secondary thickening to a slender apex. The narrow microphylls were ligulate, about 1 cm (0.4 in.) long, and borne in spiral sequence. Roots were produced from a swollen rounded base.

Reproduction of the Lepidodendrales. In most Lepidodendrales the strobili, varying from 1 to 3.5 cm (0.4 to 1.4 in.) in width and 5 to 40 cm (2 to 16 in.) in length, were terminal on the branchlets. A few homosporous cones are known, but most, even some of the earliest (e.g., *Cyclostigma* from the Upper Devonian), were heterosporous, resembling the cones of *Selaginella* in general features and the placement of the sporangia in relation to the ligule. In *Chaloneria* the fertile regions were either confined to the tip of the axis or, in some forms, fertile and sterile regions alternated in the upper part of the stem.

Although generally the sporangia of the Lepidodendrales were borne in a manner similar to those of *Selaginella,* there are forms in which the wings of the sporophyll were inflexed, the margins coming tightly together above the megasporangium. Representative of this kind of development is *Lepidocarpon* (Fig. 6.29). An even more elaborate form is *Miadesmia,* also from the Upper Carboniferous, which showed a distinct micropylelike opening, surrounded by hairs, at the distal end of the enclosing megasporophyll. There were also lateral membranous outgrowths which may have assisted dispersal of the entire reproductive structure. Since *Miadesmia* reached a length of only about 3 mm (0.12 in.), it may have been borne by a herbaceous lycophyte. In both *Lepidocarpon* and *Miadesmia* the mature megasporangium contained only one functional megaspore.

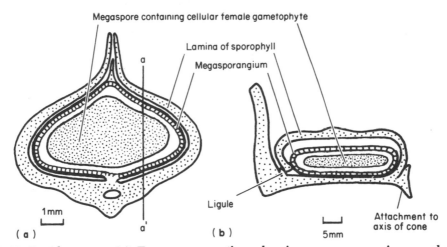

Figure 6.29. *Lepidocarpon.* **(a) Transverse section showing megasporangium enclosed in folded sporophyll. (After Arnold. 1947.** *An Introduction to Paleobotany.* **McGraw-Hill, New York.) (b) Vertical section of region indicated by a . . . a′ in (a). (After Hoskins and Cross, from Arnold. 1947. In** *Introduction to Paleobotany.* **McGraw-Hill, New York.)**

Germinating megaspores and microspores are encountered in petrified material. Megagametophytes were wholly or partly endosporic, the exposed portion often bearing rhizoids at its margin and archegonia in the central cushion. Possible mitotic figures have been seen in germinating microspores. Undifferentiated embryos are known in gametophytes as old as the Lower Carboniferous. In all essential respects the reproduction of the free-sporing Lepidodendrales resembled that of *Selaginella* and *Isoetes. Lepidocarpon* and *Miadesmia* are outstanding in that the single megaspore was retained within the sporangium. In *Lepidocarpon* germination was certainly *in situ,* leading to rupture of the sporangial wall. Some specimens of *Lepidocarpon* have yielded remarkable examples of differentiated embryos lying within the gametophyte, the whole enclosed within the megasporophyll. It would be difficult to deny such structures the status of seeds. When detached, they may have floated, and provided a means of distribution of the plant in the Carboniferous swamps.

Although the Lepidodendrales appear to have died out at the end of the Paleozoic, a few forms possibly derived from them have been found in Mesozoic rocks. *Pleuromeia* (Fig. 6.30), for example, known from Triassic sandstones is suggestive of an intermediate between the Lepidodendrales and the Isoetales. A single trunk, little more than 1 m (39 in.) high, with spirally arranged leaf scars, rose from a rounded or lobed base which produced roots in the same manner as Stigmarian axes. It terminated above in a crown of narrow ligulate leaves, and, when fertile, a single erect cone with curious obtuse sporophylls. *Pleuromeia* was heterosporous and may occasionally have been dioecious. Ecologically, it seems to have been maritime, some species possibly forming coastal thickets.

The Jurassic and Cretaceous periods, and Tertiary era yield evidence of herbaceous forms very similar to *Lycopodium* and *Selaginella*. *Selaginellites hallei,* for example, was a small, heterophyllous plant, with leaves faintly denticulate at the margin, like those of many species of *Selaginella*. There were four megaspores in each megasporangium, and each megaspore reached a diameter of about 500 μm, about ten times the size of the microspores. A ligule, however, has not be demonstrated. *Nathorstiana*, from Cretaceous rocks of Germany, recalls at once the Isoetales. An upright stock, made irregular by leaf scars, reached a height of about 12 cm (4.8 in.) and bore a crown of needlelike leaves. The base was divided into a number of vertical lobes from which the roots emerged. Nothing is yet known of the sporophylls.

The phylogeny of the Lycophyta. It seems likely that the origin of the Lycophyta is to be sought in the Devonian among plants of the kind assigned to the Zosterophyllophyta, particularly those with radially symmetrical axes. *Asteroxylon*, for example, is very suggestive of a lycophyte. Once they become recognizable the fossil record of the lycophytes is impressive. It is clear that the living representatives are relicts of a component of the Plant Kingdom which reached the peak of its morphological complexity and floristic success in the Paleozoic, about 200 million years ago. Although this period was marked by the spectacular Lepidodendrales, it seems very likely that herbaceous forms also existed throughout the evolution of the division and that they closely resembled those living today. One form from the Upper Carboniferous, for example, is considered to be so close to *Selaginella* as to be confidently placed in that genus (*S. fraipontii*). Following the discovery of fossils of small *Isoetes*-like plants in Triassic rocks even such a reasonable hypothesis as the derivation of the Isoetales from arborescent antecedents by way of such forms as *Pleuromeia* (Fig. 6.30) and *Nathorstiana* now seems improbable. Many of the living homosporous lycophytes have high chromosome numbers (reaching 2n = 134 in some species), suggesting that they may be ancient allopolyploids behaving cytologically as diploids. Nevertheless, isozyme analysis shows no evidence of the heterogeneity normally associated with species of known allopolyploid origin. Hybridization may therefore have played little role in their evolution.

Besides the antiquity of the lycophyte kind of construction, the fossil record also indicates that the Lycophyta always had a distinctly heteromorphic life-cycle. As in the living representatives, there is a conspicuous absence of morphological and anatomical similarities between the sporophytic and gametophytic phases. The gametophytes appear never to have been other than wholly parenchymatous plants of lowly status. Indications, as in *Psilotum,* of a possible ancestral isomorphic cycle are wholly lacking.

The reduction of the gametophytes in the heterosporous Lycophyta to short-lived, almost wholly endosporic thalli can be regarded as a major step in the direction of become independent of humid conditions for the survival of this vulnerable phase. Had the lycophytes acquired this valuable adaptation to terrestrial life earlier and proceeded from *Lepidocarpon* to the evolution of fully integumented seeds, their spectacular decline at the end of the Paleozoic might never have occurred.

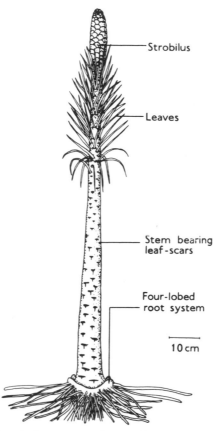

Figure 6.30. *Pleuromeia sternbergi.* **Reconstruction. (After Hirmer, from Andrews, Jr. 1961.** *Studies in Paleobotany.* **Wiley, New York)**

SPHENOPHYTA

As with the Lycophyta, the division Sphenophyta also contains both living and fossil representatives. All share a readily recognizable morphology. The general features of the division can be summarized as follows:

> *Sporophyte consisting of a monopodial branch system, some axes rhizomatous and bearing roots. Leaves microphyllous, borne in whorls. Vascular tissue of tracheids and phloem. Sporangia borne on sporangiophores, aggregated in terminal strobili. Gametophyte (in living forms) terrestrial. Spermatozoids multiflagellate. Embryogeny exoscopic.*

The only living sphenophyte is *Equisetum*, but the division has a rich fossil record. Four orders are recognized: the **Equisetales** (containing *Equisetum* and a number of fossil genera), and the extinct **Calamitales, Sphenophyllales,** and **Pseudoborniales**.

The Equisetales. The striking feature of the Equisetales is the jointed structure of the stem, and, in the regions of uniform diameter, the regular alternations of the microphylls in the successive whorls. The stems are often conspicuously ridged, each ridge being in line with the leaf above. Consequently, provided the number of leaves in a series of whorls remain the same, the ridges also show regular alternation from one inter-node to the next.

Equisetum (Fig. 6.31a), the horsetail, is a familiar sight in parts of the north temperate zone, but is rarer in the tropics and Southern Hemisphere, being absent altogether from Australia and New Zealand. About 15 species are now living, but others are known as Mesozoic fossils. The genus, or a form very closely similar, was wide-spread in Cretaceous times, and some species appear to have formed dense stands at the fringes of Cretaceous lakes. Moist habitats, such as river banks and marshy ground, are also favored by most living species, but some are able to thrive in much drier places. Equisetum arvense, for example, often flourishes on well-drained railway embank-ments. All species have a similar growth form. A perennial underground rhizome gives rise to green aerial shoots and occasionally also to perennating tubers packed with starch. In temperate and arctic regions the aerial shoots die back at the end of the growing season and new shoots emerge in the following spring.

Although the height of the aerial system of Equisetum varies from a few centimeters in arctic and alpine species to as much as 10 m (33 ft.) in the tropical E. giganteum, its morphology is strikingly uniform. Branches appear only at the nodes, and where several are present they too are whorled. The branch primordia are not, however, axillary, but they arise between the microphylls (Fig. 6.31b) and eventually break through the sheath formed by their congenitally fused bases. In subterranean axes roots emerge from directly below the sites of branch primordia. Branch and root primordia are in fact present at every node, but they develop only in appropriate environmental conditions, sometimes reproducible in the laboratory. In E. arvense, for example, green branches can be made to grow from the nodes of an etiolated unbranched fertile shoot if they are enclosed in a moist chamber.

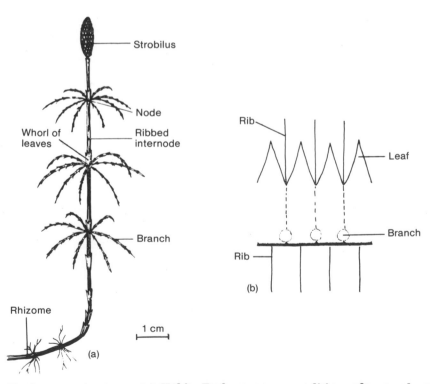

Figure 6.31. *Equisetum sylvaticum.* (a) Habit. Early summer condition, after production of vegetative branches, but before shedding of cone. (b) Diagram showing arrange-ment of leaves and ribs at a node at which there is exact alternation.

The structure of the axes of *Equisetum* also shows little variation. The mature stems have a large central cavity surrounded by a ring of vascular bundles (Fig. 6.32). These bundles are of the same number as the ribs on the outside of the internode (and hence as the leaves of the node above), and are also co-radial with them. An endodermis can usually be distinguished, either encircling the stele on the outside alone (as in *E. arvense*), or forming two continuous cylinders, one inside and one outside the ring of bundles (as in *E. hyemale*), or surrounding each bundle individually (as in *E. fluviatile*). In some species the position of the endodermis in the rhizome is different from that in the aerial stem. Alternating with the vascular bundles and lying between the endodermis and the periphery are large longitudinal air chambers known as vallecular canals. The continuity of the central cavity is interrupted at the nodes by parenchymatous septa.

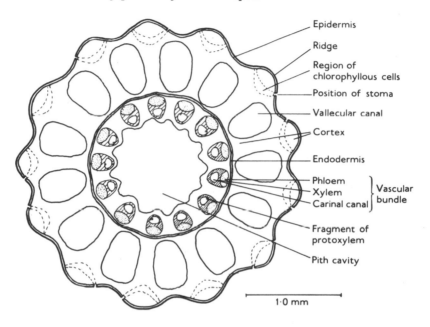

Epidermis

Ridge

Region of chlorophyllous cells

Position of stoma

Vallecular canal

Cortex

Endodermis

Phloem
Xylem } Vascular
Carinal canal } bundle

Fragment of protoxylem

Pith cavity

1·0 mm

Figure 6.32. *Equisetum arvense.* **Transverse section of young stem.**

The vascular bundles themselves contain very little lignified tissue. In a differentiated internode the protoxylem is represented solely by fragments of tracheids adhering to the sides of a cavity, termed the carinal canal, present on the adaxial side of each bundle. The metaxylem differentiates as two groups of tracheids, one placed tangentially on each side of the phloem. The cells between the phloem and the carinal canal frequently develop labyrinthine walls. These may regulate the flow of solutes from the symplast of the sieve cells to the apoplast of the canal. The vascular anatomy of the nodes is highly peculiar. Here the tracheids, resembling those of the metaxylem, run horizontally and form a ring linking the bundles of the adjacent internodes. In *E. hyemale* the phloem close to the node is notable in lacking well-defined sieve cells. The root, branch, and leaf traces also originate at the level of the node, the leaf trace departing immediately above the entry of the bundle from the internode below.

The strength of the *Equisetum* stem depends principally upon the cortical ridges. These consist of sclerenchymatous cells reinforced by deposits of silica. The nodal septa have also been shown to add to the rigidity of the axis. The support this form of construction can provide is evidently limited and stems frequently buckle under mechanical stress. The taller species of *Equisetum* usually grow in groves and, since the rough siliceous stems and branches do not readily slide over each other, the plants hold each other up.

The microphylls of *Equisetum* soon become scarious, and photosynthesis takes place predominantly in the surface layers of the stem. Apart from a few curious "water stomata" or hydathodes in the adaxial epidermis of the tip of the microphyll, the stomata are confined to the valleys of the internodes, and thus lie above the vallecular canals. The stomata are often deeply immersed, merely a pore being visible externally. Each guard cell is flanked by a subsidiary cell, and each subsidiary cell bears transverse bars of cutin on its exposed surfaces (Fig. 6.33).

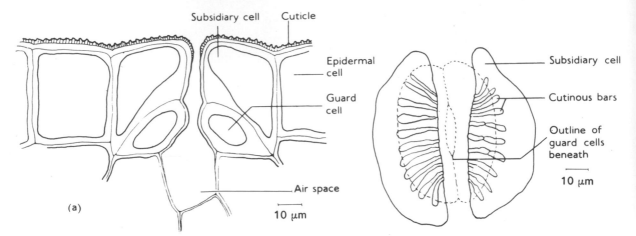

Figure 6.33. *Equisetum* **sp. Structure of stomata. (a) In vertical section. (b) In surface view partially macerated, showing the bars of cutin on the subsidiary cells. (After R. L. Hauke. 1957.** *Bulletin of the Torrey Botanical Club* **84.)**

The growth of the axis of *Equisetum* provides a striking example of coordinated differentiation. The axis is surmounted by a single tetrahedral apical cell. The extent to which this particular cell undergoes division is disputed, but adjacent to its three posterior faces, daughter cells are cut off in a regular, clockwise sequence. While the stem is growing, there is no pause in the meristematic activity of the apical initials and their products are recognizable as three tiers of cells in the extreme apex. The formation of a whorl of leaves is first indicated by the elevation of a ring of meristematic tissue at the base of the apical cone. While this ring is being superseded by another, leaf teeth initials become visible around the upper margin of the first. These initials are of limited growth and give rise to the free part of the microphylls, the basal ring meanwhile forming the sheath of fused bases.

The upper three or four nodal primordia remain close together, but the cells between them become organized as an intercalary meristem which surrounds, but does not cut across, the procambial strands. This meristem becomes active between the fourth and fifth nodes and cuts off cells acropetally which go to form the internode. The activity continues while four or five new nodes are initiated at the apex. In consequence, the mature length of the internode is reached by about the ninth or tenth node. The secretion of silica on to the surface of the internode begins at the top and proceeds downward, becoming completed as the intercalary growth ceases. The metabolic pathway of the silica and the manner of its secretion are not yet understood.

The procambial tissue, the disposition of which foreshadows that of the vascular bundles in the mature axis, advances continuously, in step with the advance of the apex, and is always to be found at the level of the uppermost leaf whorl primordia. Differentiation of the procambial strand is not, however, a simple acropetal process. Protoxylem begins to appear at about the fourth node, and differentiation extends acropetally into the leaf and basipetally into the node below. Since this differentiation occurs during the time of maximum extension, most of the tracheids first formed in the internode are ruptured, and the area of weakness provided by the differentiating cells is pulled apart by the radial and tangential expansion of the stem, so yielding the carinal canal. Metaxylem begins to appear at the fifth node and also differentiates basipetally. The rate of differentiation is such that the descending strands do not fuse with the metaxylem of the node beneath until the tenth node, when extension of the internodes is ceasing. The internodal metaxylem thus escapes rupture, although occasional small lacunae indicate that it is subjected to some stress during the concluding phases of differentiation. The differentiation of the phloem follows more or less the same course as that of the metaxylem.

In temperate and arctic habitats the aerial branches perish in the winter, and, in this instance, the growth of the shoot is a little different from that just described. Almost the whole of the next year's shoot overwinters in primordial form as a subterranean bud. In spring, elongation of the internodes and further differentiation proceeds acropetally, so generating an aerial system of limited growth.

Reproduction of the Equisetales. In all species of *Equisetum* the sporangiophores are aggregated into a terminal strobilus (Fig. 6.34) which terminates either a vegetative axis (as in *E. palustre*) or a specialized axis lacking pigmentation which appears early in the growth season (as in *E. arvense*). An intermediate state is shown by *E. sylvaticum,* in which the fertile shoots are at first colorless, but after release of the spores become green and branch. The vascular system inside the cone recalls that of a node, since it consists of a cylinder of metaxylem. Fine traces depart to the sporangiophores, and the cylinder is broken here and there by parenchymatous performations which bear no evident relation to the departing traces. The sporangiophores, which are not necessarily arranged in whorls, are peltate, and are tightly packed so that heads acquire a polygonal outline. About 20 sporangia are pendent from the margin of the head, and so lie more or less radially in the intact cone.

A sporangium develops from a group of cells, in which a central archesporial tissue, surrounded by a wall several layers thick, can soon be distinguished. The inner layer of

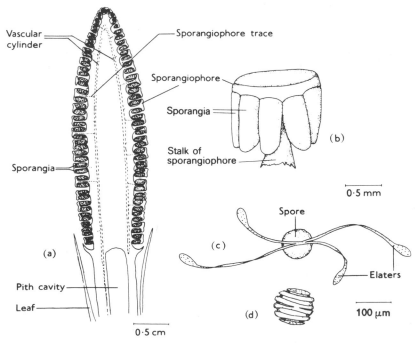

Figure 6.34. *Equisetum maximum.* **(a) Median longitudinal section of strobilus. (b) Peltate sporangiophore. (c) and (d) Spores with elaters in dry and moist condition respectively.**

the wall functions as a tapetum, together with about a third of the archesporial tissue, the remainder being sporogenous.

Groups of sporogenous cells tend to become separated by nonsporogenous tissue, possibly accounting for some lack of synchrony in meiosis within a sporangium. The spore output is nevertheless very high. The manner in which the spore wall is formed is complex and unique in the Pteridophyta. After the laying down of the exine, four spirally arranged elaters are added as a further outer layer. Each consists of two parts: an inner band of cellulose microfibrils (parallel to arrays of microtubules in the adjacent plasmodial tapetum) and an outer covering of homogeneous material secreted from the tapetum. The elaters are not acetolysis-resistant and can therefore be taken to lack sporopollenin. When extended, the elaters form an X, remaining attached to the spore only at the intersection. The mature spore retains no triradiate scar.

The mature sporangium opens along a longitudinal stomium as the result of tensions arising from the drying out of spirally thickened cells elsewhere in the wall. When the spores are freed, the elaters lift and respond to changes of humidity with jerky movements which assist in the distribution of the spores. The entanglement of the elaters also ensures that the spores are dispersed in groups.

Equisetum is usually considered to be homosporous. In *E. arvense* a large sample of spores falls into two classes whose mean diameters differ by about 10 μm, but this is probably an effect of drying. The spores, which contain chlorophyll, soon lose their viability if stored. On germination, the cellulosic inner layer of the spore wall is continuous with the wall of the emerging filament. A high concentration of potassium ions has been demonstrated at the site of germination. This probably leads to a local increase in the plasticity of the wall, since potassium ions are known to reduce the amount of cross-linking in cellulose microfibrils. The initial filament is transformed by division in a number of planes into a cushion of cells anchored by rhizoids. Subsequent growth is from a marginal meristem which forms a number of obliquely ascending lobes (Fig. 6.35) on the upper surface of which the sex organs appear. Growth in size is, however, limited, and the gametophytes rarely exceed 1 cm (0.4 in.) in diameter and 3 mm (0.12 in.) in height, although much larger forms have sometimes been found.

Culture experiments have shown that there are two kinds of gametophytes. About half the gametophytes of a mass sowing of spores remain small and produce only antheridia; very few, if isolated, continue to grow and produce archegonia. The remaining gametophytes are larger and longer lived. They first produce archegonia, and then, if none is fertilized, a crop of antheridia, followed by another of archegonia. Such gametophytes may last at least two years and in favorable circumstances produce several

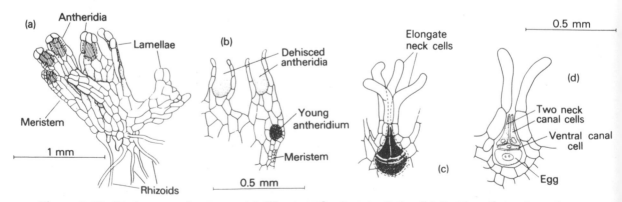

Figure 6.35. *Equisetum sylvaticum.* **(a) Young male gametophyte. (b) Section through male branch. (c,d)** *Equisetum palustre,* **archegonia. (c) General view. (d) Longitudinal section. (After J. E. Duckett. 1968. Ph.D. dissertation, Cambridge University.)**

sporophytes. There is thus some evidence that *Equisetum* is heterothallic, but, since the proportions of the two kinds of gametophyte áre related to the density of sowing and other cultural conditions, the differences in behavior cannot have a simple Mendelian basis. Chance variations in the cytoplasmic complements of the spores, leading to differences in competitiveness, may influence the growth and subsequent gametogenesis.

The antheridia of *Equisetum*, like those of the Lycopodiales, contain numerous spermatocytes. They open by the parting of the cover cells and the spermatocytes are discharged. Each is furnished with a number of long fibrils, an intriguing parallel to the elaters attached to the spore. The flagella of the enclosed spermatozoids soon become active and the spermatocytes break open, releasing the gametes. Each is about 20 μm long and is twisted into a left-handed screw of three gyres, with an apical crown of about 100 flagella. The ultrastructure is similar to that of other archegoniate male gametes, and a ribbon of microtubules adjacent to the nucleus is a prominent feature. The spermatozoid swims at about 300 μm per second in a helicoid path. The archegonia are formed mostly between the aerial lobes of the gametophyte. The necks project, and at maturity the four distal cells become elongated and reflexed. Fertilization, dependent, as in the Lycopsida, upon the presence of a film of water, is followed by the division of the zygote in a plane perpendicular to the axis of the archegonium. The subsequent embryogeny is exoscopic, and the products of the outer cell give rise to the stem apex, the primordium of the first whorl of leaves, and, in a variable lateral position, a root apex.

The development of the young sporophyte follows a curious course (Fig. 6.36), without parallel in other living archegoniate plants. The first axis, which contains a simple protostele, is of determinate growth and never increases in diameter. In *E. arvense*, for example, it produces about six whorls, each with about three leaves. As this shoot ceases to grow, a bud grows out from below the first whorl of leaves. This also produces an upright axis of slightly greater diameter than the first and containing a protostele that shows a tendency towards medullation. This process is repeated two or three times until

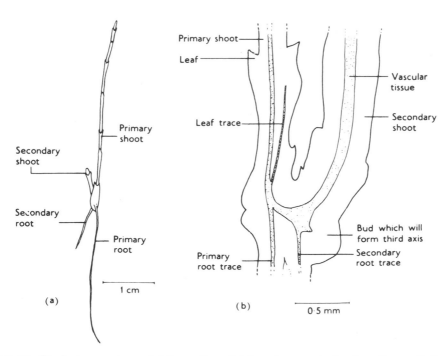

Figure 6.36. *Equisetum arvense.* **(a) Sporeling. (b) Development of mature form of plant. (After K. Barratt. 1920.** *Annals of Botany* **34.)**

the axis reaches the diameter and structure characteristic of the mature plant. The rhizomatous growth habit is then initiated. The primary root of the embryo persists only a short time, and roots are produced freely from the nodes at and below the soil surface as the young plant becomes established.

The early Sphenophyta. The fossil record of the Sphenophyta parallels that of the Lycophyta, beginning early in the Paleozoic and expanding in the Carboniferous, when the Sphenophyta must have formed a large part of the earth's vegetation. From the end of the Paleozoic to the present time they have been of diminishing importance; today only a single genus remains.

The fossil sphenophytes which show the closest resemblance to *Equisetum* are the **Calamitales**. One of the earliest may have been *Archaeocalamites* from the Upper Devonian and Lower Carboniferous. This was an arborescent plant with secondary vascular tissue, the mature stems reaching about 1.5 cm (0.6 in.) in diameter. The branches were whorled and the stem ribbed, but the ribs did not alternate as in living *Equisetum*. The upper branches bore leaves, also in whorls, which dichotomized several times, sometimes reaching a length of 10 cm (4 in.). *Calamites*, which flourished in the Carboniferous forests, is altogether better known. These plants had a creeping rhizome from which arose massive aerial stems bearing whorls of branches. In some forms the main axis reached a diameter of about 15 cm (6 in.) and a height of at least 20 m (65.5 ft.). The calamites probably formed a second story beneath the Lepidodendrales.

The stems of *Calamites* were conspicuously ribbed, the ribs in some form alternating from node to node, and in others colinear. Although *Calamites* is often represented only by pith casts, petrifactions are also know which have revealed details of the anatomy. There was a basic similarity to *Equisetum,* and the protoxylem was associated with a canal. Subsequently, a cambium arose between the primary xylem and phloem and contributed substantial amounts of secondary vascular tissue to the stem. Another difference from *Equisetum* was that no air canals were present in the cortex. The axis did, however, grow, as in *Equisetum,* from a single apical cell, beautifully preserved in some specimens. The leaves of *Calamites* were simple with a single median vein. The subsidiary cells of the stomata bore transverse bars of cutin (see Fig. 6.33).

The strobili of the Calamitales, in which terminated lateral branches, consisted of alternate whorls of sporangiophores and bracts, although the cone of one early form appears to have contained peltate sporangiophores alone. Since cones are often found detached, they are placed in form genera, defined by the relative arrangements of the sporangiophores and bracts. The two form genera most widely represented are *Calamostachys* (Fig. 6.37) and *Palaeostachys*. The cones assigned to these form genera may, of course, have been produced by plants differing widely vegetatively. Both homosporous and heterosporous cones have been described, and in at least on (*Calamocarpon*) the megasporangium appears to have contained only one functional megaspore. The evolution of heterospory in the Calamitales appears to have paralleled that in the Lepidodendrales, but there is no evidence of its having reached the seedlike formations of *Lepidocarpon* and *Miadesmia*. In *Elaterites* (possibly the microsporangiate cone of a heterosporous calamite) the spores had *Equisetum*-like elaters.

Apart from the Cretaceous plants referred to earlier which closely resemble living *Equisetum,* there is evidence of herbaceous forms having also existed in the Palaeozoic. *Equisetites hemingwayi,* for example, from the Upper Carboniferous, is the remains of a fertile shoot very like *Equisetum* in the structure of the cone and order of size.

The **Sphenophyllales,** another order of the sphenopsids, appeared in the late Devonian and were prominent in Carboniferous floras. The share little with the Equisetales and Calamitales except the whorled arrangement of the leaves. They were probably scrambling plants, supporting themselves on other vegetation in the manner of the familiar *Galium aparine* (cleavers or goose-grass). The leaves were wedge-shaped,

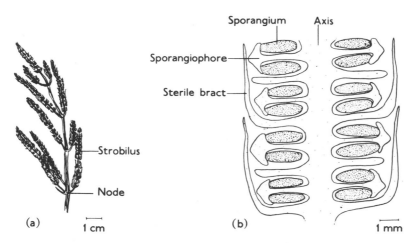

Figure 6.37. *Calamostachys.* **(a)** *C. ludwigi.* **Fertile shoot (from a compression). (After Weiss, from Andrews, Jr. 1961.** *Studies in Paleobotany.* **(b)** *C. binneyana.* **Longitudinal section of cone. (After Andrews, Jr. 1961.** *Studies in Paleobotany,* **Wiley, New York.)**

with dichotomously branching venation, and were usually borne in multiples of three (Fig. 6.38). The stems contained a solid core of primary xylem, triangular in transverse section, with protoxylem at the vertices. The primary xylem was surrounded by secondary in radial files, resulting in a vascular system strikingly reminiscent of that of a root. These stems of *Sphenophyllum* provide the finest example in the Palaeozoic of a spore-bearing plant in which well-defined secondary vascular tissue, both xylem and phloem, was regularly produced from a bifacial cambium. The strobili were terminal and frequently consisted of whorls of sterile bracts, each forming a cuplike sheath around the axis, on the adaxial side of which were attached the sporangiophores (Fig. 6.38b). The distal portion of the sporangiophore, which was free, branched, and recurved, bore a number of sporangia. In other forms of cone the sterile part was less well developed and the sporangiophore branching more complex. Most of the cones were homosporous, but distinct heterospory has been detected in one form. The Sphenophyllales disappeared at the beginning of the Triassic period.

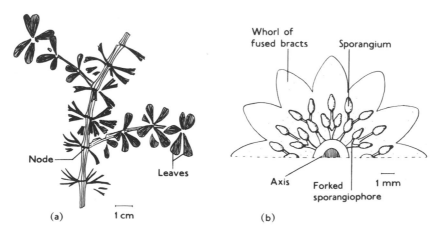

Figure 6.38. *Sphenophyllum.* **(a)** *S. verticillatum.* **Habit of vegetative shoot (from a compression). (After Potonié. 1899.** *Lehrbuch der Pflanzenpalaeontologie.* **Dümmler, Berlin.) (b)** *S. (Bowmanites) dawsoni.* **Fertile whorl viewed from above. (After Hirmer. 1927.** *Handbuch der Palaeobotanik.* **Oldenbourg, Munich.)**

The **Pseudoborniales** are based upon remains from the Upper Devonian of Bear Island and Alaska. They appear to have been sizeable trees, many branches reaching a diameter of 10 cm (4 in.) and the main axes as much as 60 cm (24 in.). The leaves, borne in whorls of four, were branched once or twice. The upper branches terminated in loose cones, up to 30 cm (12 in.) in length, consisting of whorled bracts and sporangiophores. These branched several times, each ultimately producing about 30 sporangia. Although their general features are in accord with the Pseudoborniales being placed with the Sphenophyta, little is known about them. They appear to have been a minor evolutionary line which disappeared relatively quickly.

Origin of the Sphenophyta. The bearing of the leaves or branches in whorls and the articulate structure of the stem, although characteristic features of the Sphenophyta, are not, of course, confined to this division of the Plant Kingdom. They are present in some of the algae (e.g., *Draparnaldia, Batrachosperum*) and in the flowering plants (e.g., *Galium, Casuarina*), and have clearly arisen a number of times in plant evolution. Nevertheless, the Sphenophyta appear to have specialized in this organization of the plant body from their beginnings, and they may even have had a common origin in the remote past.

Unfortunately, no fossils are known which show convincingly how the Sphenophyta might have arisen. *Calamophyton* (Middle Devonian), consisting of axes bearing lateral appendages which forked twice in planes at right angles to each other, was once thought to be an early articulate plant. This interpretation in now discredited; the transverse markings on the axes, formerly taken as an indication of nodes, appear to be nothing more then regular cross-fractures developing during fossilization. The affinities of *Calamophyton* probably lie elsewhere (Chapter 7). A possibly more satisfactory candidate for the primordial sphenophyte is *Hyenia*. The Middle Devonian plant had a rhizome reaching about 5 cm (2 in.) in diameter and producing aerial branches in close succession. Some of these bore sterile appendages which forked once or twice and probably functioned as leaves. Other branches bore complex sporangiophores, each with fine sterile extensions and six reflexed sporangia. Both sterile and fertile appendages were borne in a shallow spiral. If the sphenophytes came from a *Hyenia*-like ancestor, the shallow spiral must have given way to a series of whorls separated by distinct internodes. Consolidation of this form of development with corresponding changes in the vascular anatomy might in time have led to plants with well-defined articulate morphology.

Although in some of the Carboniferous sphenophytes, sporangia were produced in association with bracts, these complexes seem to have no affinity with the lycophyte sporophyll, a reproductive structure, as we have seen, well established in early Devonian times. The relationship between the sphenophyte and lycophyte lines of evolution, despite the parallels, seems altogether remote.

Although the Sphenophyta themselves may have had a common origin, the relationships between the Calamitales and Equisetales are clearly closer than between the other orders. There seems to have been a general tendency towards the sporangiophore becoming peltate, and this perhaps hindered the evolution of seedlike structures. The almost total elimination of the sphenophytes points to their limitations, which possibly lay principally in the reproductive mechanism. It is difficult to account for the survival of the Equisetales in preference to the other orders. It may have been because only this order contained herbaceous forms sufficiently adaptable to meet the fluctuating conditions of the Mesozoic.

___7___
THE SUBKINGDOM TRACHEOPHYTA
Part 2

SPORE-BEARING TRACHEOPHYTES WITH LIVING REPRESENTATIVES: THE FERNS

The division FILICOPHYTA encompasses both wholly fossil groups and also living representatives, many with a long fossil history. The general features of the division can be summarized as follows:

> Sporophyte herbaceous or arborescent, in many forms rhizomatous. Leaves often compound (megaphylls). Vascular system of tracheids and phloem, usually lacking clearly defined secondary tissue. Stele often divided into meristeles. Leaf traces often complex, leaving a parenchymatous gap in the stele at their origin. Sporangia borne on leaves, but never on the adaxial surface of a microphyll. Mostly homosporous; a few (living and fossil) heterosporous. Gametophytes (known only in living forms) simple, usually autotrophic, lacking vascular tissue. Archegoniate. Spermatozoids multiflagellate. Embryogeny typically endoscopic.

The living filicophytes fall into four orders, namely the **Ophioglossales, Marattiales, Filicales,** and **Hydropteridales,** referred to collectively as ferns. The ferns are an important element of the world's flora, numbering about 10,000 species and being particularly conspicuous in warm humid regions. They show the greatest range of growth forms among the vascular archegoniates. Although largely herbaceous, a number of ferns of the tropics and subtropics (mostly belonging to the family Cyatheaceae) achieve the form and stature of simple trees. *Cyathea* in the Kermadec Islands, for example, has a palmlike habit and reaches a height of 20 m(65.5 ft.). At the other extreme are minute epiphytes hardly bigger than leafy liverworts (with which they commonly grow). Some families of ferns have a rich fossil record, their distinctive characteristics being recognizable as far back as the Carboniferous. These ferns are among the most ancient of living plants.

The Ophioglossales

These ferns form a small and morphologically peculiar order of living ferns which since they have no well-established fossil record, are of obscure origin. In all members the fertile region of the frond takes the form of a spike or pinnately branched structure, clearly set off from the vegetative portion. A feature that separates the Ophioglossales from other living ferns is that the fronds, instead of expanding from a closely coiled immature state (a condition known as "circinate vernation"), grow marginally from a more or less flat primordium. That of *Botrychium lunaria* when young shows a distinct kind of folding, the upper margins of the pinnules being covered by the lower margins of the pinnules above.

Of the three genera of the order, *Botrychium* (moonwort) and *Ophioglossum* (adder's tongue)(Fig. 7.1) are fairly widespread, the former mainly in the north temperate zone and the latter chiefly in the tropics. Both genera include species native to the British Isles. The third genus, *Helminthostachys* (Fig. 7.2), is restricted to the Polynesian Islands in the South Pacific and a few regions in the Asian tropics, but is often locally abundant along roadsides. In parts of rural India the young frond is used as a vegetable.

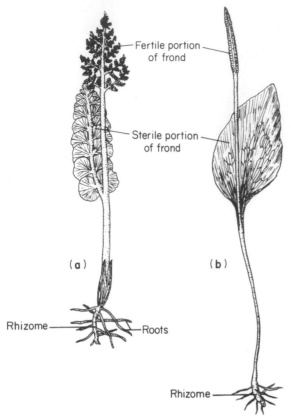

Figure 7.1. (a) *Botrychium lunaria*. Habit. (b) *Ophioglossum vulgatum*. Habit. Both after Lowe. 1874–78. *Our Native Ferns*. Groombridge, London.)

Figure 7.2. *Helminthostachys*. **Upper part of plant showing fertile spike. Photographed in Sri Lanka by R. D. E. Jayesekera. Scale bar 5 cm.**

In *Botrychium* (Fig. 7.1), where the frond is annual, the vegetative and fertile parts are pinnately branched. *Helminthostachys* is basically similar, but the branches of the fertile part of the frond are very contracted (Fig. 7.2). In *Ophioglossum* (Fig. 7.5) the sterile part of the frond, which is reticulately veined, is elliptical and entire or, in a few epiphytic forms, dichotomously lobed. The fertile part is never anything other than a simple spike.

The vascular system of the rhizome. The Ophioglossales are rhizomatous, growth taking place from a single apical cell. In *Botrychium* the rhizome of the young plant contains a medullated protostele (Fig. 7.3), but in the stele of older plants a parenchymatous area perforates the xylem anterior to the departing leaf trace. The endodermis remains wholly exterior. We thus arrive at a stele intermediate between a protostele and a solenostele, often referred to as a siphonostele (see Fig. 7.14). A rudimentary solenostele does, in fact, arise in some species of *Botrychium* as a result of an endodermis appearing on the inside of the xylem cylinder. The stele shows a number of points of anatomical interest. The metaxylem tracheids, for example, bear bordered circular pits, found outside the Ophioglossales only in the gymnosperms and angiosperms. There is also limited cambial activity leading to secondary vascular tissue, otherwise unknown in living ferns. Apart from this feature, the anatomy of *Ophioglossum* and *Helminthostachys* resembles that of *Botrychium*.

The roots of all Ophioglossales tend to be fleshy. The central vascular strand is either diarch, or in the larger roots polyarch. Many species of *Ophioglossum* build up colonies by means of buds formed on the roots.

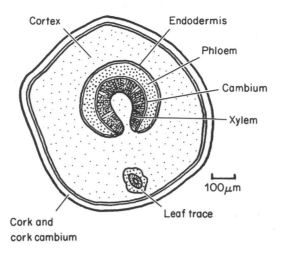

Figure 7.3. *Botrychium lunaria.* **Transverse section of rhizome.**

Reproduction of the Ophioglossales. The spherical sporangia of *Botrychium,* eusporangiate in origin, rise in two ranks on the ultimate branches of the fertile part of the frond. The spore mother cells are enclosed in a tapetum, several cells in thickness, in whose disintegration products the spores mature. Even at maturity the wall of the sporangium is massive, and stomata interrupt its outer layer. The spores, a few thousand in number, are released by transverse dehiscence. *Botrychium,* like the Ophioglossales as a whole, is homosporous.

The gametophyte of *Botrychium* is a flattened tuberous prothallus, subterranean and invested with an endophytic fungus, presumably in mycorrhizal relationship (Fig. 7.4). Gametophytes have also been raised in pure culture, but germination is poor unless the spore walls are first abraded by shaking with sand. Although the gametophytes will grow in the light, the spores must pass through a period of darkness before they will germinate. Chlorophyll remains absent and sugar is essential for successful growth. The antheridia are sunken, and each yields over a thousand multiflagellate spermatozoids. The archegonia, of quite normal construction, are partially immersed. The embryogeny of *Botrychium* is somewhat variable: in some species there is a suspensor and development

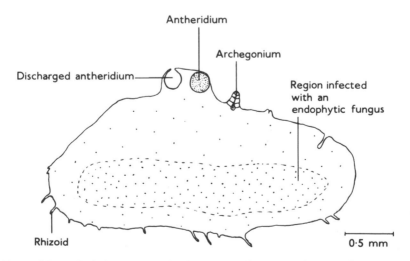

Figure 7.4. *Botrychium virginianum.* **Vertical section of gametophyte. (After Campbell. 1905.** *The Structure and Development of Mosses and Ferns.* **Macmillan, New York.)**

is endoscopic, in others the suspensor is lacking and the development exoscopic. The first organ to emerge is the root, infected from the first with the same endophytic fungus as the gametophyte. The young plant may remain subterranean in an immature condition for several years. Until the first leaf appears above ground the nutrition is presumably supplied principally by the mycorrhiza. Despite the gametophyte being subterranean, interspecific hybrids of *Botrychium* are known and sometimes locally frequent.

The reproduction of *Helminthostachys* is similar to that of *Botrychium,* but the dehiscence of the sporangia is longitudinal, and the embryogeny is regularly endoscopic. In *Ophioglossum* the sporangia, which occur as two rows partially embedded in the spike (Fig. 7.5), open by transverse clefts. Each contains numerous spores, in some species of the order of 15,000. The gametophyte of *Ophioglossum* is subterranean and cylindrical, sometimes approaching 5 cm (2 in.) in length. Both antheridia and archegonia are sunken. The embryogeny is exoscopic.

Phylogeny of the Ophioglossales. The Ophioglossales have no close relatives, and the evidence of distribution and comparative anatomy, particularly in relation to the massive eusporangiate sporangia and the stele, points to their being the relics of an ancient lineage. The chromosome numbers of *Ophioglossum* are remarkably high, exceeding 1,000 in one species. They may therefore be ancient polypoids. The negligible fossil record of the Ophioglossales as a whole suggests that they were never very numerous.

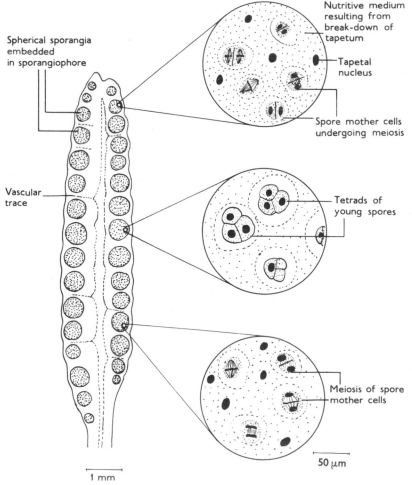

Figure 7.5. *Ophioglossum vulgatum.* **Longitudinal section of young fertile spike.**

The Marattiales

This small order of ferns is wholly tropical. Although not conspicuous in contemporary vegetation, they have a remarkably rich fossil record, and marattiaceous ferns have been recognized with certainty as far back as the Carboniferous.

Morphology and anatomy of the Marattiales. Of the living genera, the most common are *Angiopteris* and *Marattia*. Both have short upright trunks bearing large, pinnately branched and rather fleshy fronds (Fig. 7.6), sometimes reaching a length of 5 m (16.5 ft.), and showing circinate vernation. At the base of the petiole are two prominent stipules which persist after the leaf has fallen. *Christensenia*, a monotypic genus of the Indo-Malayan region, has a creeping rhizome with palmately divided fronds, which have the distinction of containing the largest stomata known in the Plant Kingdom. *Danaea*, a small genus confined to tropical America, has one species with a simple, ovate frond, and another with a small pinnate frond in which the lamina is pellucid and filmy. These forms, although revealing the diversity in the fronds of the Marattiales, are nevertheless unusual, and a massive angular construction is characteristic of the fronds of the Marattiales as a whole. The laminae normally show differentiation into palisade and mesophyll, with the stomata confined to the lower surface.

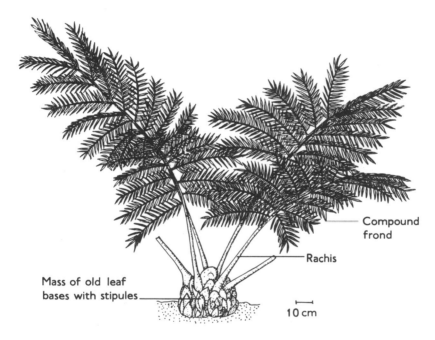

Mass of old leaf
bases with stipules

Compound
frond

Rachis

10 cm

Figure 7.6. *Angiopteris teysmanniana.* **Habit of young plant. (After Bitter, in Engler and Prantl. 1902.** *Die natürlichen Pflanzenfamilien 1:4.* **Engelmann, Leipzig.)**

The stems of the upright forms grow not from a single apical cell, but from a more massive meristem, and in this they are unique among the living ferns. The leaves form an apical crown, and since each receives an extensive trace, consisting of several strands of vascular tissue, the form of the stele is highly complex. A transverse section of the stem shows a number of concentric cycles of partial steles (**meristeles**), and dissection reveals that the meristeles of each cycle anastomose freely, occasional anastomoses also occurring between adjacent cycles. Leaf traces originate from the outer cycle of meristeles. Root traces, which may arise at any depth, pass out obliquely into the cortex. An endodermis, although present in young plants, is usually absent from the stelar regions of the older.

The stems of the Marattiales also contain little if any sclerenchyma, but there is an abundance of mucilage canals and tannin sacs, as elsewhere in the plant (Fig. 7.7). These indicate a particular kind of carbon metabolism that seems to have been widespread among the ancient ferns.

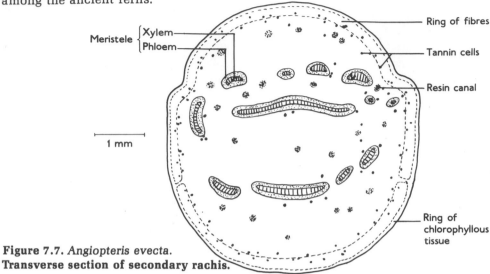

Figure 7.7. *Angiopteris evecta.*
Transverse section of secondary rachis.

Reproduction of the Marattiales. The fertile fronds resemble the sterile in most genera, and the sporangia, always eusporangiate in origin, are confined to the lower surface. In *Angiopteris* they arise in two ranks beneath veins towards the margins of the pinnules (Figs. 7.8a, 7.10a). The group is called a **sorus**. Dehiscence of the sporangia, along a longitudinal stomium, is directed towards the midline of the sorus (Fig. 7.10a). In *Marattia* the fertile regions are similar, but the sporangia are congenitally fused into a **synangium** (Figs. 7.8b, 7.9 and 7.10b). As the synangium matures and dries, it splits longitudinally into two valves (Fig. 7.10b), and each compartment dehisces by a pore in the inner face. The number of spores in each sporangium (or synangial compartment) in the Marattiales reaches a thousand or more.

Given warmth, moisture, and light, the spores germinate rapidly, and after passing through a brief filamentous phase, generate a green thalloid gametophyte with apical growth, resembling a thallose liverwort such as *Pellia*. Although autotrophic, the lower cells contain an endophytic, and presumably mycorrhizal, fungus. The gametophyte can be long-lived and old specimens reach a length of 3 cm (1.4 in.) or more. The antheridia are sunken and occur on both surfaces, but the archegonia, also sunken, are confined to the median region of the ventral surface. The protruding neck cells form little more than a cap over the ventral canal (Fig. 7.11).

The first division of the zygote is by a wall transverse to the longitudinal axis of the archegonium. The subsequent embryogeny is endoscopic, and the embryo often emerges from the upper side of the gametophyte. A suspensor has been reported in *Danaea*, but is elsewhere lacking.

The fossil history of the Marattiales. The Marattiales are represented in the Carboniferous period by both vegetative and fertile material. *Psaronius,* for example, is the remains of a trunk surrounded by a mantle of descending roots (Fig. 7.12). The vascular tissue, which was wholly primary, formed a polycyclic array of anastomosing, bandlike meristeles. Morphologically and anatomically *Psaronius* is so suggestive of an arborescent *Angiopteris* that there seems little doubt of its affinity. Fertile material is represented by *Scolecopteris* (Fig. 7.13) and *Eoangiopteris,* the sporangia of which were very similar to those of *Angiopteris,* although there were minor differences in the sorus. Fertile fronds of Marattiales, resembling those of various modern genera, are also found throughout the Mesozoic.

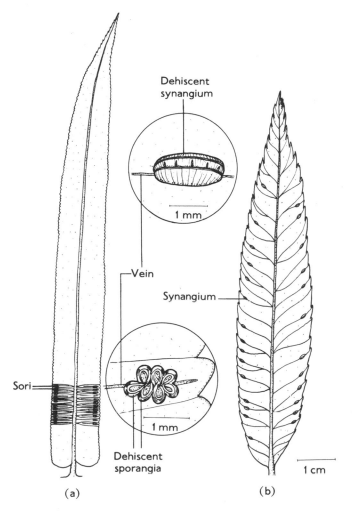

Figure 7.8. Fertile pinnules of the Marattiales. (a) *Angiopteris*. (b) *Marattia*.

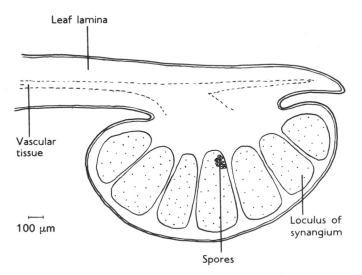

Figure 7.9. *Marattia*. Vertical section of synangium.

Figure 7.10. (a) *Angiopteris evecta*. The abaxial surface of a fertile pinnule showing the grouping of the sporangia into sori, and the arrangement of the sori. The sporangia have dehisced. Scale bar 2.5 mm. (b) *Marattia fraxinea*. The abaxial surface of a fertile pinnule showing the arrangement of the synangia. The synangia have dehisced. bar Scale 2.5 mm.

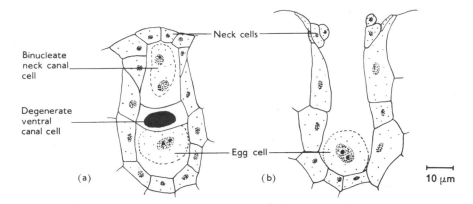

Figure 7.11. *Angiopteris evecta*. Vertical section of archegonium. (a) Immature. (b) Prior to fertilization. (After Haupt, from Foster and Gifford. 1959. *Comparative Morphology of Vascular Plants*. Freeman, San Francisco. © 1959.)

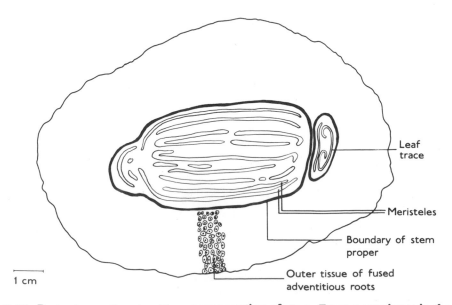

Figure 7.12. *Psaronius conjugatus*. Transverse section of stem. From a specimen in the Oliver Collection.

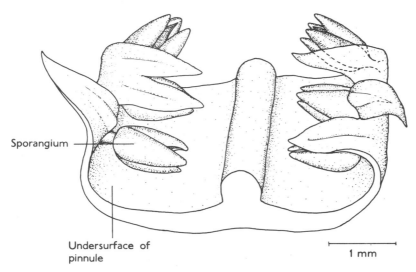

Figure 7.13. *Scolecopteris incisifolia.* **Reconstruction of portion of fertile pinnule. (After Mamay, from Andrews, Jr. 1961.** *Studies in Paleobotany.* **Wiley, New York.)**

The Filicales

These, the largest order of ferns, include (with the exception of *Botrychium* and *Ophioglossum*) all the homosporous ferns of temperate regions. The Filicales reach their greatest representation in the humid tropics and subtropics. Despite the diversity of the Filicales, there are two features which distinguish them consistently from the remainder of the living ferns. These are, first, the origin of the sporangium, and, second, the plane of the first dividing wall of the zygote. In the Filicales the sporangium develops from a single initial cell, a condition termed **leptosporangiate,** contrasting with the eusporangiate condition of the Ophioglossales and Marattiales. Also in the Filicales, at least in those species which possess the typical heart-shaped gametophyte, the first dividing wall of the zygote is vertical or slightly oblique, parallel to the longitudinal axis of the archegonium, whereas in the Ophioglossales and Marattiales the first wall is perpendicular to the archegonial axis. The subsequent embryogeny of the Filicales is regularly endoscopic.

Growth forms and economic uses of the Filicales. In the constantly humid and warm-temperate conditions of tropical mountains the Filicales adopt a wide range of growth forms. Besides terrestrial species, both upright and rhizomatous, are found arborescent and scandent forms, and numerous epiphytes. In an epiphytic species of Southeast Asia (*Lecanopteris*) the rhizome is curiously inflated and specialized to house colonies of ants. The sporangia, the walls of which contain fat, are sought after by the ants and the spores thereby distributed. Adaptations of this kind (myrmecophily) are otherwise found only in flowering plants.

Few Filicales figure in today's economy although some (e.g., *Pteridium*) have been valued in the past as a source of fuel, thatch, and food. Young croziers of *Pteridium* are still cooked and eaten in parts of Southeast Asia, despite the presence of a carcinogen. In North America, *Matteuccia* is similarly used as a delicacy ("fiddleheads"). Secondary plant products are a notable feature of some fern fronds and lead to extracts having minor medicinal uses. In some ferns these products are toxic (in *Pteridium* they include the insect-moulting hormone ecdysone) and may provide protection from predators.

In the less stable regions of the Andes, Indians use the trunks of tree ferns for the framework of buildings. The resistance of the sclerenchyma to shattering by earthquakes is superior to that of timber. Several species of herbaceous ferns are popular horticultural plants, particularly those mutants with striking modification of the form of the frond.

The growth of the stem. The stems of the Filicales, with the occasional exception of some of the larger specimens of *Osmunda*, grow from a single, conspicuous, apical cell (Fig. 7.14), tetrahedral in shape. Daughter cells are cut off adjacent to its three posterior faces, although it is disputed whether the apical cell itself divides. Studies with polarized light have shown that the cellulose microfibrils in the tangential walls of the daughter cells lie in arrays transverse to the axis of the stem. Consequently, tangential growth is constrained, and expansion is predominantly radial and longitudinal. The apex of the filicalean ferns is evidently much less specialized than that of a flowering plant. A slice taken from the tip of the apex, no more than 0.25 mm (0.01 in.) in depth, will generate a normal stem in culture. The same result with flowering plants can be obtained only with a substantially greater explant.

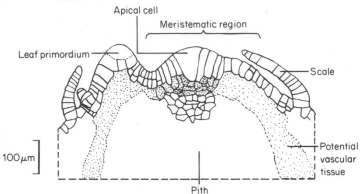

Figure 7.14. *Dryopteris aristata.* **Longitudinal section of stem apex. (After C. W. Wardlaw. 1944.** *Annals of Botany, New Series 8.*)

The meristematic activity in the apical cone diminishes towards its base. Below the apical cone, cell divisions are more generalized and variously directed. Leaf primordia arise in this region in a definite phyllotactic sequence. A leaf primordium, first visible as a slight protuberance, soon develops its own apical cell. As the leaf primordia age and become separated by the expansion of the apex, bud primordia may be formed between them, but in some species buds do not appear at all so long as the apex is actively meristematic. Development of buds beyond the stage of primordia is rarely seen in the region of developing leaves.

A curious situation is found in *Pteridium* (bracken). The rhizomes of the mature plant are arranged in layers, the lowermost (up to 30 cm/1 ft. or more beneath the surface) consisting solely of "long" shoots with extended internodes. Most of the fronds are borne on short stubby rhizomes nearer the soil surface. In the event of a "front" of bracken invading a new area, the lowermost rhizomes head the advance.

Since the apices of many ferns are comparatively broad and accessible, they provide excellent material for experimental work on phyllotaxy. The results indicate that the young leaf primordia, each a center of meristematic activity, suppress growth in their immediate vicinity. Thus, if the position in which a leaf primordium is expected to arise is isolated by radial incisions from the neighboring recently initiated primordia, then the new primordium develops with unusual vigour and outgrows the others. Similar experiments also confirm the fundamental similarity of stems and leaves in the Filicales. For example, tangential incisions on the anterior side of very young primordia that would normally yield leaves result in the production of stem buds instead. Incisions on the posterior side are without any effect. Consequently the determination of the subapical primordia appears to depend upon their being initially traversed by gradients of metabolities originating in the apical meristem. If a primordium is isolated from these gradients by an anterior incision, it yields the radially symmetrical structure of a stem instead of the dorsiventral symmetry of a leaf. Naturally in interpreting the results of microsurgery, the growth-stimulating effects of wounding must also be taken into account.

The formation and morphology of the stele. The cells which yield the vascular tissue first become recognizable as a distinctively staining ring shortly below the apical cell. The diameter of the ring increases in register with that of the apex as a whole, and beneath the leaf primordia its cells become confluent with crescentic strands of similar cells descending from them. Further down in the apex the cells of the ring are continuous with the procambium, and this in turn with the vascular tissue of the mature shoot. The position of the protoxylem is variable, but the xylem is commonly mesarch. The large tracheids of the metaxylem have scalariform pitting, and in a few instances (e.g., in the rhizome of *Pteridium*) the oblique end walls have scalariform perforations. These vessellike channels recall the situation in some species of *Selaginella*. The phloem consists of sieve cells with sieve areas confined to the oblique end walls. The vascular tissue is usually surrounded by a narrow zone of parenchyma, and then by an endodermis with a clear Casparian strip. The walls of the cortical cells adjacent to the endodermis are often thickened and made conspicuous by impregnation with phlobaphene. The endodermis and these thickened tangential walls probably together limit apoplastic transport between stele and cortex.

The form of the stele in the Filicales shows considerable variation (Fig. 7.15). In some species of *Gleichenia,* for example, the procambial tissue yields a solid core of tracheids from which the leaf traces depart without any break in the continuity of the xylem (Fig. 7.15a). Some other species of *Gleichenia* show a similar stele, but with the medullation of the tracheidal core leading to the production of a *siphonostele* (Fig. 7.15b). In *Osmunda* (Fig. 7.16) the stele is basically a siphonostele, and the phloem and endodermis remain wholly external. The continuity of the xylem, however, is broken at the departure of the leaf traces, leaving a so-called "leaf gap" which closes again anteriorly. Since, when dissected, the xylem (but not the stele as a whole) has the appearance of a cylinder of netting, *Osmunda* is said to have a dictyoxylic siphonostele.

In some ferns in which the stele is cylindrical phloem and endodermis are present both inside and outside the xylem (Fig. 7.15c). This form of stele, found principally in rhizomatous ferns, is termed a *solenostele* (or amphiphloic siphonostele). Leaf gaps are regularly present and sometimes additional perforations unrelated to the departure of leaf

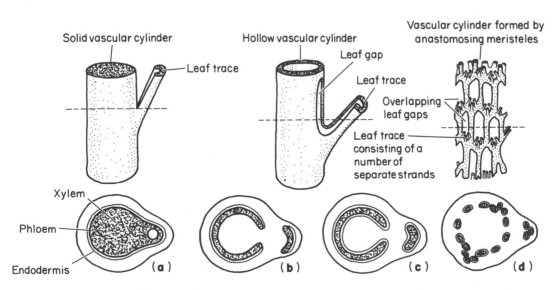

Figure 7.15. Principal forms of fern steles. (a) Protostele. (b) Siphonostele. (c) Solenostele. (d) Dictyostele.

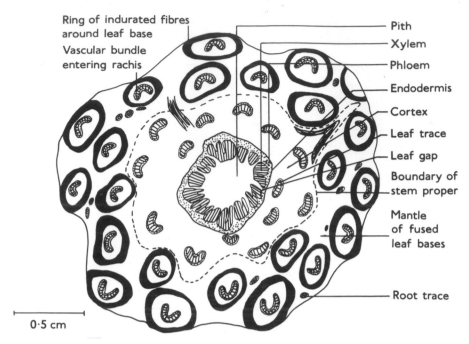

Ring of indurated fibres around leaf base

Vascular bundle entering rachis

Pith

Xylem

Phloem

Endodermis

Cortex

Leaf trace

Leaf gap

Boundary of stem proper

Mantle of fused leaf bases

Root trace

0·5 cm

Figure 7.16. *Osmunda regalis.* **Transverse section of rootstock.**

traces. The internal and external phloem and endodermis are in continuity around the margins of the gaps in the xylem. If the leaf gap and other perforations are close together, as in *Dryopteris,* the stele in section appears as a ring of anastomosing vascular bundles (meristeles) (Fig. 7.15d), each with xylem at the center and concentric phloem. This type of stele, which is of widespread occurrence, is termed a **dictyostele**. A complication, shown, for example, by the rhizome of *Pteridium* (Fig. 7.17) and the trunks of the tree fern *Cyathea,* is the presence of two or more concentric vascular systems, interconnected at intervals and usually all contributing to the leaf traces. These steles are said to be poly-cyclic. A point to be noted in passing is that steles are not always radially symmetrical. Dictyosteles in the ferns with creeping rhizomes, for example, are often markedly dor-siventral, the departure of the leaf traces being confined to the upper surface and flanks.

The experimental investigation of stelar morphology. The form of the stele in the Filicales has also been the subject of experimental investigation. In *Dryopteris,* for example, if the apical region is isolated by vertical cuts, but left in contact below, it con-tinues to grow and a solenostele differentiates behind it. As the apex expands and builds up a new crown, the stele gradually opens out to reform a dictyostele. In any one species, therefore, the size of the apex determines the form of the stele. This is also well shown in sporelings where a protostele is always present at the beginning. In protostelic species this merely increases in diameter as the plant develops, but in solenostelic and dictyostelic species the protostele of the sporeling becomes medullated, and phloem and endodermis appear within in step with the increasing girth of the apex. This relationship between size and form is clearly the consequence of physiological equilibria, but they are undoubtedly complex and have yet to be resolved.

Other anatomical features of the stem. In addition to the xylem, which is wholly primary, there are frequently bands or rods of sclerenchyma in the stem contributing to its rigidity. In the tree ferns, for example, many which reach heights of 10 m, (33 ft.) mechanical stability is dependent almost entirely upon the extremely tough girdle of sclerenchyma in the outer cortex, often in association with the leaf bases. In those Filicales which are believed, on the basis of fossil evidence, to be relicts of very ancient

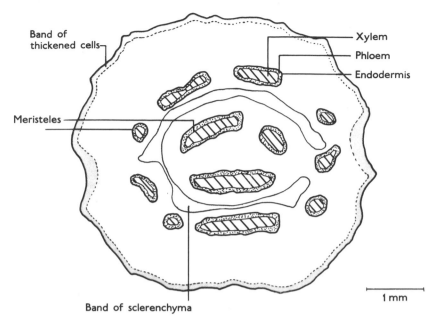

Figure 7.17. *Pteridium aquilinum.* **Transverse section of rhizome.**

groups (e.g., *Gleichenia*), the parenchymatous tissue of the stem often contains resin sacs and mucilage canals. Among the Filicales more recent in origin these features are less evident.

Roots. After the primary root all subsequent roots are adventitious, in arborescent forms often being produced even in the aerial regions and providing a mantle of stubby outgrowths between the leaf bases. Roots show a distinct apical cell, but in some instances this may be quiescent, divisions being confined to the cells at its flanks. As in other filicophytes a root cap is present, and in many species root hairs. The xylem is commonly diarch, and, in many epiphytic forms, all but the protoxylem often remains unlignified.

The morphology and anatomy of the vegetative leaves. The leaves of the Filicales retain an apical cell during their development from the primordium. Some form of pinnate branching is usually present in the mature leaves. True dichotomous branching occurs very rarely (the frond of *Rhipidopteris peltata* provides one of the few examples), but cymose branching, superficially resembling dichotomy, is shown by the leaves of several species of *Gleichenia*. In a few Filicales meristematic areas are retained in the differentiated leaf, and these subsequently grow out to form either additional leaves (as in *Trichomanes proliferum,* Fig. 7.18) or new plantlets (as in *Asplenium mannii* and *Camptosorus radicans*). These forms illustrate how in the living ferns, as in the extinct, the leaves sometimes display features suggestive of stems. All parts of the young leaves usually show circinate vernation. The extension of the rachis and the unrolling of the pinnae clearly involve considerable coordination of growth in space and time. There is evidence that this is dependent upon the diffusion and varying relative concentrations of growth-regulating substances (auxins) in the expanding leaf, but these auxins are not necessarily identical with those in seed plants. The expanding leaves of some ferns, for example, the tropical *Dryopteris decussata,* are enveloped in mucilage, possibly with some protective effect. Extrafloral nectaries are found at the points of branching of the frond of *Pteridium*. There is no evidence that they have any function apart from the secretion of unwanted metabolites.

The lamina of the leaf is commonly differentiated into palisade and mesophyll, but

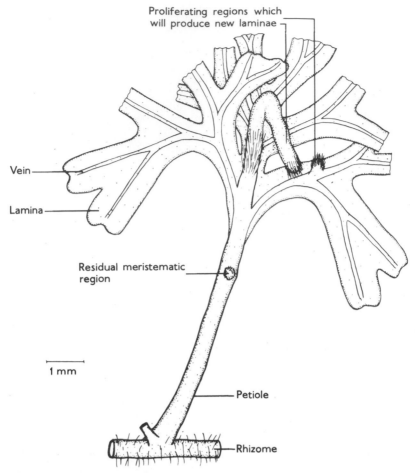

Figure 7.18. *Trichomanes proliferum.* **Frond, showing indefinite growth. (After P. R. Bell. 1960.** *New Phytologist* **59.)**

the texture is very variable and in some species a thick cuticle on the upper surface gives the leaf a surprising harshness. "Filminess," the possession of laminae only one cell in thickness, is found in *Leptopteris* and throughout the family Hymenophyllaceae. Filmy ferns are necessarily confined to situations of continuously high humidity. They are often able to thrive in irradiances far below those tolerated by flowering plants, and more akin to those of the bryophyte communities with which the epiphytic forms are frequently mixed. In *Hymenophyllum malingii,* a peculiar epiphyte of New Zealand, the leaf lacks a lamina. Instead, the axes of the pinnately branched frond bear green filaments interspersed with stellate trichomes. Some of the larger tropical epiphytes are distinguished by producing sterile leaves of two forms. In addition to the erect photosynthetic leaves are others which clasp the support. The latter soon die, but persist in a rigid scarious condition and serve as collectors of humus and moisture. *Platycerium* (Fig. 7.19) provides a striking example of this kind of habit. In other epiphytes of similar situations the leaves, borne on a short upright rootstock, are stiff and tightly overlapping, so forming a funnel which traps rain and organic matter. *Asplenium nidus* provides a typical example of these "nest ferns." The material at the base of the funnel is freely penetrated by absorptive rootlets.

   The fertile leaves and the nature of the sporangia. The fertile leaves of the Filicales are often quite similar to the sterile (as in *Dryopteris*), but dimorphy is not uncom-

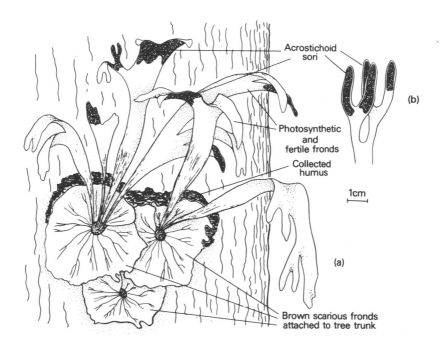

Figure 7.19. *Platycerium.* **(a) Habit, showing the two forms of leaves. (b) Lower surface of fertile portion of frond.**

mon. In *Blechnum spicant,* for example, both sterile and fertile fronds are simply pinnate, but, in the fertile, the sterile part of the lamina is very reduced. The sporangia arise from single initial cells (except in *Osmunda* where a few additional cells are involved), either at the margin or on the lower surface of the leaf. The mature sporangium has a distinct stalk, the structure of which ranges from a broad multicellular stump to a delicate and relatively long column of cells. The wall of the capsule is only one cell thick (again with the exception of *Osmunda,* where a thin inner layer may also be present), and it always contains a series of indurated cells and a well-defined stomium. In *Osmunda* the indurated cells are grouped laterally (Fig. 7.20a), and a linear stomium extends from them over the apex of the sporangium. In other Filicales the indurated cells are arranged in a single band (**annulus**) which encircles the sporangium transversely near the apex of the sporangium, as in *Anemia* (Fig. 7.21) and other Schizaeaceae; obliquely, as in *Gleichenia* (Fig. 7.20b) and the Hymenophyllaceae; or vertically, as in *Dryopteris* (Fig. 7.22) and most common temperate ferns. The annulus is interrupted by the stomium. Where the annulus is vertical, it is also interrupted by the stalk of the sporangium, the stomium then lying just in front of the stalk. The sporangium dehisces at the stomium as a consequence of tensions set up in drying. Although this process is of a general nature in *Osmunda,* it is more precise in those Filicales where the sporangia have annuli, particularly where the annulus is vertical. Here, as the cell sap in the annular cells diminishes by evaporation, asymmetrical thickening of the cell walls (Fig. 7.22) causes an increasing tangential tension which tends to reverse the curvature of the annulus. The stomium eventually breaks, and the upper part of the sporangium gradually turns back as if on a hinge (Fig. 7.22). Tension in the cells of the annulus soon reaches a critical level; at this point the water remaining in the annular cells spontaneously becomes vapor. The tension is immediately released, and the upper part of the sporangium flies back to more or less its original position (Fig. 7.22). These two movements effectively disperse the spores. It has sometimes been observed that the movements are repeated, a feature not so readily explained.

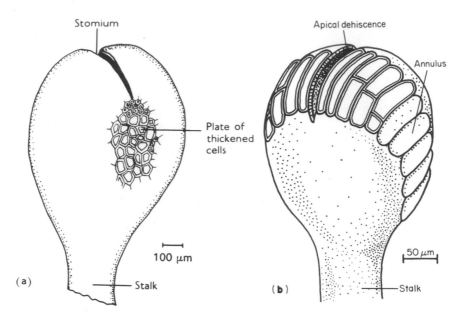

Figure 7.20. (a) *Osmunda regalis.* **Sporangium. (b)** *Gleichenia.* **Sporangium.**

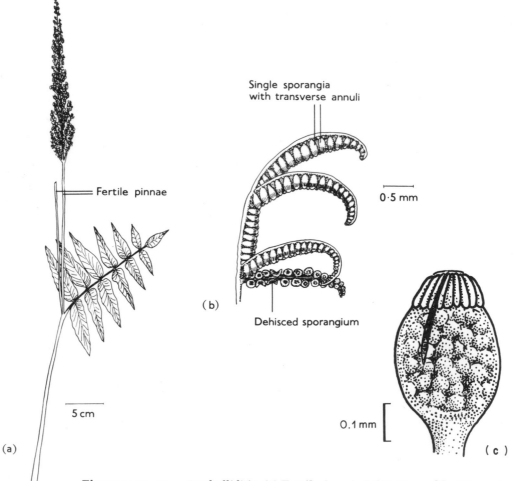

Figure 7.21. *Anemia phylliditis.* **(a) Fertile frond. (b) Portion of fertile region.**
(c) Sporangium showing transverse annulus.

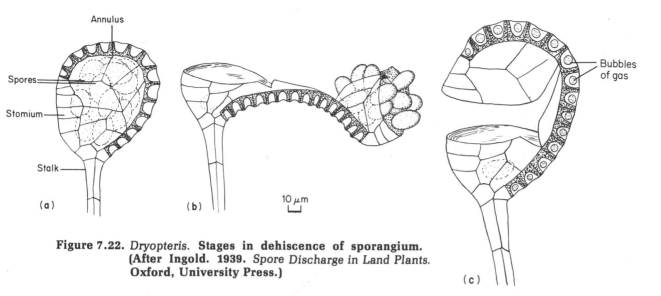

Figure 7.22. *Dryopteris.* **Stages in dehiscence of sporangium.**
(After Ingold. 1939. *Spore Discharge in Land Plants.*
Oxford, University Press.)

The arrangement of the sporangia. In most Filicales the sporangia arise in distinct sori, usually beneath veins or near their extremities, these two positions being called superficial and marginal respectively. Sometimes the sporangia are produced in a continuous line, referred to as a *coenosorus,* well shown, for example, by *Pteridium.* The sorus is often partly or wholly covered by an outgrowth of the lamina, called an indusium, which adopts a characteristic form. In *Dryopteris,* for example, the indusium is reniform (Figs. 7.23 and 7.24) and in *Polystichum,* peltate. In *Onoclea* the sori are protected by a rolling up of the fertile pinnules. In some species of *Polypodium,* where an indusium is typically absent, the sori are immersed in the lower surface of the lamina. A few ferns show the so-called "acrostichoid" condition in which the sporangia arise as a continuous felt on the lower surface of the fertile frond (e.g., *Platycerium;* see Fig. 7.19). In some ferns (e.g., *Gleichenia*) the sporangia in a sorus are all of the same age (Fig. 7.25a, b.); in others they are produced in spatial and temporal sequence on an elongated receptacle, as in the Hymenophyllaceae (Fig. 7.25c, d), yielding a so-called "gradate" sorus; and in yet others the sporangia are produced over a period but intermingled, leading to a so-called "mixed" sorus (Fig. 7.25e, f), as in *Dryopteris* and most common temperate Filicales.

The development of the sporangia and spores. In the development of the sporangium a cluster of spore mother cells becomes surrounded by a two-layered tapetum. In origin this is part of the wall tissue and not of the archesporium, as in eusporangiate sporangia. The cytological changes accompanying meiosis provide a model for sporogenesis in the land plants generally. As their nuclei enter prophase the spore mother cells become surrounded by a thickened wall, and cytoplasmic connections between them are extinguished. The presence of callose in this special wall has not yet been convincingly demonstrated. During prophase the density of the cytoplasm diminishes and there is a loss in affinity for basic stains, largely accounted for by a fall in the frequency of ribosomes and hence in the concentrations of ribonucleic acid. Spores are produced in tetrads. The first exine is secreted by the spore itself. At this stage the thickened wall of the mother cell, weakened by autolytic degradation, breaks open. The spores then separate and complete their development in the breakdown products of the tapetum.

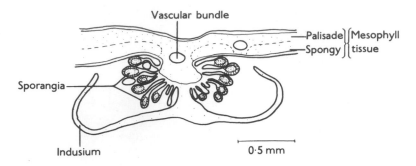

Figure 7.23. *Dryopteris filix-mas*. **Vertical section of sorus.**

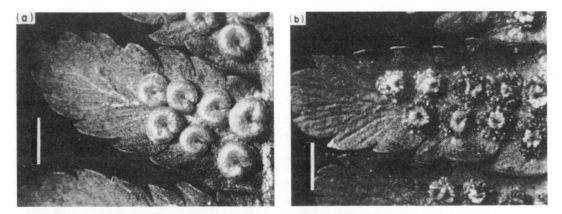

Figure 7.24. The abaxial surface of fertile pinnules of *Dryopteris filix-mas*. (a) The sori in a young condition, before dehiscence of the sporangia. (b) Three weeks later, most of the sporangia having dehisced. Scale bar 2 mm.

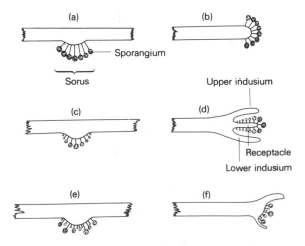

Figure 7.25. Principal forms of sorus. (a,b) Simple, the sporangia all of the same age. (a) Superficial. (b) Marginal. (c,d) Gradate. (c) Superficial. (d) Marginal. (e,f) Mixed. (e) Superficial. (f) Marginal.

Some 2^6 to 2^8 spores are produced in each sporangium, the higher numbers being characteristic of the families believed to be more primitive. The spores are usually of the order of 40 μm in diameter, but measurements of a representative sample will usually show a normal distribution with a range of about $\pm$ 10 μm about the mean. A few ferns normally included with the homosporous do, in fact, produce spores of different sizes. The most extreme example is seen in *Platyzoma,* a fern of very unusual appearance growing on deep sands in Northeast Australia. Although the mature sporangia are about the same size, some produce 16 large spores and others 32 small ones. The difference in spore size seems to result from the greater availability of nutrients and space for spore growth when the number is lower.

The outer part of the spore wall (exine), the material of which is derived from the tapetum, is sometimes deposited in a characteristic pattern of bars and ridges. This is especially true of the Schizaeaceae, and it provides a feature that has been very useful in identifying fossil forms. In some ferns, principally the more recent, the spores are monolete instead of trilete. Monolete spores often have an additional translucent investment, called a **perispore,** formed from the remains of the tapetum. In a few ferns monolete and trilete spores are found in the same sporangium.

The gametophytic phase. In the Osmundaceae and sporadically in some other families the spores are green, but normally chlorophyll is absent. Nongreen spores are often rich in lipid, amounting in some instances to 60% of the dry weight. Green spores are short-lived, but nongreen spores may remain viable for many years, particularly if buried in soil. In this way "banks" of viable spores may be formed beneath stands of ferns, germination occurring only if the spores are brought to the surface. Such reservoirs of spores provide for the re-colonization of areas cleared by cultivation or other disturbance.

Although moisture is essential for germination, the requirement for light is variable. Some spores (e.g., *Pteridium*) germinate in the dark, but others require the stimulus of red light. The phytochrome system is evidently involved, but the situation is not simple since in some instances the effect of red is reversed by both far-red and blue. After the first division the protruding daughter cell often yields a colorless rhizoid, the elongation of which may be promoted by metal ions present in the perispore. Continued divisions give rise to an algal-like filament, the growth of which is predominantly apical, furnished with a few rhizoids. Germination and production of the first rhizoid appear to be resistant to inhibitors of transcription, so it is probable that these initial steps are dependent upon long-lived messenger RNA already present in the spore.

In some Filicales (the Hymenophyllaceae, for example, and some Schizaeaceae), the gametophyte remains filamentous and the sex organs are borne on lateral cushions of cells. In most, however, the cells at the apex of the filament soon begin to divide in a number of directions and so form a cordate (heart-shaped) gametophyte (Fig. 7.26), of which *Dryopteris* and *Pteridium* provide familiar examples. There is no general agreement about the cause of this change from one-dimensional to two-dimensional growth, but it seems likely that it is a consequence of a changing balance between carbohydrate and protein metabolism. When the protein metabolism is depressed in relation to the carbohydrate (as can be done experimentally by growing the cultures in red light) the gametophytes persist indefinitely as filaments with elongated cells. In blue light, which changes the balance of the metabolism in favor of protein, the cells divide more frequently and become progressively shorter, their width remaining little changed. Ultimately the apical cell becomes broader than long. In these conditions one-dimensional growth gives way almost immediately to two-dimensional. The change in the direction of division at the apex of the filament is in accordance with the concept that the plane of the new wall in a dividing cell is transverse to the principal stress in the plasmalemma (which in turn may influence the alignment of microtubules). In cells which are longer than broad this stress, caused by the turgor of the cells, is longitudinal, and hence the new wall transverse. In cells broader than long the principal stress is transverse to the filament, and the new wall correspondingly longitudinal or almost so. Substances which affect the rate of growth, and the rapidity with which growth changes from one-dimensional to two-dimensional can be isolated from media in which spores have been cultured. The proportions of these substances vary in relation to the wavelength of the light to which the cultures have been exposed.

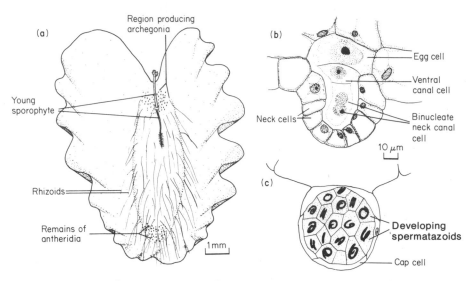

Figure 7.26. (a) *Pteris ensiformis*. Lower surface of gametophyte, bearing young sporophyte. (b) *Pteridium aquilinum*. Median longitudinal section of young archegonium. (c) *Pteridium aquilinum*. Almost mature antheridium.

Despite their delicate structure, the gametophytes of some ferns have proved surprisingly resistant to desiccation and freezing, their tolerances often far exceeding those of the sporophytes. A few gametophytes have modifications which clearly promote survival. That of the Mediterranean *Anogramma leptophylla,* for example, has a tuberous portion resistant to drought. In a few tropical species of the Schizaeaceae the gametophyte is wholly subterranean and closely similar to that of *Psilotum*.

Many cordate gametophytes produce only antheridia in the first stages of growth, but subsequently the production of antheridia declines or even ceases, and archegonia then appear in sequence on the lower surface of the gametophyte behind the apical meristem. When gametophytes are grown in pure culture substances accumulate in the medium which stimulate the production of antheridia in other young gametophytes. The chemical nature of these substances, called **antheridiogens,** varies with species but some are similar to gibberellins. As a gametophyte passes from the antheridial to the archegoniate phase it becomes insensitive to its antheridiogen, but it is still secreted into the medium. Although studied mostly in axenic conditions, there is now evidence that antheridiogens regulate sexuality in natural populations of gametophytes.

When a sample of spores is sown, a proportion of the gametophytes never develops a clearly defined meristem. These gametophytes, termed **ameristic,** continue to produce antheridia indefinitely. Although the proportion of such gametophytes is influenced by the density of sowing, the tendency to develop ameristically may also depend upon spore size. A relationship between sex expression and spore size is clearly seen in *Platyzoma*. The small spores give rise to gametophytes which remain depauperate and male, while the large spores yield complex gametophytes which are at first female and only later produce antheridia.

Antheridia usually contain 2^5 or 2^6 spermatocytes, each of which differentiates a spermatozoid. When the antheridia are mature, flooding causes the mucilage within to swell. The cap cells are then forced off and the spermatocytes extruded. The flagella of the spermatozoids become active and each breaks free from its mucilaginous shell. The motile spermatozoid is about 5 μm long and takes the form of a helix with a left-handed (anticlockwise) screw. The anterior gyre is taken up with the multilayered structure and the associated mitochondrion, and the remaining gyres with the nucleus and microtubular ribbon. The flagella, confined to the anterior gyres, emerge tangentially and are directed posteriorly. They beat with a helical wave which drives the spermatozoid forward and at the same time causes it to rotate in the sense of the screw.

The necks of the archegonia, consisting of four to six tiers of cells, usually project conspicuously and are often recurved. The maturation of the egg, completed within about 24 hours, involves in some species extensive interpenetration of nucleus and cytoplasm. The full significance of this remarkable cytology is not yet known, but it may be concerned with the establishment of sporophytic growth. The mature egg is surrounded by a conspicuous lipoidal membrane. Flooding causes mucilage within the canal of the ripe archegonium to swell. The cap is forced off and the contents of the canal ejected. A clear passage, containing watery mucilage, now runs down to the surface of the egg cell.

In ferns with cordate gametophytes the production of archegonia continues only so long as growth remains ordered and symmetrical. This, in turn, is dependent upon the activity of the apical meristem, from which growth-regulating substances stream posteriorly. Beyond a certain size the activity of the meristem declines. The gametophyte then begins to proliferate irregularly and a male phase is re-established. Some fern gametophytes appear to have lost the power to produce sex organs. Pale green thalli, resembling small liverworts and locally abundant in shaded rocky areas of parts of temperate North America, are believed to be gametophytes of *Vittaria,* but sporophytes have never been observed. The evolution of these forms may have followed from the gametophytes being more tolerant of cool conditions than the corresponding sporophytes. The filamentous gametophytes of some species of *Trichomanes* are also found without functional sex organs beyond the distribution of the sporophyte in both North America and Britain. Reproduction of these imperfect forms is solely by fragmentation and gemmae. The gametophytes of some ferns with normal cycles also multiply asexually.

Fertilization and embryogeny. Fertilization in the Filicales depends, as with most other archegoniate plants, upon the presence of water. The mucilage around the mouth of the archegonial canal, which may contain traces of malic acid (known to possess chemotactic properties), attracts the spermatozoids and effectively confines them to the region of the archegonia. Several spermatozoids commonly enter an open archegonium, and occasionally some can be seen to swim out again. Where crosses are attempted between certain genera (e.g., *Athyrium* ♀ × *Dryopteris filix-mas* ♂) the mucilage immobilizes the foreign spermatozoids and no hybrids are produced. Barriers to gametic fusion in ferns are so far little studied. It is possible that in some species self-fertilization is prevented or its chances lessened by incompatibility mechanisms in the archegonial mucilage or at the surface of the egg, but the evidence so far is inconclusive. It appears, however, to be a general rule that only one spermatozoid enters the egg, although others may be seen pressed against it surface.

Division of the zygote follows about 48 hours after fertilization. The first vertical wall is succeeded by a horizontal so that in lateral aspect the zygote appears divided into quadrants. These quadrants indicate in a general way the course of the subsequent embryogeny. The upper anterior region, for example, goes to form the apex of the new sporophyte, the lower anterior the first leaf, while the posterior regions give rise to the first root and the foot (Fig. 7.27). There is no suspensor. Following fertilization, possibly a consequence of growth-regulating substances coming from the zygote, the growth of the gametophyte diminishes and the initiation of archegonia ceases. At the same time the cells of the archegonium immediately above the fertilized egg proliferate, forming a conspicuous cap (calyptra). Experiments have shown that this calyptra, probably by exerting mechanical pressure on the developing embryo, plays an important part in determining normal embryogenesis, recalling the situation in the mosses. If the calyptra is removed from above a very young zygote, the zygote gives rise to a mass of parenchymatous tissue before producing differentiated growing points. With cordate gametophytes, an intact apical meristem, probably in consequence of the auxin it produces, is also essential for normal embryogenesis. If this meristem is destroyed, differentiation of the embryo is markedly slower and the emergence of the first root very much delayed.

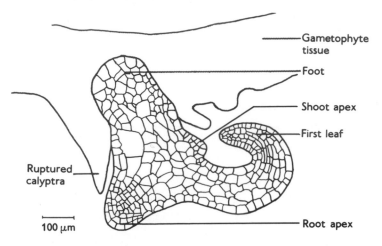

Figure 7.27. *Pteridium aquilinum.* **Vertical section of embryo.**

Aberrations of the fern life cycle are not uncommon. Apogamy, the production of a sporophyte without sexual fusion, occurs regularly in some ferns and can be induced experimentally in others. *Dryopteris affinis* (*D. borreri*) provides an example of a regular apogamous cycle. In this and similar species the final mitosis of the sporogenous cells is incomplete. Following division of the chromosomes, the nucleus reforms. The cell remains undivided, although it has grown considerably as if a normal division were about to occur. Instead, it becomes a spore mother cell. The restitution nucleus, which has twice the sporophytic number of chromosomes, enters meiosis. The four spores produced, reflecting the large size of the spore mother cell, have diameters of about 80 μm, and contain nuclei with the same number of chromosomes as the parent sporophyte. The spores germinate in the normal way and the gametophytes pass through a male phase. The spermatozoids are perfectly formed and functional, and are capable of fertilizing eggs of related sexual species. Subsequently, however, in place of archegonia, a sporophyte develops directly from the subapical region of the gametophyte. The cellular mechanisms underlying this kind of life cycle are not yet understood. The condition is however known to be genetically dominant since hybrids with sexual species have the same kind of apogamous cycle. Apogamy can sometimes be induced by withholding water from gametophytes and preventing sexual reproduction. Sporophytes have been raised in this way from gametophytes of *Dryopteris filix-mas* and *Dryopteris dilatata*. In *Pteridium* in pure-culture apogamous outgrowths are promoted by a high level of sucrose. Such experimentally produced sporophytes are often depauperate, and if they reach meiosis fail to produce viable spores. They naturally lack the pre-meiotic doubling of the chromosomes characteristic of the *Dryopteris affinis* kind of cycle.

The direct production of gametophytes from sporophytes, termed apospory, can be very readily induced in culture by placing fragments of juvenile leaves in sterile conditions on an agar medium. The gametophytes are usually sexually perfect and yield tetraploid sporophytes. This phenomenon may occur sporadically in the wild and lead to the production of natural autopolyploids. Autopolyploids may also arise by chance failure of reduction in meiosis.

The evolution of the Filicales. The evolutionary relationships of the living Filicales can be studied at two levels. First, by carrying out crosses between living species and examining the pairing behavior of the chromosomes at meiosis we can obtain evidence of the extent of the genetic identity between them. Second, by comparing the morphology and anatomy of living species with the fossil, we can obtain a general impression of the evolution of the contemporary fern flora.

Studies of chromosome homologies at meiosis, facilitated by the ease with which squash preparations can be made of developing sporangia, have shown that many familiar species are of hybrid origin. *Dryopteris filix-mas*, for example, is an allotetraploid probably derived form a hybrid between two diploid ancestors, one similar to *D. oreades* (*D. abbreviata*) and the other to *D. caucasica*. Particularly interesting is that widespread forms of some species are autotetraploids. A familiar example is the subspecies *quadrivalens* of *Asplenium trichomanes*. Other species may be tetraploid in origin, but isozyme analysis shows that they behave as diploids. Representative of this situation is the common *Pteridium* (n = 52). Selfing experiments have revealed considerable genetic variation in the progeny, possibly arising from recombination between duplicated unlinked loci (**homoeologous heterozygosity**). Hybridization and duplication of chromosome number undoubtedly account for much of the diversity in living ferns, but provide for little profound anatomical or morphological change.

The study of the fossil Filicales has given valuable information about the evolutionary status of the present-day families. It is clear, for example, that the Osmundaceae have a very long history. A transverse section of the stem of the late Permian *Thamnopteris* is very similar, even to the extent of the arrangement of the sclerenchyma and the packing of the leaf bases, to a section of *Osmunda*. The record of this family, which includes preserved fronds and sporangia, continues through the Mesozoic to the present. The Schizaeaceae also have a well-established fossil history extending back to the Paleozoic, and that of the Gleicheniaceae is similar. Other families first appear in the Mesozoic. Nevertheless, despite these evidences of antiquity, most of the living Filicales either have no fossil record, or no record extending back further than the Tertiary. This is particularly true of the large family Polypodiaceae, and we must suppose that these ferns are comparatively recent, probably having evolved towards the end of the Cretaceous period and subsequently.

Comparison of the living Filicales with the fossil, quite apart from tracing particular lineages, also reveals those features which can be regarded in a general way as primitive. Protostelic and solenostelic vascular systems, the simultaneous production of the sporangia in the sori, short and thick sporangial stalks, the indurated cells of the sporangial wall aggregated laterally or arranged in a transverse annulus, and a large number of spores in each sporangium are all features of the early Filicales. Conversely, dictyosteles, the production of mixed sori, long and delicate sporangial stalks, vertical annuli, and low spore numbers are all features of Filicales with little or no fossil record.

On the basis of these criteria, it is possible to assign the families of living ferns to three grades according to their evolutionary advancement. It is also possible to arrange them in two series, according to whether the sporangia are marginal or superficial in origin, but recent research has thrown considerable doubt upon the significance of this feature. Nevertheless, it is tentatively retained here, and the position of some of the more important families and genera in this double classification is shown in Table 7.1. The

Osmundaceae are considered assignable to both series, since the sporangia are marginal in *Osmunda*, but superficial in *Todea*. This classification does not, of course, imply that living families and genera have evolved from each other; it merely illustrates relative primitiveness. The arrangement is substantiated by the fossil record which is of greater duration in respect of the Filicales at the bottom of the table than of those at the top.

Table 7.1.   Classification of selected genera and families of living Filicales based on (1) evolutionary development and (2) origin of sporangia.

Features of evolutionary significance	Position of origin of sporangia	
	Marginal	Superficial
Steles, commonly dictyosteles. Sori mixed. Sporangia with vertical annuli interrupted at stalk	*Pteridium* *Davallia*	*Dryopteris* *Polystichum* *Asplenium* *Athyrium* *Polypodium*
Steles, dictyosteles or solenosteles. Sori gradate. Sporangia with oblique annuli	Hymenophyllaceae	Cyatheaceae
Steles, protosteles or solenosteles. Sori simultaneous. Indurated cells of sporangial wall aggregated, or in transverse or oblique annuli	Schizaeaceae *Osmunda* (Osmundaceae)	Gleicheniaceae *Todea* (Osmundaceae)

(left axis) Numbers of spores/sporangium 2^{5-6} ... 2^{7-8} — Reduction in number of cells forming sporangial stalk

The Hydropteridales

These so-called water ferns are outstanding in being heterosporous. Although mainly plants of fresh waters and swamps, some species are characteristic of sites subject to seasonal drying. Hydropterids of the family Marsileaceae have a creeping rhizome bearing subulate (Fig. 7.28), bifoliate, or quadrifoliate leaves. Those of the Salviniaceae are floating plants with short branching rhizomes. The leaves are shortly petioled or sessile. In *Salvinia* the surface of the upper leaves is made unwettable by a covering of waxy hairs; the lower leaves are submerged and much divided, taking the place of roots. In *Azolla* (Fig. 7.29) the leaves are minute and unequally two-lobed, the larger upper lobe floating and containing a mucilaginous chamber inhabited by the blue-green alga *Anabaena azollae*. The delicate roots of *Azolla*, being devoid of soil, are very suitable objects for the study of root growth. Serial sectioning reveals how the cell lineages derived from the three posterior faces of the apical cell generate the mature root.

Some hydropterids have economic value. In rice-growing areas, *Azolla*, on account of its nitrogen-fixing symbiont *Anabaena*, is used extensively as a green manure. The vegetative parts of *Marsilea* yield a sedative used in Indian medicine, and the sporocarps are a source of starch.

Reproductively, *Pilularia* (Fig. 7.28), is representative of the Marsileaceae. Fertile pinnules are produced at the base of the petiole. They curve over the developing sporangia and become concrescent and eventually hardened, so producing the sporocarp. The ridgelike placentae bear megasporangia below and microsporangia above (although in *Marsilea* the positions are reversed). In *Azolla* (Salviniaceae) the sporangia develop on the submerged lower lobe of the leaf, microsporangia and megasporangia being produced in separate sori. The sporocarp is formed by an elaborate indusium which totally encloses

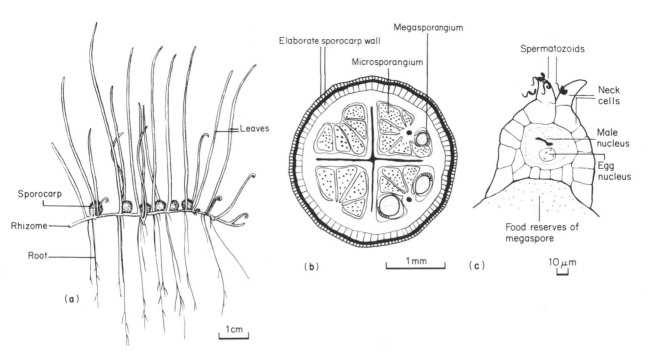

Figure 7.28. *Pilularia globulifera.* **(a)** Habit. **(After Hyde and Wade. 1940.** *Welsh Ferns.* **National Museum, Cardiff.) (b)** Transverse section of sporocarp showing the four sori. **(c)** Longitudinal section of inseminated archegonium.

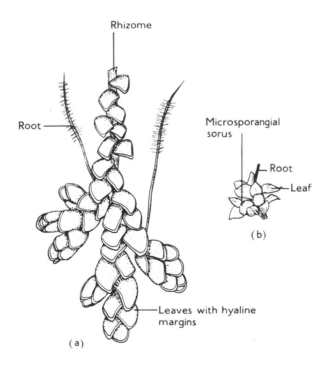

Figure 7.29. *Azolla filiculoides.* **(a)** Habit. **(b)** Lower surface of shoot showing microsporangial sori. **(After Campbell. 1905.** *The Structure and Development of Mosses and Ferns.* **Macmillan, New York.)**

the sorus. Sporogenesis in the hydropterids is similar to that in homosporous ferns, but in the megasporangia all but one of the potential megaspores are resorbed. Contrasting with the single megaspore in the mature megasporangium, the microsporangia typically contain 64 microspores.

The sporocarps of the Marsileaceae open only after the wall has weathered sufficiently to permit the entry of water. This may take years, but the viability of the spores declines only very slowly during this period of dormancy. When water eventually enters, the sori are carried out of the ruptured sporocarp by a wormlike expansion of the gelatinous remains of the fertile pinnules. The liberated spores germinate at once; the megaspore rapidly gives rise to a single archegonium surrounded by a few somatic cells (which may develop chlorophyll), while the microspores each produce a single antheridium containing 16 spermatocytes. The sporocarps of the Salviniaceae open more readily. In *Azolla* the megaspore is surrounded by four frothy massulae (Fig. 7.30) formed from the tapetum, and these give the liberated megaspore buoyancy, but it is doubtful whether they keep it afloat indefinitely. A variable number of massulae are formed in the microsporangium (Fig. 7.31). Each includes a number of microspores at its periphery and is furnished externally with peculiar anchorlike glochidia. These male massulae hook themselves to the female, and the complex then sinks. Germination of the spores, in the main similar to that of the spores of the Marsileaceae, then follows. Embryogenesis is much more rapid in the heterosporous ferns than in the homosporous. Fertilization occurs within about 12 hours of the germination of the spores, and an embryo emerges from the archegonium on the following day.

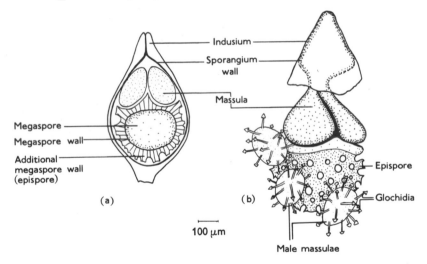

Figure 7.30. *Azolla filiculoides.* **(a) Longitudinal section of megasporangium. (b) Liberated megaspore with male massulae attached. (After Strasburger. 1873.** *Ueber Azolla.* **Abel, Leipzig.)**

The origins of the hydropterid ferns are altogether obscure. Although the fossil record of the Marsileaceae is meager, remains of reproductive parts resembling those of *Azolla* are identifiable as far back as the Lower Cretaceous.

Precursors of the Ferns

Although it is generally agreed that no true ferns have yet been detected in the Devonian, fossils are known which may have belonged to the complex of forms from which the ferns evolved.

Conspicuous among the Devonian floras are the **Cladoxylales,** represented by a number of genera including *Cladoxylon* and *Calamophyton.* They have been found in the

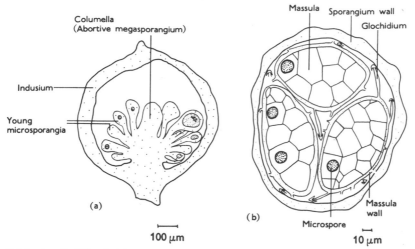

Figure 7.31. *Azolla filiculoides.* **(a) Longitudinal section of young microsporangial sorus. (After Campbell. 1905.** *The Structure and Development of Mosses and Ferns.* **Macmillan, New York.) (b) Transverse secton of mature microsporangium. (After Smith. 1955.** *Cryptogamic Botany,* **2. McGraw-Hill, New York.)**

Devonian and Lower Carboniferous of Europe and North America. The axes of *Cladoxylon* (Fig. 7.32), the larger reaching diameters of the order of 1.5 cm (0.6 in.), branched irregularly, and bore small dichotomously forking leaves, rarely reaching 2 cm (0.75 in.) in length. Some of the terminal shoots were fertile, and here small fan-shaped sporangiophores took the place of leaves, each segment of the sporangiophore terminating in a sporangium. The plant was probably homosporous.

The vascular system of the stem consisted of ascending, anastomosing plates of xylem, each in life probably surrounded by phloem and endodermis. Since the system has the appearance of falling into a number of independent subsystems, it is said to be polystelic. The vascular supply of the branches was compound in origin, departing from several of the adjacent ascending plates. A curious anatomical feature was the presence of small parenchymatous areas within the xylem, especially towards the periphery of the stem. Some species of *Cladoxylon* may have had rudimentary secondary thickening. *Calamophyton* was similar to *Cladoxylon* but the fertile appendages had sterile segments in addition to tassels of sporangia. *Pseudosporochnus* from the Middle Devonian of Belgium was possibly an allied plant. It appears to have been a small tree with an upright trunk. The main axis reached about 8 cm (3.13 in.) in diameter and bore a crown of primary branches. They underwent repeated dichotomies and some of the ultimate branchlets were fertile. The anatomy of the stem was cladoxyloid, and the parenchyma was characterized by numerous nests of sclerotic cells ("stone cells").

The **Zygopteridales,** an order mostly confined to the Carboniferous and regarded as truly fernlike, may have been represented in the Devonian by the remarkable *Rhacophyton.* This was probably a large bushy plant reaching a height of about 2 m (6.5 ft.). The main stem, about 2 cm (0.75 in.) in diameter, bore lateral branch systems of two kinds. The sterile branches were complanately branched leading to a spray of branchlets appropriately referred to as a frond; the fertile branches had a much more complicated morphology. At each node of the axis of the frond (rachis) there were two sterile branches subtended by two densely branched tassels of sporangia. The branching of the fertile frond was clearly not in one plane, but quadriseriate. *Rhacophyton* was probably homosporous.

The quadriseriate branching of the frond is characteristic of the Carboniferous zygopterids, of which *Metaclepsydropsis* is an example (Fig. 7.33). The vascular system

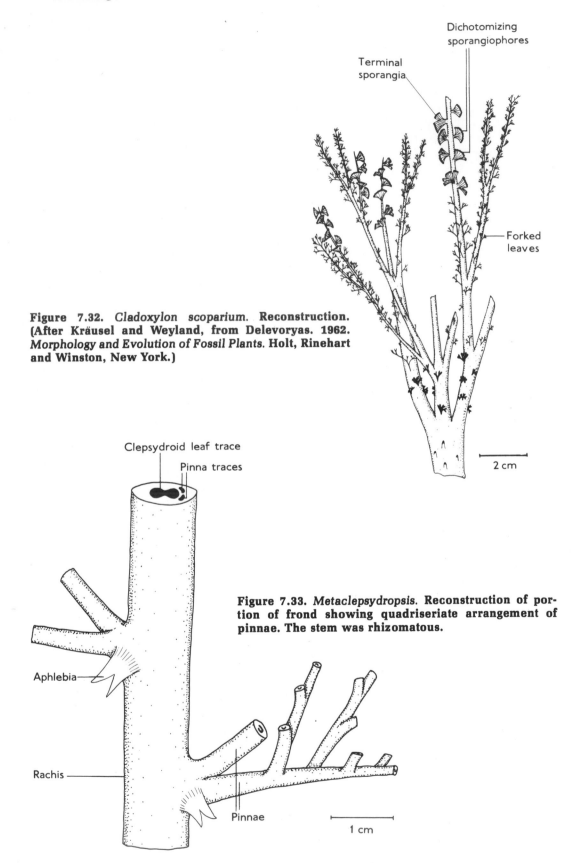

Figure 7.32. *Cladoxylon scoparium.* **Reconstruction. (After Kräusel and Weyland, from Delevoryas. 1962.** *Morphology and Evolution of Fossil Plants.* **Holt, Rinehart and Winston, New York.)**

Figure 7.33. *Metaclepsydropsis.* **Reconstruction of portion of frond showing quadriseriate arrangement of pinnae. The stem was rhizomatous.**

of the stem of the zygopterids was commonly a protostele, often with a parenchymatous medulla. The fronds, which were usually in a recognizable phyllotactic spiral, received a single vascular trace which assumed in the petiole a definite and characteristic symmetry. Consequently, the genera and species of these ferns are largely based upon the profile of the leaf trace in transverse section. In *Metaclepsydropsis*, for example, the section of the trace was hour-glass–shaped (clepsydroid). The protoxylem lay towards each pole, each group associated with an island of parenchyma. Pinna traces arose from each pole in alternate pairs, in register with the quadriseriate branching. In *Stauropteris*, where the frond was similarly constructed, the rachis contained four groups of tracheids ascending in parallel. Well-defined phloem with elliptical sieve areas has been seen in several of these petrified leaf traces. In several zygopterid fronds there were small-branched emergences, each with an exiguous vascular supply, at the base of the rachis and at the sites of branching of the fronds. These are referred to as aphlebiae (Fig. 7.33) and occur in some living ferns (e.g., *Hemitelia* and other tree ferns).

Although metaxylem tracheids in radial rows are occasionally seen in zygopterid stems, it is doubtful whether these were produced by cambial activity. As in living ferns secondary activity seems to have been absent.

The sporangia of the zygopterids were massive (often reaching or exceeding 2.5 mm (0.1 in.) in length) and presumably eusporangiate in origin. In many forms (but not in *Rhacophyton*) a distinct annulus of thickened cells interrupted by a thin-walled stomium is visible. Homospory appears to have been general, but a possible example of heterospory is provided by the Carboniferous *Stauropteris burntislandica*. The sporangia (Fig. 7.34) of this species (*Bensonites*), which were parenchymatous at the base, produced only one tetrad, consisting of two large and presumably functional spores about 200 μm in diameter, with two small, possibly abortive, spores lying between them. In a related

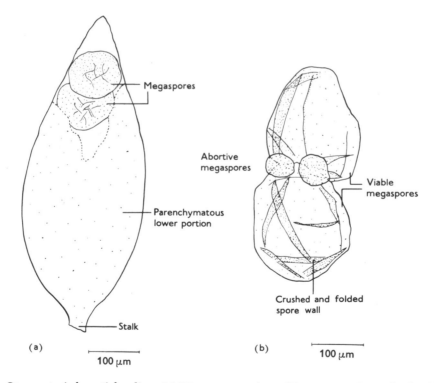

(a)   100 μm

(b)   100 μm

Figure 7.34. *Stauropteris burntislandica.* (a) Megasporangium. (From a specimen in the Oliver collection.) (b) An isolated tetrad (*Bensonites*). (After W. G. Chaloner. 1958. *Annals of Botany, New Series 22.*)

species, believed to be homosporous, spores germinating in a manner typical of living ferns have been found petrified within sporangia. Apart from a few isolated instances of this kind, nothing further is known of the gametophytic phase of the zygopterids.

With the exception of those forms clearly referable to families still existing, the remainder of the Carboniferous ferns are placed in the **Coenopteridales.**

As in the zygopterids, the fronds of the coenopterids were often highly branched and spreading, but the symmetry remained complanate as in the frond of a living fern. Genera are again defined by the shape of the leaf trace in section. In *Botryopteris,* for example, which probably formed large sprawling clumps, the trace was ω-shaped, the curvature being abaxial and the protoxylem lying at the tips of the adaxial extensions (Fig. 7.35). In *Tubicaulis,* some of whose species were simple epiphytes, the trace was C-shaped, the open part of the curvature being abaxial. Where known the vascular system of the axis appears to have been a protostele lacking secondary thickening.

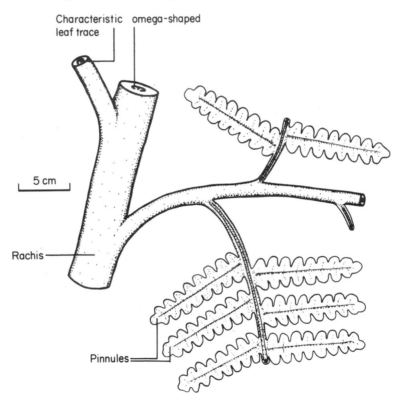

Figure 7.35. *Botryopteris.* **Reconstruction of portion of frond. (After Delevoryas and Morgan, from Delevoryas. 1962.** *Morphology and Evoluton of Fossil Plants.* **Holt, Rinehart and Winston, New York.)**

The sporangia of the coenopterids were less massive than those of the zygopterids and may have been leptosporangiate in origin. The wall was a single cell thick and furnished with an opening mechanism of the *Osmunda* kind. They were often attached to pinnules, either near the margin or superficially. The spores were trilete, often with finely decorated exines, 40–70 μm in diameter. There have been no indications of heterospory.

The zygopterids and coenopterids persisted until the close of the Paleozoic era, but so far as is known are absent from the Mesozoic.

The Origin of the Filicophyta and the Morphological Nature of the Megaphyll

Since the Cladoxylales, Zygopteridales, Coenopteridales, Marattiales, Filicales, and possibly the Ophioglossales, were all in existence together towards the close of the

Paleozoic era, it seems clear that these orders represent parallel lines of evolution within the Filicophyta, only three of which have survived. There are sufficient similarities between the orders, however, to suggest a common ancestor in some earlier period. A significant feature of many Filicophyta is the close resemblance, both in appearance and behavior, between leaves and branch systems. In species of *Botryopteris,* for example, the decumbent fronds gave rise here and there to root and shoot buds. These evidently played a major role in vegetative propagation. Similar behavior is seen in some living ferns. In *Stromatopteris,* a rare fern of New Caledonia with a *Psilotum*-like gametophyte, it is even more difficult to distinguish between frond and axis. Fronds and branches arise from the creeping rhizome in an almost identical manner. Moreover, the frond, which consists of a rachis with a row of pinnules on each side, is morphologically similar to the shoot of *Tmesipteris.*

These stemlike properties of the megaphyll have given rise to the view that it is fundamentally different from the microphyll of the lycophytes. Its origin is probably to be sought in the lateral sprays of branchlets produced by some of the earliest axial forms. It is significant that some trimerophytes from the Lower Devonian of both Norway and North America did bear sprays of lateral branches suggestive of the fronds of early zygopterids. Pinnules presumably arose by the ultimate branchlets becoming flattened and developing a lamina.

The filicophytes distinguished themselves from other plants acquiring megaphylls by remaining largely homosporous and failing to develop extensive secondary tissues. The absence of secondary thickening limited the stature attained by the filicophytes, but they were nevertheless an important component of Paleozoic floras.

___ 8 ___
THE SUBKINGDOM TRACHEOPHYTA
Part 3

PRIMITIVE OVULATE PLANTS AND THEIR PRECURSORS

PROGYMNOSPERMOPHYTA

The division Progymnospermophyta contains only fossil plants. The concept of progymnospermy, a stage at which plants with coniferlike anatomy and morphology were still reproducing by spores, followed the surprising discovery that certain well-preserved trees of the Upper Devonian produced frondlike sprays of branches, some of which bore clusters of sporangia. *Callixylon* (*Archaeopteridales*), the first progymnosperm to be recognized, provides a splendid example of this stage of evolution (Fig. 8.1a). The trunks reached a diameter of 1.5 m (5 ft.) and a length of 8 m (26 ft.) or more. Permineralized remains reveal fine details of the anatomy. A central pith was surrounded by mesarch primary xylem. Outside this lay a considerable thickness of well-developed secondary xylem traversed by narrow rays. The pits in the radial walls of the tracheids were frequently grouped, the groups aligned horizontally and in register with tracheids in the rays. Dense wood of this kind, also characteristic of modern conifers, is termed **pycnoxylic**.

Although *Callixylon* was known for many years as the trunk of a late Devonian tree, only much later were discovered specimens in organic connection with frondlike branches. These were already known a *Archaeopteris* and had been assumed to be the fronds of ferns. *Archaeopteris* is known both sterile and fertile (Fig. 8.1b). The sporangia were spindle-shaped, up to 3.5 mm (0.14 in.) in length, and occasionally with stomata in the epidermis. Heterospory occurred in some species, the microspores being 30–70 μm and the megaspores 150–500 μm in diameter.

Among other plants assigned to the progymnosperms are *Aneurophyton* (**Aneurophytales**), known from the Middle and Upper Devonian of Europe and North America, and *Protopitys* (**Rotopityales**) from the Lower Carboniferous of Scotland. The axes of *Aneurophyton* contained a core of primary xylem, triangular in section, surrounded by secondary wood. The spirally arranged branches dichotomized two or three times, some terminating in tassels of sporangia. *Aneurophyton* appears to have been homosporous. The abundant pycnoxylic wood of *Protopitys* indicates that it is the remains of a tree. The branches were distichously arranged and sporangia were borne in small terminal clusters. The diameter of the spores lay between 80 and 160 μm, but too little is known of the plant to be certain that it was heterosporous.

The progymnosperms, which first become recognizable in the Middle Devonian and extend to the Lower Carboniferous (and possibly later), represent a significant intermediate stage in the evolution of the land flora. The notion that the frond originated from a lateral branch system is clearly supported by *Callixylon* and *Archaeopteris*. Anatomically the pycnoxylic wood of the progymnosperms is strikingly modern in appearance, and some forms show clear examples of a bifacial cambium producing files of secondary phloem at its outer face. Further, although seeds are known almost as old as the progymnosperms, gymnospermous reproduction presumably arose from heterosporous of the kind seen in many progymnosperms.

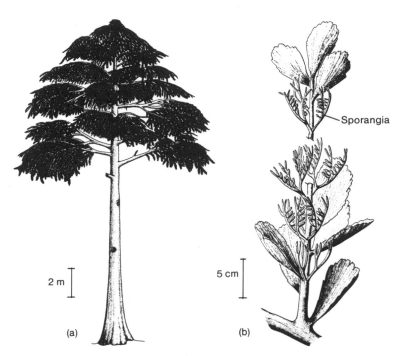

Figure 8.1. (a) *Callixylon (Archaeopteris)*. **Reconstruction. (b)** *A. halliana.* **Reconstruction of branch with sterile leaves and fertile appendages. (Both after Phillips, Andrews and Gensel, from Stewart. 1983.** *Paleobotany and the Evolution of Plants.* **Cambridge University Press.)**

GYMNOSPERMOPHYTA

The seed habit, the formation and retention of the embryo within an integumented megasporangium, is evidently of great antiquity. In paleobotany the term "seed" is also used for fossilized ovules, the later stages of seed formation being inferred. In general, the early seeds of the Paleozoic terminated axes and were either bilaterally symmetrical (*platyspermic*) or radial (*radiospermic*). Examples of each are known from the Late Devonian. Similar seeds are borne by some living plants. Together the living and fossil forms constitute the gymnosperms, a name which implies that the seeds are naked, unenclosed in any carpellary structure. The fossil record indicates clearly that gymnospermy is the most primitive form of the seed habit.

Although a diverse division with possibly more than one origin from the early land plants, the general characteristics of the gymnosperms (not all being represented in the fossil forms) can be summarized as follows:

Sporophyte usually arborescent; branching and leaves various. Secondary vascular tissue always present, consisting of tracheids (in a few forms also of vessels) and sieve cells. Sporangia borne on specialized structures, probably of axial origin. Heterospory general. Megasporogenesis occurring within a specialized tissue (nucellus), this in turn surrounded by a distinctive sheath (integument), the whole termed the ovule. Neither male nor female gametophytes autotrophic. Fertilization by multiflagellate spermatozoids, or by male cells with no specialized means of locomotion, occurring within the ovule, either before or after its being shed. Embryogeny endoscopic, the embryo remaining contained within the seed developed from the ovule.

One of the most fully investigated of the early gymnospermous seeds is *Archaeosperma* (Fig. 8.2) from the Upper Devonian of Pennsylvania. Here the nucellus (megasporangium) was surrounded by an integument. This was deeply lobed above, revealing radial symmetry. The single functional megaspore was surmounted by three abortive spores. Their triradiate markings show that the tetrad had tetrahedral symmetry. Megasporogenesis in *Archaeosperma* thus resembled that in a heterosporous fern such a *Pilularia*.

Little is yet known of the plants which bore the Devonian seeds. Knowledge of the early Carboniferous gymnosperms, although fuller than that of the Late Devonian representatives, also remains limited. It seems likely, however, that some were impressive trees. *Pitus (Pitys)*, for example, almost certainly a gymnosperm, had enormous trunks, reaching in some instances a diameter of 2 m (6.5 ft.). Later in the Carboniferous, remains become sufficiently numerous for the recognition of distinct orders, of which two (Cycadales, Coniferales) have descendants living today.

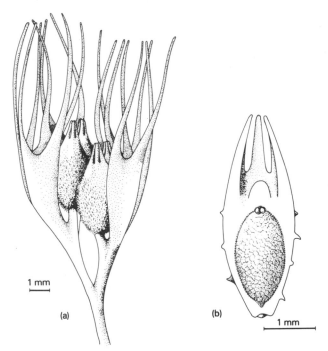

1 mm

(a)

(b)

1 mm

Figure 8.2. *Archeosperma arnoldii.* **(a) The two-seeded cupules were borne in pairs, the two pairs of seeds facing each other. (b) Reconstruction of an individual seed showing the three abortive megaspores at the top of the one which is functional. The integument is deeply lobed above. (After J. Pettitt and C. B. Beck. 1968.** *Contributions from the Museum of Paleontology* **[University of Michigan] 22.)**

Lyginoptendales

Investigators of the fernlike fronds found in Carboniferous rocks soon became aware that not all these were, in fact, referable to ferns. Some were undoubtedly associated with seeds, and others with stems in which there were secondary thickening and other anatomical features rarely found in the Filicophyta. Nevertheless, the habit of these plants was probably something like that of the Marattiales. The leaves were megaphyllous, compound and pinnately branched, and borne on stems of varying height. Together they comprise the Lyginoptendales (Cycadofilicales), frequently referred to as pteridosperms and representative of the earliest Pteridospermopsida.

The features of the pteridosperm stem are well shown by the Carboniferous *Lyginopteris*. These well-preserved axes ranged in diameter from about 0.5 to 4 cm (0.2 to

1.6 in.) and occasionally branched. There was a central pith, which contained nodules of thick-walled cells (similar to the groups of stone cells in the flesh of a pear), surrounded by a ring of primary xylem strands (Fig. 8.3). Exterior to these was a relatively large amount of secondary xylem. The tracheids of the secondary xylem, like those of the metaxylem, were furnished with bordered pits, but in the somewhat smaller tracheids of the secondary xylem they were absent from the tangential walls. A girdle of phloem, rarely well preserved, lay outside the xylem. A characteristic feature of *Lyginopteris* was the anatomy of the outer cortex. This contained radially elongated bands of fibers which anastomosed freely and clearly gave considerable mechanical support to the stem. The outer surface of the stem was furnished with peculiar multicellular glands.

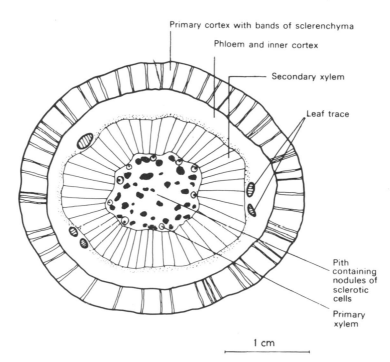

Figure 8.3. *Lyginopteris oldhamia.* **Transverse section of stem. The glandular epidermis has been lost. The leaf trace divides into two strands in the cortex, but these reunite to form a V-shaped trace at the base of the petiole.**

The leaves of *Lyginopteris* (originally described as *Sphenopteris*) were borne in a 2/5 phyllotaxy and when young showed circinate vernation. In mature leaves, which sometimes reached a length of 50 cm (19 in.), the rachis dichotomized at about half its length, but the remainder of the branching was pinnate and the ultimate segments were narrow pinnules. All surfaces of the leaf bore glands similar to those of the stem.

The female reproductive organs of *Lyginopteris* are also known in detail. The ovules (originally described as *Lagenostoma*) terminated axes which were probably branches of otherwise normal leaves. Each ovule (or seed) was partially enclosed in a cup formed by a number of glandular and basally fused bracts (Fig. 8.4a). This structure, called a cupule, in some forms contained more than one ovule. The ovule itself was an upright, radially symmetrical structure, about 0.5 cm (0.2 in.) long and a little less broad. The central part (Fig. 8.4b), the nucellus, possibly a specialized archesporial tissue, was surrounded by an integument of two layers, the outer of which contained a sclerenchymatous sheath. A single vascular bundle entered the base of the ovule and divided symmetrically into nine parts which ascended the inner fleshy part of the integument. The upper part of the integument around the micropyle was shallowly lobed, the lobing corresponding to the intervals between the ascending veins. The nucellus was fused to the integument except at its summit. Here the apical portion ascended as a column, surrounded by a sheath of similar tissue (Fig. 8.4b). Pollen grains have been observed in the space between the sheath and the column (the "pollen chamber"). Germination may have occurred here, but no details are yet known. A female gametophyte, surrounded by distinct membrane and bearing archegonia at its upper surface, lay within the nucellus. This gametophyte probably developed from a megaspore, formed as one of a tetrad within the nucellus, the remaining megaspores degenerating.

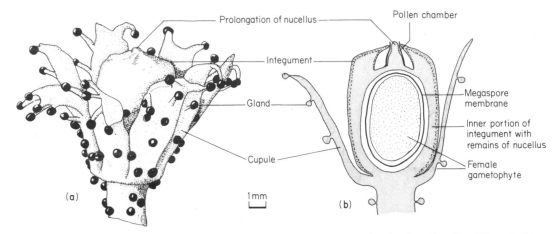

Figure 8.4. *Lagenostoma lomaxi.* **(a) Reconstruction of seed. (After a drawing by Oliver.) (b) Longitudinal section of seed. (After Walton. 1940.** *An Introduction to the Study of Fossil Plants.* **A. and C. Black, London.)**

The male reproductive organ of *Lyginopteris* is not yet known with certainty. It is, however, very probable that it consisted of a small ovate plate, about 2 mm (0.08 in.) in length, terminating a branchlet and bearing about six bilocular sporangia (Fig. 8.5a). Each sporanguim was about 3 mm (0.13 in.) long and 1.5 mm (0.06 in.) wide. The microsporangiophores were borne on a branch system (the whole being known as *Crossotheca*) which may have formed part of a *Lyginopteris* leaf.

The pollen grains (Fig. 8.5b), which bore triradiate scars, were presumably distributed by wind. By analogy with the pollination of living gymnospermous ovules, it is thought the grains were trapped by a drop of sugary fluid which protruded from the micropyle, and that subsequent absorption of this drop drew the grains down into the pollen chamber. Germination of the grains is thought to have been proximal (i.e., at the site of the triradiate scar, as with the spores of ferns and bryophytes). This contrasts with the regular distal germination of conifer and angiosperm pollen. Pteridosperm pollen is accordingly often referred to as "prepollen."

Embryos have only rarely been found in pteridosperm seeds, and their absence is not readily explained. Megagametophyes with archegonia are not uncommon and have been seen in the earliest seeds of the Late Devonian.

The pteridosperms probably had their origin in axial plants resembling the progymnosperms, the fronds again being derived from lateral sprays of branchlets. The radial symmetry of many pteridosperm seeds suggests that they may have evolved from tassels of megasporangia (resembling the male organ *Crossotheca*) in which a central megasporangium was closely surrounded by a ring of similar megasporangia. Sterilization of the outer ring, but not of the center, would then have resulted in an integumented megasporangium. Although no intermediate stages in such a transformation have yet come to light, it is significant that in one of the earliest pteridosperm seeds from the Lower

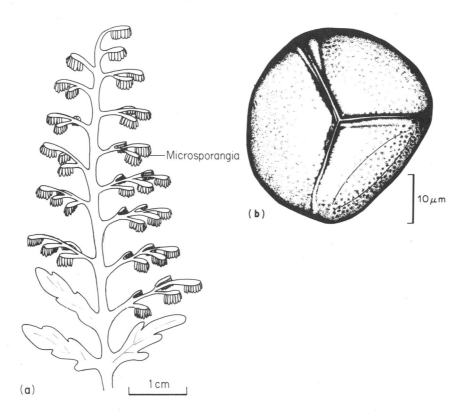

Microsporangia

(a)

1 cm

(b)

10 μm

Figure 8.5. Crossotheca. (a) Reconstruction of fertile shoot. (After Andrews. 1961. Studies in Paleobotany, Wiley, New York.) (b) Microspore. The deep cleavage at the site of the triradiate scar indicates proximal germination. (From a preparation by R. Kidston. photographed by W. G. Chaloner.)

Carboniferous (Salpingostoma) the integument consists of a ring of fingerlike processes, each containing a vein, fused only in the basal region. Progressive fusion of components in this manner might then have led to the entire integuments of the pteridosperm seeds of the Upper Carboniferous, the shallow lobing often seen in the micropylar region of these integuments being the only remaining indication of their compound origin. Experiments with models have shown that the fingerlike extensions of the integument seen in some of the early seeds may, by setting up local turbulence, have assisted in trapping wind-borne pollen above the nucellus. The distinctive pollen chamber of most pteridosperm seeds is found at this site.

The development of ovules containing single megaspores from progymnospermous megasporangia is unlikely to have been a smooth progression. The regular failure of three of the spores in the tetrad to develop (seen also in Lepidocarpon) indicates precisely controlled events within the tetrad. This may have been the consequence of a relatively simple mutation, but clearly one of great significance in plant evolution.

The later pteridosperms. Lyginopteris is representative of a wide range of Carboniferous pteridosperms, but towards the end of the Paleozoic other, more complex forms (**Medullosales,** of which Medullosa is representative) became prominent. They were more like tree-ferns in habit, sometimes possibly scrambling. The leaves had pinnules conspicuously larger than those of the earlier pteridosperms. The seeds, about six times the size of those of Lyginopteris, were more clearly in association with foliar organs, and the microspores, which were monolete, were produced in large cup- or spindle-shaped synangia. The grains of some forms had lateral air bladders and may have germinated distally. Clear indications of nuclear divisions have been seen within some appear to have produced pollen tubes. The stems often had complex vascular systems with highly dissected steles and leaf traces.

The descendants of the pteridosperms. It seems beyond doubt that some pteridosperms persisted into the early Mesozoic, although the remains become much less frequent after the close of the Paleozoic era and are of an unfamiliar form. There is also evidence that the evolution of the pteridosperms in the Northern and Southern hemispheres diverged at this time, but the two floras remained in contact in certain regions of Africa. The early Mesozoic rocks of these regions have yielded a number of puzzling plants, almost certainly derived from the pteridosperms and probably of very great importance in the evolution of the later seed plants. Some of them had leaves with reticulate venation and an appearance strikingly like that of the leaves of some modern flowering plants. Others had seeds in partially closed cupules, a development possibly leading to the remarkable fruiting body of the Caytoniales and the angiospermous carpel (Chapter 9).

The Cycadales

This order contains both living and fossil representatives. The upright, naked ovules resemble in general features and size those of the later pteridosperms. The fossil evidence indicates that the cycads came into prominence at the beginning of the Mesozoic era.

Although in Mesozoic times the cycads were distributed as far north as Siberia and Greenland, they are today confined to tropical and sub-tropical regions in both the Old and New worlds. There are nine genera in all, the most common being *Zamia, Macrozamia, Cycas, Encephalartos, Dioon,* and *Ceratozomia,* but only *Cycas,* extending eastwards from Madagascar into Polynesia, has anything approaching a wide distribution. *Encephalartos woodii* of Africa is represented only by male plants. The female is presumable extinct.

The stems of the cycads are either short, stocky trunks, often with a large portion below ground (Fig. 8.6) or are much taller, reaching heights of up to 15 m (50 ft.). Below ground is a massive taproot which bears, together with normal roots, others which are

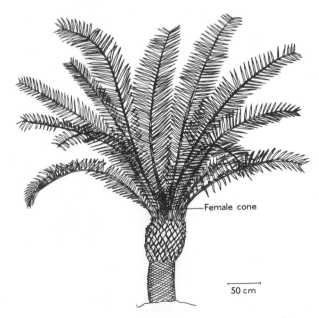

Female cone

50 cm

Figure 8.6. *Encephalartos hildebrandtii.* **Habit of plant bearing female cone. (After Eichler, from Eichler, in Engler and Prantl. 1889.** *Die natürlichen Pflanzenfamilien, 2:1.* **Engelmann, Leipzig.)**

negatively gravitropic and which break up at the soil surface into coralloid masses. These contain endophytic fungi and blue-green algae.

Apart from stature, all cycads have a similar growth form. The thick stem, usually unbranched, bears and apical rosette of large pinnate leaves. The rate of growth is very slow and, although there are no recognizable annual rings in the stem, the age of any specimen can be calculated approximately from the rate of leaf production and the number of leaf bases. A specimen of *Dioon* only 2 m (6.5 ft.) high was estimated in this way to be about 1000 years old. *Bowenia,* confined to North Queensland, is anomalous in possessing a tuberous rootstock bearing only one or two leaves at a time. The leaves of cycads contain a variety of metabolic products, some poisonous and others of possible therapeutic value.

The stem grows from a massive apex in which there is generalized meristematic activity, and considerable centrifugal expansion, as well as growth in length. Behind the apical initials a core of central tissue, which soon becomes distinguishable from the peripheral, differentiates into the vascular tissue and pith. The peripheral zone becomes cortex. The primary vascular tissue consists of a ring of bundles with endarch xylem, and these surround an extensive pith. The secondary xylem is traversed by wide parenchymatous rays and the radial walls of its tracheids are furnished with several series of circular bordered pits (except in *Zamia* and *Stangeria* where the pits are of the narrower kind characteristic in ferns). The first cambium is of limited activity and is followed by others which arise successively outside the vascular cylinder. These cambia are of diminishing activity, the last producing merely a few concentric bundles lying out in the cortex. The stele is thus highly parenchymatous, and the main mechanical support of the stem comes from its armour of sclerenchymatous leaf bases. A stem with soft spongy xylem of this kind is termed **manoxylic**. Mucilage canals, tannin cells, and cells containing crystals of calcium oxalate occur in the pith and cortex of mature stems.

The leaves are of two kinds, foliage leaves (or fronds) and scale leaves, sequences of each other in a definite phyllotactic arrangement. The scale leaves cover the apex and the upper part of the stem (Fig. 8.7), often disintegrating to form a fibrous sheath below. The

Figure 8.7. *Zamia* **sp. Young plant with female cone. Although produced terminally, the female cone becomes pushed to one side by the sympodial growth and appears to be lateral. (Photograph by F. White.) Scale 5 cm.**

foliage leaves in some genera show circinate vernation (as in *Cycas*), but in others a vernation similar to that seen in the fern *Botrychium*. In all cycads the leaves are pinnately
branched, but are twice pinnate only in *Bowenia*. The leaf trace, which is horseshoe-
shaped in section in the petiole, has a complex origin in the stem. Some strands arise
opposite the insertion of the leaf and girdle the stem obliquely upwards into the leaf base.
In an individual bundle of the trace much of the metaxylem is adaxial to the protoxylem,
but characteristically a few tracheids, often separated by parenchyma, lie on the abaxial
side adjacent to the phloem ("centrifugal xylem").

The venation of the pinnae is various, but any branching is dichotomous and open.
In sections of the veins small patches of transfusion tissue (anatomically intermediate
between parchenyma and tracheids) are often seen on each side of the xylem. In *Cycas*,
where each pinna has only a midrib, a sheet of similar cells extends from the midrib to the
margin ("accessory transfusion tissue").

The pinnae have a leathery texture. A conspicuous cuticle is usually present, and an
epidermis (often accompanied by a hypodermis), palisade, and mesophyll are well differentiated. The cell walls of the lower epidermis are straight or slightly sinuose, and the
stomata, although usually sunken, are surrounded by a simple ring of subsidiary cells
(*haplocheilic* stomata). These epidermal features, which remain clearly evident in fossil
material, are of great value in distinguishing extinct Mesozoic forms from superficially
similar contemporary plants (see Fig. 8.14).

Reproductive structures of the cycads. The mega- and microsporangiophores of
the cycads are aggregated into separate strobili borne on different plants. The female cone
either terminates the main axis (in which case subsequent growth is sympodial (Fig. 8.7) or
it is lateral, according to the genus. The situation in *Cycas* is exceptional for here the main
axis, having given rise to a sequence of megasporangiophores, continues to be active and
reverts to the production of normal vegetative leaves.

The female cones vary in compactness and in the number of ovules borne on each
megasporangiophore. At one extreme stands *Cycas*, in which the female cone consists of
a loose aggregate of megasporangiophores, the distal, sterile portions of which are more
extensive than in any other cycad. Several pairs of ovules are attached in the proximal
region (Fig. 8.8a), the micropyles of the ovules being directed obliquely outwards. At the
other extreme are *Zamia* and *Encephalartos*, in which small, peltate sporangiophores
(Fig. 8.8b) are tightly packed in a distinct ovoid cone. Each sporangiophore bears two

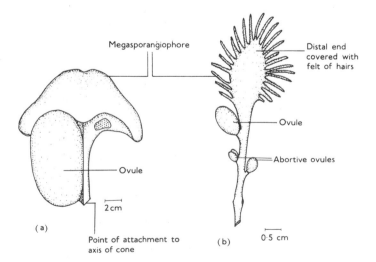

Figure 8.8. (a) *Encephalartos hildebrandtii*, **megasporangiophore. (b)** *Cycas revoluta*, **megasporangiophore.**

ovules, the micropyles of which are directed towards the center of the cone. Both the cones and ovules in the cycads generally are of extraordinary size. In *Encephalartos* female cones have been recorded weighing as much as 45 kg (90 lb.) and in *Macrozamia* the ovules reach a length of 6 cm (2.4 in.). The whole of the female reproductive system is thus on a much larger scale then in any other living plants.

The male cones of the cycads (Fig. 8.9) are also either terminal or lateral. Where terminal, subsequent growth is always (including that of *Cycas*) sympodial. Where lateral, growth is monopodial and the cones may be present in considerable numbers. In *Macrozamia,* for example, 20–40 cones may be produced in rapid sequence around the lower part of the apex. There is much more uniformity in the structure of the male cones than in that of the female, although again considerable variation in size. In some species of *Encephalartos* the male cones reach a length of 50 cm (19 in.), but in *Zamia* only 5 cm (2 in.). The microsporangiophores are in the form of scales, closely appressed during growth and, when mature, covered on their lower surfaces with several hundred microsporangia (Fig. 8.9b). The sporangia, about 1 mm (0.04 in.) in length and structurally resembling those of the fern *Angiopteris,* are grouped in sori, each consisting of three to four sporangia. Their origin is eusporangiate and their formation almost simultaneous. Each sporangium produces hundreds of spores. In *Encephalartos* temperatures some 15°C above the ambient have been recorded in male cones at the time of meiosis, a consequence of the intense respiratory activity throughout the cone at this stage of development. The microspores of the cycads as shed (and partially dehydrated) are characteristically boat-shaped as a consequence of a broad germination furrow (*colpus*) in the distal face. The exine on the proximal face shows only an indistinct tetrad scar. Germination of the microspores begins in the sporangium, and each spore when shed already contains three cells, namely, a single prothallial cell, a generative cell, and tube nucleus (Fig. 8.11).

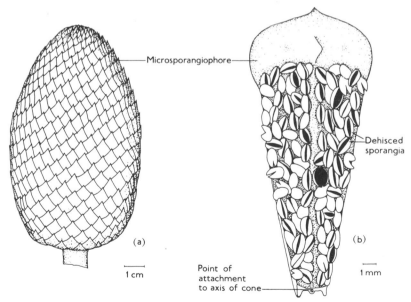

Figure 8.9. *Cycas* sp. (a) Male cone. (b) Microsporangia, viewed from below.

Development of the gametophytes and fertilization in the cycads. The ovules of the cycads are distinctly radiospermic (although platyspermic forms may have occurred in the Paleozoic), and the integument is differentiated into sclerenchymatous and fleshy layers. In the immature ovule the megaspore mother cell, which is deeply embedded in the nucellus, undergoes meiosis and gives rise to four megaspores in a linear tetrad. The three outer megaspores degenerate, but the inner (chalazal) megaspore, which inherits most of the cytoplasm of the mother cell, remains viable and develops a layered wall of surprising complexity. It germinates in situ and begins to form the female gametophyte, the initial development of which consists of a sequence of free nuclear divisions (Fig. 8.10a). The gametophyte enlarges with the expanding ovule and eventually a vacuole forms at its center. Wall formation then begins at the periphery of the gametophyte and continues towards the center. The archegonia (Fig. 8.10b) appear at the micropylar end of the gametophyte where the cells are comparatively small. A distinct and thickened boundary, known as a "megaspore membrane," persists between the haploid gametophyte and the diploid nucellus.

Although at first sight unfamiliar, the archegonia of the cycads can be seen from their development to be quite similar to those of the lower archegoniate plants. A single initial cell divides into an outer primary neck cell (which subsequently gives rise to one tier of neck cells) and an inner central cell. The latter rapidly expands, and then divides to form the egg and a small superficial ventral canal cell (Fig. 8.10c), which degenerates as the egg becomes ready for fertilization. Maturation of the egg involves considerable cytological activity. The nucleus enlarges and becomes weakly staining, and in Zamia, small bodies, which appear in the light microscope as refractive droplets, are seen to stream away from its surface into the cytoplasm. The cycads possess the largest egg cells known among land plants. The diameter may reach or exceed 3 mm (0.13 in.) or more, and even that of the nucleus may be as much as 0.5 mm (0.02 in.).

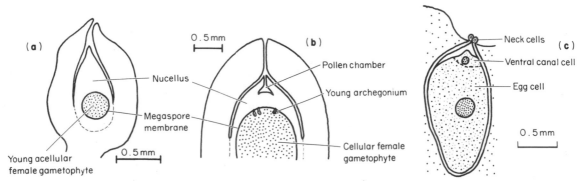

Figure 8.10. Female reproductive system of cycads. (a) Longitudinal section of young ovule. (b) Completion of development of female gametophyte. (c) Archegonium containing newly formed egg. (All after McLuckie and McKee. 1958. *Australian and New Zealand Botany.* **Horwitz, Sydney.)**

Pollination, which occurs during the closing stages of the growth of the female gametophyte, is brought about by the microspores, which are distributed by wind (and in some species possibly by insects), being caught in a sugary "pollination drop" at the orifice of the micropyle. This fluid, probably secreted by the cells at the tip of the nucellus, is now withdrawn, carrying the microspores with it. These now become lodged in a shallow pollen chamber formed by autolysis at the tip of the nucellus (Fig. 8.10b). Here each grain puts out a tube from the distal part of the grain (i.e., on the side away from the center of the original tetrad) laterally into the nucellus. This tube, which is of limited growth, has a purely haustorial function, and only the tube nucleus enters it. In the presence of the pollen, and possibly accelerated by enzymes secreted by it, the upper part of the nucellus continues to break down until all that remains above the mature archegonia is a small pool of fluid.

The development of the male gametophyte meanwhile continues. The generative cell, the only cell to show further activity, divides, giving rise to a spermatogenous (body) cell and a sterile (stalk) cell which surrounds the prothallial cell (Fig. 8.11b). The proximal part of the male gametophyte now bends over the archegonia, and the nucleus of the body cell ultimately divides to form two coiled, multiflagellate spermatozoids (Fig. 8.11c). In *Microcycas* more than one division occurs and up to 16 spermatozoids may be formed. The spermatozoids are constructed on the same principles as those of the lower archegoniates, but are much larger and in some species may reach 300 μm in diameter. They are finally released from the proximal part of the tube close to the ruptured exine directly into the fluid above the archegonia. One or more eggs become fertilized. The penetrating spermatozoid sheds its cytoplasm (including the flagella) in the cytoplasm of the egg, and its nucleus enter and disperses in the large female nucleus. Because of the nature of the male gamete, the cycads are termed *zooidogamous*.

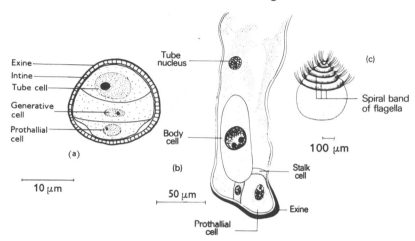

Figure 8.11. *Cycas* sp. **Development of the male gametophyte. (a) Vertical section of microspore at liberation. (b) Longitudinal section of pollen tube. ([a], [b] After B. G. L. Swamy. 1948.** *American Journal of Botany* **35.) (c) Spermatozoid. Modified after B. G. L. Swamy. 1948.** *American Journal of Botany* **35.)**

This account of the development of the male gametophyte and fertilization is based on the events in *Macrozamia, Dioon, Zamia* and *Cycas*. The distal germination of the microspores in these genera coupled with the proximal release of the gametes suggests a position intermediate between the prepollen of most of the pteridosperms and the pollen of the conifers and angiosperms. Many features of sexual reproduction in the cycads however need confirmation and reinvestigation by modern techniques.

Embryogenesis in the cycads. Although the development of the male and female gametophytes and the interval between pollination and fertilization is prolonged and may extend over months, the formation of the proembryo follows immediately after fertilization. After a period of free nuclear division, in which as many as 256 (2^8) or even more nuclei may be formed, the proembryo becomes partly or wholly cellular. Further growth takes place at the chalazal end, and embryogeny is evidently endoscopic. At the extreme base of the proembryo is a group of small meristematic cells which develop into the embryo proper, in some species protected on the outside by a layer of cap cells which later degenerate. Above the embryonic cells are a number of elongating cells which form a conspicuous suspensor. The mature suspensor may reach several centimeters in length, but the resistance it meets in driving the young embryo into the nutritive tissue of the female gametophyte causes it to be highly twisted and coiled (Fig. 8.12).

The embryo grows and differentiates at the expense of the food reserves, including starch and fats, prominent in the cells of the female gametophyte ("pseudoendosperm"). The mature embryo has two or more cotyledons (depending upon the species) directed away from the micropyle and enclosing the stem apex (plumule). Although a short axis is

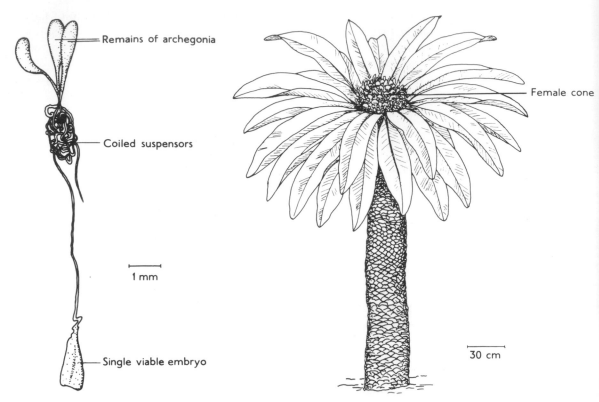

Figure 8.12. *Cycas* **sp. Embryo and suspensor. (After B. G. L. Swamy. 1948.** *American Journal of Botany* **35.)**

Figure 8.13. *Bjuvia simplex.* **Reconstruction of plant with female cone. (After Florin, from Arnold. 1947.** *An Introduction of Paleobotany.* **McGraw-Hill, New York.)**

present below the cotyledons (hypocotyl), a root is still lacking at this stage. The whole is surrounded by the exhausted remains of the female gametophyte and nucellus, and externally by the integument. Germination occurs as soon as conditions are favorable and the seed has imbibed sufficient water. The hypocotyl pushes its way through the micropylar end of the seed and then begins to develop a strong taproot which persists throughout the life of the plant. The cotyledons remain partially enclosed in the seed, but the plumule emerges and gives rise to scale and mature leaves, the first of which have only a few pinnae.

    The fossil history and possible origin of the cycads. The first cycads are found in the late Paleozoic. Both petrified male cones (resembling those of *Stangeria*) and megasporangiophores of the *Cycas* kind have been discovered in North American rocks. A cycad of the Mesozoic era is *Bjuvia simplex* from the Rhaetic of Sweden, the remains being sufficient to attempt a reconstruction of the plant (Fig. 8.13). There was a general resemblance to *Cycas,* but the leaves were entire instead of pinnate. The megasporangiophores, which showed little development of the distal sterile region, formed a loose terminal cone. *Beania* is a female cycad cone from the Jurassic, resembling that of *Zamia,* but less compact. The same beds yield what are almost certainly male cycad cones and also compressions of leaves, the epidermal features of which are quite similar to those of living cycads.

    In their radiospermic seeds and in the complexity of their stelar structure the cycads so closely resemble the later pteridosperms that it seems beyond doubt that they had their origin in some common stock. If so, both leaves and sporangiophores would have been derived from lateral branch systems. If we compare the megasporangiophore of *Cycas revoluta* and its pinnate distal portion with that of *Zamia,* where the sterile distal portion is lacking, we may see the process by which an ovuliferous megaphyll lost its sterile region and became wholly reproductive. This specialization seems to have been accomplished earlier in the male inflorescences, since in all cycads, even those of Paleozoic age, the microsporangiophores have little sterile tissue.

The Bennettitales

This order, represented solely by fossils, extends from the Triassic to the Cretaceous periods. The frequency of their remains is such that they were probably a more conspicuous element of the Mesozoic floras than the Cycadales. In habit, there was a general resemblance to the cycads. Some Bennettitales were upright, sparingly branched plants, while others were squat, bearing a crown of leaves near the soil surface.

The leaves of the Bennettitales were entire or pinnate and very similar to those of the cycads. They were not, in fact, easily distinguishable from these until it was discovered that the epidermal features were quite different (Fig. 8.14). In the Bennettitales the walls of the epidermal cells were highly sinuose, and the guard cells and subsidiary cells appear to have had their origin from the same mother cell, giving rise to so-called **syndetocheilic** stomata of characteristic form.

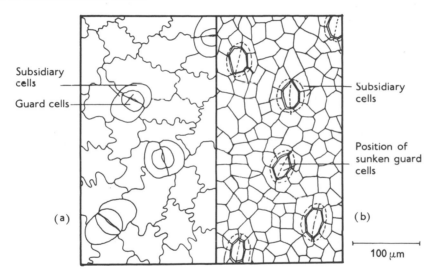

Figure 8.14. Fossil cuticles of (a) the Bennettitalean type and (b) the Cycadalean type. (From a preparation by W. G. Chaloner.)

The stem structure, so far as it is known, was similar to that of the cycads, except that girdling leaf traces appear to have been absent.

The female cone of the Bennettitales consisted of an axis bearing upright ovules interspersed with sterile scales. The male inflorescence was a whorl of microsporangiophores which produced either microsporangia or large complex synangia. The pollen grains were monocolpate, similar to those of the cycads and later pteridosperms. In most forms the inflorescence was bisexual, the male portion being below the female (Fig. 8.15a). In *Cycadeoidea* the ovules are often well preserved, and this genus provides the oldest known example of megaspores produced in a linear tetrad. The seeds were similar in structure and symmetry to those of cycads and pteridosperms, although the existence of a second integument has been suspected in some forms. Unlike the pteridosperms, embryos are sometimes well preserved. A massive hypocotyl was directed towards the micropyle, resembling the situation in the cycads.

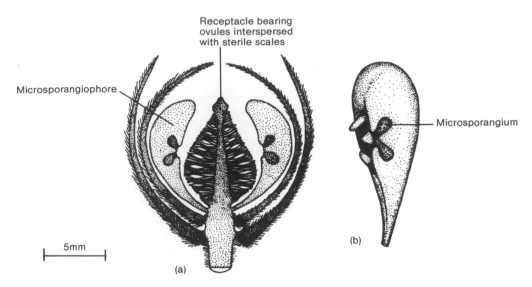

Figure 8.15. *Williamsoniella coronata.* **(a) Longitudinal section of fertile shoot. (b) Micro-sporangiophore showing partially enclosed microsporangia. (After Harris, from Stewart. 1983.** *Paleobotany and the Evolution of Plants.* **Cambridge University Press.)**

The relationships and origins of the Bennettitales. The general anatomical and reproductive features of the Bennettitales indicate that they also had their origin in some pteridospermous stock, and in some respects the resemblance is closer than with the cycads. The microsporangiophores (Fig. 8.15b), for example, can be readily envisaged as derived from condensed microsporangiate pteridosperm fronds. The fossil record indicates that the Bennettitales evolved parallel to the cycads. There is no obvious explanation of why they should have become extinct whereas the Cycadales survived. It is noteworthy, however, that the microsporangiophores of the Bennettitalean cone commonly ascended close to the ovules, and self-pollination may have been the rule. This would have limited genetic recombination and consequent adaptability.

The Caytoniales

This order, Jurassic in age (but possibly extending into the Lower Cretaceous), also had radially symmetrical seeds. They lay within a spherical cupule, almost totally closed except for a small opening adjacent to the stalk. Since the Caytoniales are particularly important in relation to the problem of the origin of the angiospermous carpel they are considered further in this context (Chapter 9).

The Cordaitales

Although vegetatively resembling some modern conifers, the Cordaitales were in other ways distinct and must have been among the most impressive of the seed-bearing plants of the later Paleozoic. So far as is known, they were all arborescent with columnar trunks, many probably reaching heights of 30 m (98 ft.) and diameters of 1 m (39 in.). The leaves, confined to the upper branches, were spirally arranged and strap-shaped (Fig. 8.16a). In some forms they were as much as 1 m (39 in.) in length and 15 cm in width. There was regular parallel venation interspersed with longitudinal bands of hypodermal fibers, a structure not dissimilar to that of the leaves of the modern conifer *Araucaria araucana*.

In general, the vascular tissue of the cordaitalean trunks consisted of a large amount of secondary xylem, typically pycnoxylic, traversed by narrow parenchymatous rays.

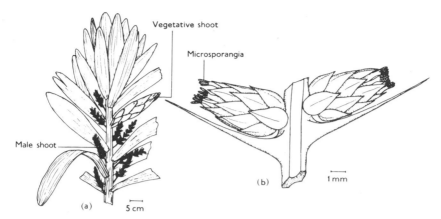

Figure 8.16. (a) Reconstruction of a Cordaitalean shoot. (After Grand'Eury, from Andrews, Jr. 1961. *Studies in Paleobotany,* **Wiley, New York.) (b)** *Cordaianthus concinnus.* **Reconstruction of two male shoots. (After T. Delevoryas. 1953.** *American Journal of Botany 40.***)**

The secondary tissues surrounded a medullated primary stele. The primary xylem tended to diminish in later forms, leaving a ring of mesarch bundles bordering an extensive pith, often broken up into lenticular diaphragms. The secondary tracheids showed several series of circular bordered pits on their radial wall and were closely similar to those of living *Araucaria.* The leaf traces, which passed outwards from the primary xylem, were simple in origin and commonly consisted of two parallel strands. The roots of the Cordaitales are quite well known, since they often became petrified while they were penetrating decaying remains of other plants. They show a triarch stelar structure and a distinct root cap at the growing tip.

Reproductive structures of the Cordaitales. The reproductive organs of the Cordaitales (known as *Cordaianthus*) were borne on slender branches. Although male and female were separate, they possibly occurred on the same tree. Each reproductive region was basically an axis, from 10 to 30 cm (4 in. to 12 in.) in length (Fig. 8.16b), bearing two rows of bracts in a complanate distichous arrangement. The male and female shoots occurred singly in the axils of the bracts.

The individual male shoots (Fig 8.16b) were about 1 cm (0.4 in.) long. Each consisted of a short, stout axis bearing a large number of linear-lanceolate scales, each with a single vein, in a close spiral. The lower scales were sterile and acute or obtuse at their apices, but the upper scales were emarginate and terminated in several (usually six) cylindrical microsporangia. Since both sterile and fertile scales lay in one spiral, they appear to have been of similar morphological nature. The pollen grains were surrounded by air bladders formed by the separation of the layers of the wall. The two layers remained in contact, however, in one region, possibly the site of liberation of the gametes. This thin area was opposite the small triradiate scar at the proximal pole of the grain. It thus appears that the Cordaitales had true pollen with distal germination.

The female shoots were of similar organization, but the fertile scales terminated in ovules instead of microsporangia. In earlier forms (e.g., *Cordaianthjus pseudofluitans,* Fig. 8.17a) the fertile scales (megasporangiophores) projected conspicuously from the shoot, branched, and carried more than one seed. In the later forms, however, the megasporangiophores were shorter and unbranched, terminating in only one seed concealed among the sterile scales.

The seeds of the Cordaitales were not radially symmetrical, but bilateral, the margin of the seed often being extended as a wing. Because of their characteristic flattened appearance, these seeds are termed platyspermic, and they are readily distinguishable from the predominantly radiospermic seeds of the pteridosperms. The integument of a cordaitalean seed (Fig. 8.17b) has the appearance of having been formed by two valves, each containing a vein, coming together and enclosing a megasporangium. This view is supported by the occasional occurrence in female shoots of what are interpreted as abortive megasporangia subtended by two unfused lobes. Each seed would thus have three components, the two outer which form the integument possibly having been derived from sterilized sporangia. It is noteworthy in this connection that the microsproangia were commonly produced in multiples of three.

Platyspermic seeds, similar to those seen in cordaitalean inflorescences, are frequently found detached in Carboniferous deposits and their structure is now well known (Fig 8.17b). They are about 1 cm (0.4 in.) in height and only a little less in their major transverse diameter; the minor is of the order of 0.5 cm (0.2 in.). The integument was differentiated into one or more layers, at least one of which was sclerenchymatous. The nucellus, except for the basal region, appears to have been separate from the integument. A female gametophyte, surrounded by a distinct "megaspore membrane," developed within the nucellus, and archegonia were produced on its upper surface, the nucellus above this region becoming differentiated as a pollen chamber. Reticulate markings on the pollen grains lying in this chamber were once regarded as indicating endosporic germination of the grain, but this interpretation is now questioned. The pattern is more probably a relic of the sculpturing of the wall than of an internal structure. Gametes were possibly liberated into fluid above the archegonia. As with the pteridosperms, embryos are conspicuously absent from cordaitalean seeds.

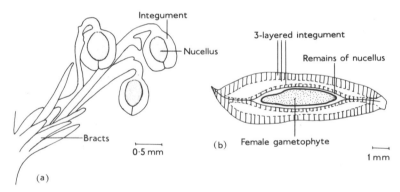

Figure 8.17. (a) *Cordaianthus pseudofluitans*. **Reconstruction of female shoot. (After Florin, from Andrews, Jr. 1961.** *Studies in Paleobotany.* **Wiley, New York.) (b)** *Kamaraspermum leeanum.* **Transverse section of seed. (After Kern and Andrews, Jr. from Andrews, Jr. 1961.** *Studies in Paleobotany.* **Wiley, New York.)**

Origin and fossil history of the Cordaitales. Little is known of the origin of the Cordaitales, but it possibly lay well before the Carboniferous period. Remains of substantial woody plants, the xylem of which showed araucarian pitting, have been found as early as the Middle Devonian. In the leaves of some of the Cordaitales of the Lower Carboniferous the nerves branched as they approached the tip, possibly indicating an origin in a fan-shaped structure. Axes bearing leaves of the kind envisaged are in fact known from the Middle Devonian of Bohemia, but the relationship of these fossils (placed in the genus *Barrandeina*) to the Cordaitales is quite unproven. Nevertheless, the impression is that the Cordaitales were derived from axial heterosporous forms in much the same way as the pteridosperms, but that in the Cordaitales the megaphyll condensed into the characteristic strap-shaped leaf and the integument of the seed evolved in a slightly different way. The earliest platyspermic seeds come from the later Devonian.

The Cordaitales probably persisted into the beginning of the Mesozoic, but then became extinct.

The Coniferales and Taxales

The conifers are the most widespread of all the groups of gymnosperms, and they form the climax vegetation at high altitudes and in the colder regions of the temperate zones, particularly the north. They are much less common in the tropics, where they are usually confined to mountains and are often mixed with angiospermous trees. Of all the vascular plants discussed so far, the conifers are the first of significant economic importance. They are almost all arborescent and the wood is used extensively as timber and as a source of pulp for paper-making and related industries.

Growth form. The growth form of a conifer is frequently pyramidal, the conspicuous main axis being the principal source of the valuable timber. A few conifers of this form attain remarkable sizes and ages. Specimens of *Sequoia,* for example, in California frequently exceed 100 m (300 ft.) in height, their trunks reaching diameters of several meters and showing over 2000 growth rings. *Pseudotsuga* in the forest of the Olympic Peninsula of the Pacific Northwest may attain even greater heights (but not girth). The oldest living conifers are probably specimens of *Pinus aristata* at high altitudes on the arid White Mountains of the California–Nevada border. Modern techniques of dating show that some of these are almost 5000 years old.

Some conifers (such as the junipers) are bushy, and a few (confined to Australasia) are dwarf, heatherlike shrubs of boggy alpine situations. Occasionally the growth form is markedly influenced by the habitat. *Pinus montana*, for example, is a pyramidal tree when growing in acid situations on lower hills, but a straggling shrub with no evident main axis (*Krummholz*) when on limestone at higher altitudes. Most conifers tend to be surface rooted, and many species produce stubby rootlets in the humus layer which are associated with mycorrhizal fungi. *Taxodium distichum* (swamp cypress) which grows in swamps in the warmer parts of eastern North America, is outstanding among conifers in producing negatively gravitropic aerophores which rise above the surface of the water. These specialized roots are, however, rarely produced by specimens planted outside the native habitat.

Leaf form. The leaves of the conifers take a variety of forms (Fig. 8.18), but they are nearly always small and simple in shape ranging from needlelike structures several centimeters in length (*Pinus*, Fig. 8.18a) to closely appressed scales reaching only a few

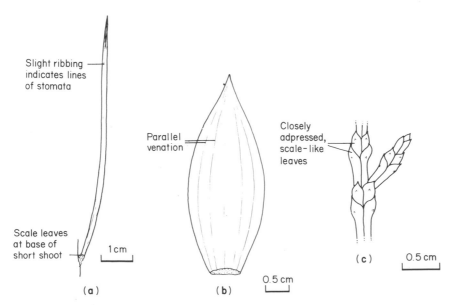

Slight ribbing indicates lines of stomata

Parallel venation

Closely adpressed, scale-like leaves

Scale leaves at base of short shoot

1 cm

0.5 cm

0.5 cm

(a)   (b)   (c)

Figure 8.18. Forms of conifer leaf. (a) *Pinus monophylla.* **(b)** *Araucaria araucana.* **(c)** *Chamaecyparis obtusa.*

millimeters (as in Cupressaceae, Fig. 8.18c). *Araucaria araucana* is unusual in having broadly lanceolate leaves 5 cm (1.4 in.) or more in length (Fig. 8.18b). In the Cupressaceae the plant frequently passes through a juvenile phase in which it produces needlelike leaves. Cutting or grafts of the juvenile phase sometimes go on producing needlelike leaves indefinitely, and these so-called *Retinospora* forms are common in gardens. The venation of conifer leaves is never reticulate. There are either a number of parallel veins (as in some species of the Araucariaceae, Fig. 8.18b) or a single median vein, often showing a double structure (as in *Pinus*). In some conifers (e.g., *Pinus*) the leaves are borne wholly or principally on short shoots. The leaves of most conifers persist for several seasons; in only a few genera (e.g., *Larix*) are the leaves truly deciduous. In *Taxodium* and *Metasequoia* the leaves are confined to the ultimate branchlets, and branchlets (phyllomorphs) are shed as a whole at the end of the growing season.

Stems. the stems of conifers grow from a group of meristematic cells. In some genera, notably *Araucaria* and *Juniperus,* the apex is organized into a distinct tunica, in which divisions are principally anticlinal, and a central corpus, where divisions are in several planes. The latter gives rise to the pith and primary vascular tissue. The mature stems of conifers are mostly secondary wood, the pith and primary xylem being relatively inconspicuous (Fig. 8.19). At the outside, the phloem, cortex, and periderm form a comparatively narrow band. The dense secondary xylem consists of radial files of tracheids, traversed by narrow parenchymatous rays. Wood parenchyma is not conspicuous. In *Pinus* it is confined to the epithelium of the resin canals, and it is entirely lacking in *Taxus*. The tracheids are usually differentiated in distinct annual rings, those formed towards the end of a season's growth being narrower than those at its beginning (Fig. 8.20). In *Pinus* the tracheids, which rarely exceed 4 mm (0.17 in.) in length, bear bordered pits, usually in a single row, on the radial walls. The central part of the pit membrane is thickened and forms the torus (Fig. 8.20). Thickenings often present along the margins of the pits are termed the "Rims of Sanio." In *Araucaria* the pits are similar, but in three to four rows, the pits of adjacent rows alternating. The tracheids of *Araucaria,* but not of *Pinus,* occasionally have small trabeculae (initially of cellulose, but subsequently lignified) extending across the lumen. These so-called "Bars of Sanio" also occur in a number of other genera.

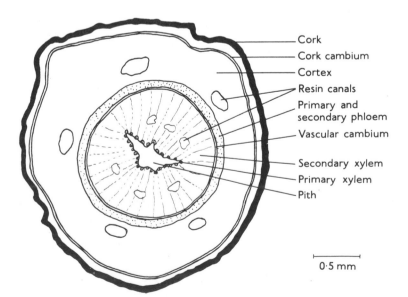

Cork
Cork cambium
Cortex
Resin canals
Primary and secondary phloem
Vascular cambium
Secondary xylem
Primary xylem
Pith

0·5 mm

Figure 8.19. *Pinus sylvestris.* **Transverse section of young stem with only one season's secondary vascular tissue.**

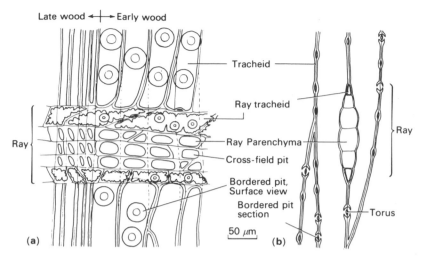

Figure 8.20. *Pinus sylvestris.* **(a) Radial longitudinal section of secondary xylem at junction of early and late wood. (b) Tangential longitudinal section of a ray similar to that shown in (a).**

Considerable differentiation is sometimes present in the parenchymatous rays of conifer woods. In *Pinus,* for example, the cells of the upper and lower margins in the xylem portion of the ray may form radially oriented tracheids (Fig. 8.20), and cells in a similar position in the phloem closely apply themselves to the sieve cells and become conspicuously rich in cytoplasm. The rays provide a means of transporting materials laterally in the growing stem.

The resin canals of the conifers, which are schizogenous in origin and interconnected, run longitudinally in the leaves (Fig. 8.21) and xylem, and also transversely in some of the larger rays. The resin itself (a complex acidic substance containing oxidized phenols and terpenes) is synthesized in the epithelium of the canals, probably most in the younger tissues, but the actual site of synthesis in the cells is not yet exactly known. The

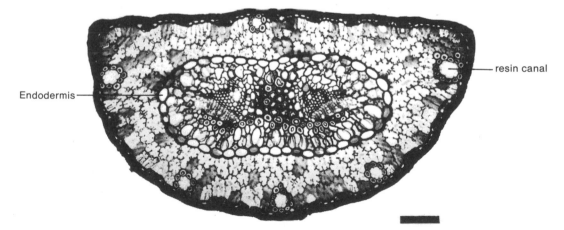

Figure 8.21. *Pinus sylvestris.* **Photomicrograph of a transverse section of a leaf, the flat adaxial side uppermost, midway along its length. The epidermal cells are lignified and the stomata obscured, but the cells of the hypodermis are visible and adjacent to them the resin canals. Two vascular bundles, separated from the mesophyll by a common endodermis, lie at the center of the leaf. Thickened and lignified cells are found between these vascular bundles and below the phloem, but the cells elsewhere within the endodermis are largely transfusion tissue. Scale bar 0.1 mm.**

resin system can be tapped by driving a gutter-shaped steel wedge into the xylem near the base of the tree (Fig. 8.22), and from some species considerable quantities of commercially valuable resin can be collected. Pine resin, for example, is the source of turpentine and colophony, both widely used in the paint and varnish industry (although less so than formerly). The male bark beetle of *Pinus ponderosa* transforms a component of the resin into the sex pheromone of the species.

Figure 8.22. *Pinus pinaster* **(maritime pine). The trunk of this tree (photographed in Portugal) has been tapped for resin. This species of pine is the principal source of natural resin in Europe. Scale bar 5 cm.**

Leaves. Many features of anatomical and physiological interest are presented by conifer leaves (Fig. 8.21). The cuticles, for example, are often furnished, especially in the region of the stomata, with distinctive patterns of tubercles and ridges. Palisade and spongy mesophyll are commonly present, and in *Pinus* the walls of the mesophyll cells have ridges projecting into the cell (Fig. 8.21). A well-defined hypodermis, the cells of which may be lignified, is present in many leaves. The vascular bundles are often surrounded by transfusion tissue. Resin canals are frequent, and in some leaves (as of *Thuja*) a prominent gland on the back of the leaf contains fragrant oil. The leaves of conifers at high altitudes and in artic regions are remarkably resistant to frost damage. Tissue water in twigs of species of *Pinus* and *Pseudotsuga*, for example, can withstand cooling to −40°C (−40°F) without the formation of ice crystals. These resistant conifers can also carry out photosynthesis at unusually low temperatures. Additionally the pyramidal form and the diagravitropic branches of many northern conifers are particularly suitable for trapping the low-angle light characteristic of high latitudes.

Roots. The roots of conifers have a simple primary structure, similar to that found in the ferns. The apical meristem is protected by a root cap, and root hairs are produced from a zone immediately behind it. Secondary vascular tissue begins to be formed at a very early stage, often before the primary tissues are fully differentiated (Fig 8.23). Resin canals are abundant in the secondary xylem, rays, and cortex.

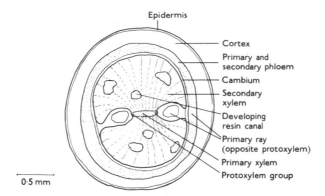

Figure 8.23. *Pinus sylvestris.* **Transverse section of young root.**

Reproductive structures. As the name of the order implies, the male and female reproductive organs of the conifers are commonly borne in cones (Figs. 8.24 and 8.27). Most conifers are monoecious but diclinous, the male and female cones being produced in different regions. In *Pinus,* for example, the female cones are produced near the apex of the tree and occupy the positions of main lateral buds, while the male cones are produced on the lower branches, usually in groups, each cone occupying the position of a short shoot. A few conifers (e.g., *Taxus* and *Juniperus communis*) are dioecious. The reproductive cones are usually compact, but in the Podocarpaceae the female cones are either lax or reduced, and in the Taxales the female reproductive region is not conelike at all. Nevertheless, the general affinities of the Taxales are clearly with the conifers.

The male cones are fairly uniform in structure, although they range widely in size. Those of *Taxus,* and of the Podocarpaceae and Cupressaceae are globose, hardly reaching 0.5 cm (0.2 in.) in diameter, but those of other conifers are commonly elongated, and in *Araucaria* they may exceed 20 cm (8 in.) in length and 3 cm (1.19 in.) in width. All, however, consist of a central axis bearing regularly arranged microsporangiophores (Fig. 8.24). These take the form of scales, somewhat peltate in shape, a variable number of pollen sacs being attached to the head and lying parallel to the stalk. The pollen grains of many species are winged and readily identifiable. The grains of *Pinus* (Fig. 8.25), for

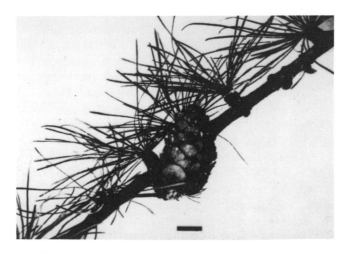

Figure 8.24. *Larix decidua* (larch). **Portion of a shoot showing the short shoots and the female cone in its second year. Note that the female cone is negatively gravitropic and that it terminates a lateral axis having a position equivalent to that of a short shoot. Scale bar 1 cm.**

example, have two asymmetrically placed air bladders (formed by local separation of the layers of the exine) between which the pollen tube emerges. Other grains have characteristic ornamentation; those of *Cryptomeria,* for example, possess a peculiar cuticular hook on one side. The pollen grains often begin to develop internally before shed. In *Pinus* (Fig. 8.30) the pollen grain when liberated contains two degenerating prothallial cells, a tube cell and a generative cell. The grains of the Taxales, Taxodiaceae, and Cupressaceae, however, lack prothallial cells and are uninucleate when shed, whereas the mature grains of the Araucariaceae contain up to 15 prothallial cells. The pollen grains of all conifers germinate distally (i.e., away from the center of the original tetrad).

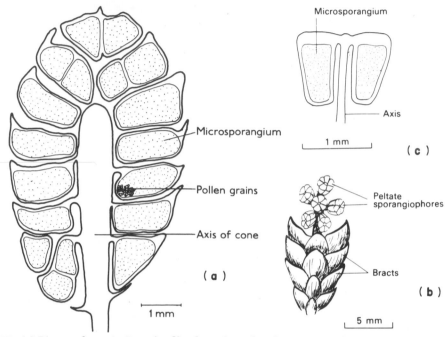

Figure 8.25. **(a)** *Pinus sylvestris.* **Longitudinal section of male cone. Each microsporangiophore bears two pollen sacs. (b,c)** *Taxus baccata.* **Mature male cone, each microsporangiophore bearing six to eight pollen sacs, and longitudinal section of microsporangiophore.**

Of the female reproductive regions, that of *Taxus* is considered first as it facilitates an understanding of the more complex situation in *Pinus* and other conifers. In *Taxus* the ovule terminates a short shoot bearing three pairs of decussate bracts. The ovule itself is upright and bilaterally symmetrical (Fig. 8.26). The single integument contains a sclerenchymatous layer, and two vascular bundles, diametrically opposed, ascend in the fleshy portion adjacent to the nucellus. A female gametophyte arises in the nucellus as in other gymnosperm ovules, and when mature it bears immersed archegonia in its micropylar surface. The minute short shoot terminating in the ovule is auxiliary to another short shoot furnished with spirally arranged scale leaves. This whole complex is itself borne in the axil of a normal foliage leaf. Both short shoots of the female reproductive system in *Taxus* are highly condensed and can be seen only by a careful dissection (Fig. 8.27).

In the female cone of *Pinus* (Fig. 8.28) we are again concerned with an axis bearing spirally arranged scales in the axils of which are ovuliferous structures (Fig. 8.28b and c). In *Pinus,* however, the ovuliferous structure is also scalelike, and it is largely fused with and ultimately projects beyond the bract scale in whose axil it arises (Fig. 8.28c). This, however, is not always the situation, even in the Pinaceae. In *Abies,* for example, the bract and ovuliferous scales remain separate, and in some species (e.g., *A. venusta*) the

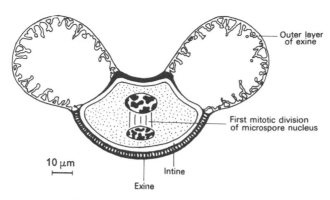

Figure 8.26. *Pinus banksiana.* **Median section of pollen grain showing the first division of the microspore nucleus and the nature of the bladders.**

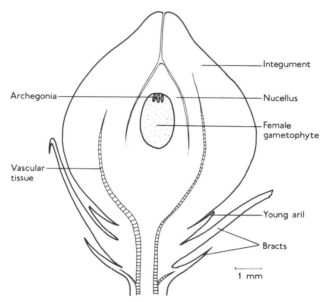

Figure 8.27. *Taxus baccata.* **Longitudinal section of young ovule. When fully mature, the nucellus bears the pollination drop at its tip.**

bract scale projects far beyond the ovuliferous. In the Pinaceae the ovuliferous scale bears two inverted ovules near its base (Fig. 8.28b), but in other families the number of ovules and their orientation vary. In *Araucaria* (Fig. 8.29) the ovuliferous scale produces and ultimately entirely surrounds a single inverted ovule. A specialized ovuliferous scale of this kind, also found in some Podocarpaceae, is termed an epimatium. Ovules throughout the conifers are regularly bilaterally symmetrical and the seeds are often winged.

The morphological and anatomical evidence, now supported by the paleobotanical, points to the ovuliferous scale being a highly modified shoot. The vascular supply to the bract scale, for example, consists of vascular bundles of which the xylem is adaxial, the orientation normal for a leaf trace. The bundles passing to the ovuliferous scale, however, are not only similar in position to those entering an auxiliary shoot, but the xylem of each is also abaxial, an orientation often seen at the base of a shoot trace. The female region of *Taxus* can be readily related to the cone of *Pinus* if the primary axis with its spirally arranged bracts is regarded as a cone, only one scale of which is fertile. The secondary axis with its decussate bracts and terminal ovule is then equivalent to an ovuliferous scale (Fig. 8.30).

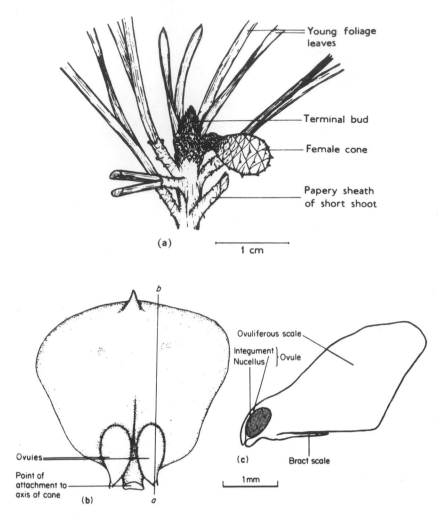

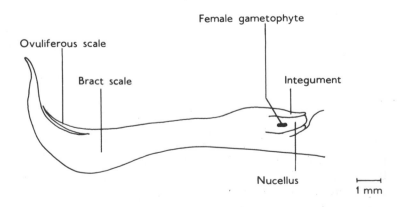

Figure 8.28. *Pinus sylvestris.* **(a) Female cone in the early summer of its first year. (b) Single scale from cone in (a), viewed from above. (c) Longitudinal section along line** *a–b.*

Figure 8.29. *Araucaria araucana.* **Longitudinal section of young ovule. (After M. Hirmer. 1936.** *Bibliotheca Botanica* **28.)**

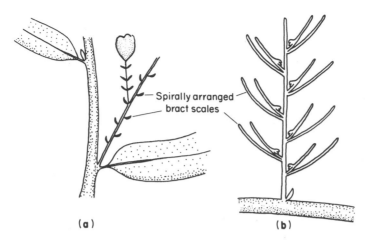

Spirally arranged
bract scales

(a)   (b)

Figure 8.30. Diagrammatic representations of female reproductive regions of (a) Taxus **and (b) Pinaceae, showing how** Taxus **can be regarded as bearing a cone only one scale of which is fertile.**

Pollination and fertilization. Pollination, which in temperate climates occurs in the spring, involves in most species a pollination-drop mechanism of the usual kind. The development of the female cone is so coordinated with that of the male that, at the time of release of the pollen, the axis of the female cone undergoes general elongation, thus opening the scales and allowing penetration of the pollen. Wind tunnel experiments have shown that air eddies around the cone promote the accumulation of wind-borne grains between the scales. Following pollination, rapid growth of the scales causes them to be tightly packed once again.

In *Pinus syvestris* the germination and development of the male gametophyte within the female cone are very slow and extend over a whole season (Fig. 8.31), coinciding with meiosis in the nucellus and the initiation of the female gametophyte. A pollen tube emerges from the grain in this first season's growth (Fig. 8.31e and h), and the generative cell divides into a sterile ("stalk") cell and a spermatogenous ("body") cell. Little further occurs in the winter, but in the following spring development is resumed. After a period of free nuclear division the female gametophyte becomes cellular (Fig. 8.31i), and one to six archegonia are formed in its upper surface (Fig. 8.31j), each surrounded by conspicuous jacket cells. The pollen tubes grow towards the archegonia, and, when a tube has come to within a short distance of an archegonium, the spermatogenous cell, which has moved into its tip, divides into two male ("sperm") cells of unequal size. The tube eventually penetrates the archegonium and the two sperm cells are liberated into the egg cytoplasm. The larger sperm nucleus passes into the egg nucleus, while the smaller sperm cell, the sterile cell, and the tube nucleus all degenerate. Several archegonia in one ovule may be penetrated by pollen tubes, and this can result in the formation of several zygotes and subsequent polyembryony.

Reproduction in *Pinus* is representative of that of the conifers generally. Among the principal variations is the "dehiscent" or "explosive" pollen of the Cupressaceae. Following hydration the grains swell and the exine is shed. In some conifers (e.g., *Larix*, larch) the pollen is trapped by a stigmatic flap of the integument covered with a sticky secretion before reaching the micropylar canal. In *Araucaria* the pollination drop mechanism is entirely absent. The pollen germinates between the scales of the female cone forming a freely branching, multinucleate, myceliumlike weft, many of the branches penetrating the nucellus. In the Taxodiaceae the female gametophyte often produces many archegonia, up to 60 being present in *Sequoia*. Pollen tubes may discharge gametes, which in this family are similar in size, above adjacent archegonia, each then being fertilized. In *Taxus* the male gametes are naked nuclei of equal size.

In all conifers (including *Taxus*) the male gametes lack any specialized means of locomotion. The delivery of nonmotile male gametes into the vicinity of the female gamete by means of a pollen tube is termed **siphonogamy.** It contrasts sharply with zooidogamy.

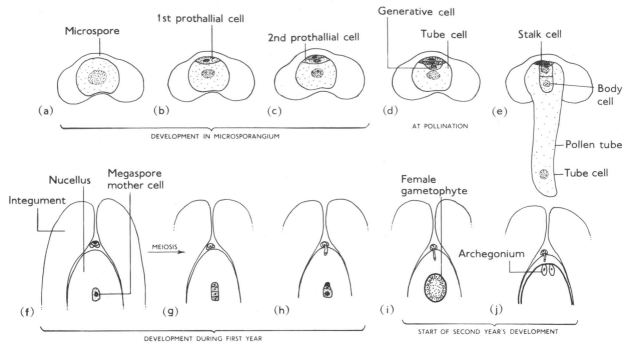

Figure 8.31. *Pinus* sp. **Development of the pollen and ovule. Diagrammatic, not to scale.**

Embryogenesis. Germination of the zygote frequently involves free nuclear division, but in *Pinus* this is not extensive, only four nuclei being so formed (Fig. 8.32a). These move to the bottom of the archegonium and form a plate, walls then being laid down between them. These cells divide longitudinally, the cells of each column behaving synchronously. This leads to the formation of a suspensor, tetragonal in section, terminating below in four groups of embryonic initials, each capable of yielding an embryo (Fig. 8.32c and d). This so-called cleavage polyembryony may be further complicated by additional embryos budding off from the basal suspensor cells (rosette cells). Usually only one of these many potential embryos reaches maturity.

The development of the zygote in other conifers differs only in detail from that seen in *Pinus*. In *Sequoia*, for example, there is no initial free nuclear division, and in the Podocarpaceae the cells of the proembryo pass through a binucleate stage, a feature believed peculiar to this family. In other conifers polyembryony seems less common than in *Pinus*. The embryos of many conifers have several cotyledons; as many as 12 may be present in *Pinus*.

In those conifers in which the male gamete retains its cytoplasm (e.g., *Pinus, Larix*) the male cytoplasm enters the zygote. During karyogamy this cytoplasm surrounds the zygotic nucleus, forming a "neocytoplasm" which displaces the original egg cytoplasm. Correlated with this unusual cytology is the discovery that in both *Pinus* and *Larix*, although the inheritance of mitochondrial DNA is maternal, that of the plastids is paternal. The proplastids of the egg cell appear to degenerate during maturation of the female gamete.

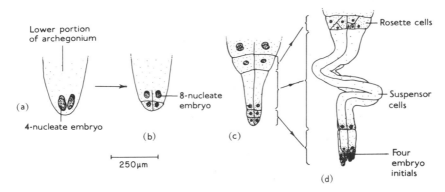

Figure 8.32. *Pinus* **sp. Stages in the development of the embryo. (After Bucholz, from Foster and Gifford. 1959.** *Comparative Morphology of Vascular Plants.* **Freeman, San Francisco. © 1959.)**

The formation and liberation of the seeds. The mature embryo lies in the remains of the female gametophyte and nucellus, and is surrounded by a hard seed coat formed from the integument. In some conifers (e.g., *Sequoia, Pinus sylvestris*) this is expanded as a conspicuous wing assisting the distribution of the seeds by wind (Fig. 8.33). The female cone often becomes dry and woody during the formation of the seeds and sometimes does not open until a long period after the seeds are mature. In *P. sylvestris* the cone opens and releases the seeds in the second year after pollination (the whole process of reproduction thus extends over three years), but in the "closed cone" pines of the Pacific coast of North America the cones remain closed indefinitely and the seeds are released by decay of the scales or as a consequence of the singeing of the cones by a forest fire. The cones of some pines are extraordinarily large; those of the Californian *P. coulteri,* for example, may reach 40 cm (16 in.) in length and 2 kg (4 lb.) in weight. In some Podocarpaceae and in *Juniperus* the ovuliferous scales become fleshy in fruit, the "berries" of *J. communis* being used to flavor gin. In some other podocarps and in *Taxus* the seed becomes surrounded by aril which grows up from the base (Fig. 8.34). In *Taxus* this becomes bright red and succu-

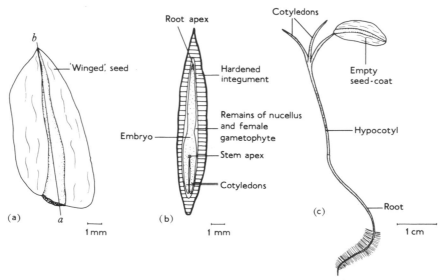

Figure 8.33. *Sequoia gigantea (Sequoiadendron giganteum).* **(a) Seed. (b) Longitudinal section (at a–b) of well-soaked seed. (c) Young seedling.**

lent. Although the seed is poisonous, the aril is wholesome and sought after by birds, probably an aid to dispersal of the seeds. The nonwinged seeds of some species of *Pinus* in both Europe and North America ("nut pines") are edible and prized by both people and wild life.

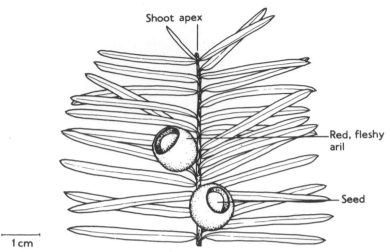

Figure 8.34. *Taxus baccata.* **Shoot bearing seeds surrounded by mature arils.**

Germination. In most conifers germination is initiated by the root pole of the embryo elongating and breaking through the seed coat. The vigorous primary root soon anchors the seedling, and the elongating hypocotyl raises the remains of the seed, from which the cotyledons are rapidly withdrawn (Fig. 8.33c). All conifer seedlings, so far as is known, become green in the dark, a remarkable property that distinguishes them from the seedlings of most angiosperms.

The seeds of *Araucaria* often germinate in the cone before it falls apart. In some species the hypocotl swells to form a tuber, and the seedling is capable of "resting" in this condition for several months. It was this curious feature that facilitated the transmission of the first specimens of *A. araucana* from Chile to Europe in the eighteenth century.

The evolution and origin of the conifers. The current geographical distribution of the conifers presents a number of features of evolutionary significance. Floristically, for example, the conifers of the Northern Hemisphere are strikingly different from those of the Southern, and some families (notably the Pinaceae in the north and the Araucariaceae in the south) hardly cross the equator. Fossils of Quaternary and Tertiary age give no indication that this is a recent segregation, but they do reveal that the distribution of some families was formerly much more extensive. *Sequoia*, for example, now confined to the west coast of North America, was once widespread in the Northern Hemisphere, and in the Tertiary era *Taxodium* swamps occurred in Spitzbergen. The distribution of the conifers has clearly contracted with the rise of the angiospermous forests.

The fossil record of the conifers extends back well into the Carboniferous, and the early forms show a close relationship, both vegetative and reproductive, with the Cordaitales. Representative of the Upper Carboniferous forms is *Walchia* (*Lebachia*), placed in the order **Voltziales** (Fig. 8.35). The female cone consisted of an axis, about 8 cm (3.13 in.) long, bearing spirally arranged, bifid bracts. In the axil of each was a radially symmetrical, but flattened, dwarf shoot bearing small bracts and a fertile scale (Fig. 8.35). The fertile scale bore a single ovule. The orientation of the ovule (which was platyspermic) in life is not yet clear, but it was either terminal and upright, or inverted and adjacent to the side of the fertile scale away from the axis of the cone. This fertile short shoot can reasonably be regarded as morphologically intermediate between the female flower of the Cordaitales (Fig. 8.17) and the ovuliferous scale of a modern conifer. There is in fact a series of fossils of late Paleozoic and early Mesozoic age in which the vegetative and fertile parts of the female short shoot become progressively less distinct, leading ultimately to a structure extremely similar to an ovuliferous scale. Some families became distinct earlier than others. The curious lax cone of some Podocarpaceae, for example, is recognizable as far back as the Lower Triassic, and the characteristic female shoot of *Taxus* seems to have evolved by the end of the same period.

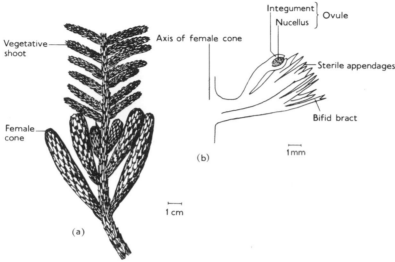

Figure 8.35. *Walchia (Lebachia)* **sp. (a) Reconstruction of a shoot bearing upright female cones. (b) Longitudinal section of ovuliferous structure. The orientation of the ovule is uncertain. (Both after Florin, from Delevoryas. 1962.** *Morphology and Evolution of Fossil Plants.* **Holt, Rinehart and Winston, New York.)**

The male cones of the early conifers closely resembled those of the modern, and thus differed sharply from those of the Cordaitales, although the pollen grains of the Cordaitales and early conifers were quite similar. Nothing is known of fertilization in the early conifers. Spermatozoids may have been produced, but it is also possible that the siphonogamous reproduction seen in modern conifers evolved quite early. This, by disposing of the necessity for free fluid at the time of fertilization, reduces the hazards of copulation. Perhaps the advantages of siphonogamy saved the Coniferales from the extinction that befell the Cordaitales.

There are sufficient similarities between the Coniferales and Cordaitales to warrant the assumption of a common origin from progymnospermous ancestors, the two orders possibly becoming distinct in the early Carboniferous.

The Ginkgoales

The order Ginkgoales is represented today by a single genus and species, *Ginkgo biloba* (maidenhair tree). This remarkable tree, with a striking pagodalike arrangement of the main branches, was unknown to the Western world until the seventeenth century. It was first discovered in Japan and subsequently in China, but always in cultivation. Suggestions that wild stands of *Ginkgo* may occur in remote parts of China, although not improbable, have never been confirmed. *Ginkgo* is now common in cultivation in all parts of the world.

Fully grown specimens of *Ginkgo* are tall, deciduous trees reaching a height of 30 m (98 ft.) or more. The lateral branches bear both long and short shoots (Fig. 8.36), and leaves occur on each. Damage to a long shoot will cause one or more adjacent short shoots to behave as long shoots, indicating that their manner of growth is not irreversible, and that the maintenance of the dwarf condition probably depends upon the presence of growth-regulating substances produced by the meristem of the long shoot. Anatomically, the apices of the long and short shoots are similar and show well-defined zonation, although no distinct tunica and corpus are present. Growth takes place from a superficial group of apical initial cells. A large proportion of a mature stem consists of secondary xylem, penetrated by narrow parenchymatous rays. The tracheids have bordered pits, usually in a single row, on their radial walls.

The leaves of *Ginkgo* are fan-shaped, usually with a distal notch (hence the specific name). Two vascular bundles ascend the petiole and dichotomize in the lamina, with occasional anastomoses. Short resin ducts may lie between the veins. The distal margin of the leaf is usually irregular, a feature much more marked in juvenile leaves where the distal part of the leaf may even be segmented.

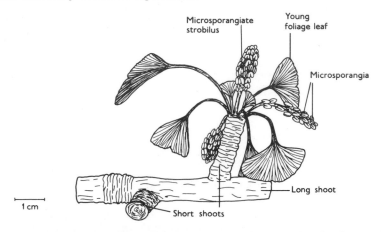

Figure 8.36. *Ginkgo biloba*. **Short shoot bearing male strobili. The leaves have not reached mature size.**

Reproduction. *Ginkgo* is dioecious, and sex determination appears to be chromosomal since the male possesses a heteromorphic pair of chromosomes. The male reproductive structures (Fig. 8.36) consist of a small strobili, resembling catkins, which arise in the axils of scale leaves of the short shoot. The axis of the strobilus bears a number of microsporangiophores arranged in a loose and irregular spiral. Each microsporangiophore is slightly peltate and the sporangia, usually two, are attached beneath the head. The pollen grains, which have a characteristic furrow in the wall, contain four nuclei when shed, two of the nuclei being associated with rudimentary prothallial cells, and the others identified as the generative and tube nuclei. The first prothallial cell soon degenerates.

The ovules are usually borne in pairs, two sessile ovules being symmetrically attached at the end of a stalklike sporangiophore. Not infrequently, however, the sporangiophore branches irregularly and bears more than two ovules. The sporangiophore itself arises, as in the male, in the axil of a scale or a leaf on a short shoot. The ovules (Fig. 8.37) are about 0.5 cm (0.2 in.) long and about as broad, and are surrounded at the base by a cushionlike swelling of the broad end of the sporangiophore. They possess a single integument into which two diametrically opposed bundles ascend. This bilateral symmetry is reflected in the micropyle which is slightly two-lipped at its tip. A megaspore is formed within the nucellus and this yields a female gametophyte bounded by a conspicuous membrane. The gametophyte is unique among seed plants in containing chlorophyll. Sufficient light reaches the gametophyte to permit some photosynthesis,

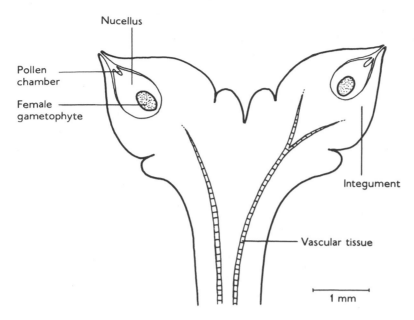

Figure 8.37. *Ginkgo biloba.* **Longitudinal section of female shoot with young ovules.**

supplementing the supply of photosynthates reaching the gametophyte from the sporophyte. At maturity, two archegonia arise at the micropylar end of the gametophyte and the nucellus above develops a pollen chamber.

Pollination is assisted by a "pollination drop" at the micropyle, and the pollen chamber, having received the pollen, then becomes closed above. The cavity of the nucellus progressively deepens, carrying the pollen with it, until it reaches the female gametophyte, the center of which is extended upwards to form a so-called "tentpole" (a feature seen also in many fossil seeds). The germinating pollen forms a tube, but, as in *Cycas,* this has a haustorial function. The tube penetrates the nucellus and there branches freely, growing through the spaces at the interstices of the cells. The proximal part of the grain forms a sac hanging in the chamber above the archegonia. As the archegonia mature, the generative cells of the male gametophytes divide. Each yields a spermatogenous and a sterile cell, the latter wrapped around the surviving prothallial cell. The spermatogenous cell yields two multiflagellate spermatozoids whose major diameters are of the order of 100 μm. These spermatozoids are released into the fluid above the archegonia and bring about fertilization.

The zygote, which may not be formed until after the ovule has been shed, begins its development by free nuclear division, leading to a proembryo containing about 256 nuclei. Walls then differentiate and a flask-shaped proembryo is formed, the lower part of which becomes the embryo proper. A clearly defined suspensor is thus absent. The mature embryo has two cotyledons. Usually only one of the paired ovules on the sporangiophore develops into a seed. In the mature seed (Fig. 8.38), which reaches a diameter of about 2 cm (0.75 in.), the outer layer of the integument becomes fleshy and resinous, and the inner hard. The formation of the seed is completed in one season.

The fossil history and possible origin of the Ginkgoales. There is fossil evidence of the Ginkgoales having existed in the Mesozoic and at that time having been much more widely distributed than today. The leaves of the extinct Ginkgoales were usually much more like the juvenile than the mature leaves of *G. biloba.* Although reproductively at about the same level of evolution as the cycads, the origin of *Ginkgo* and the Ginkgoales is obscure. They may have arisen as an offshoot of a cordaitalean stock. The bilateral symmetry of the seed and the dense secondary xylem are indications against affinities with the cycads and pteridosperms.

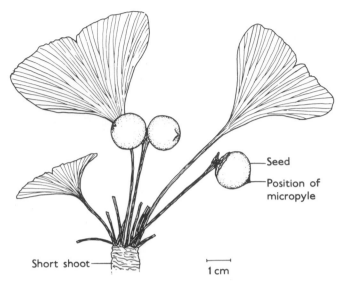

Figure 8.38. *Ginkgo biloba.* **Tip of short shoot bearing ripe seeds.**

The Gnetales

This order consists of three genera: *Ephedra, Gnetum,* and *Welwitschia.* Not only are the Gnetales very different from other gymnosperms, but the genera also differ so markedly among themselves that many have considered each to be worthy of independent classificatory rank. They have been studied extensively by morphologists because of certain features which make them appear intermediate between gymnosperms and angiosperms. However, although the Gnetales indicate how certain characteristics of angiosperms may have arisen, they themselves appear to be specialized offshoots from the main evolutionary trends. Unfortunately the fossil record of the Gnetales is very fragmentary and does not extend further back than the Cretaceous.

The genus Ephedra. The genus *Ephedra* is widely but discontinuously distributed. Some 35 species occur in the Mediterranean region, Asia, and in the Americas. They are typical "switch plants," consisting of densely branched axes, the younger of which are green and photosynthetic (Fig. 8.39). The leaves consist of whorls of small scales which soon become scarious. Many species grow in extremely arid situations, such as sand dunes and scree slopes, and these not unexpectedly have an extensive root system. The young twigs of some species have medicinal uses, and the genus is the source of the alkaloid ephedrine.

The stem of *Ephedra* grows from a group of meristematic cells, and a distinct tunica and corpus are recognizable in the apex. The primary vascular system consists of a number of bundles symmetrically placed around a central pith (Fig. 8.40), the bundles being linked at the nodes by a transverse vascular ring, as in *Equisetum.* The primary xylem becomes surrounded by secondary, traversed by broad parenchymatous rays. The tracheids have bordered pits on their radial walls, well-developed tori also being present. Many of the tracheids are arranged in columns, and the end walls are so extensively perforated that they can be legitimately regarded as vessel segments with foraminate perforation plates. The phloem consists of sieve cells and parenchyma, the sieve cells, like those of the conifers, having highly inclined end walls.

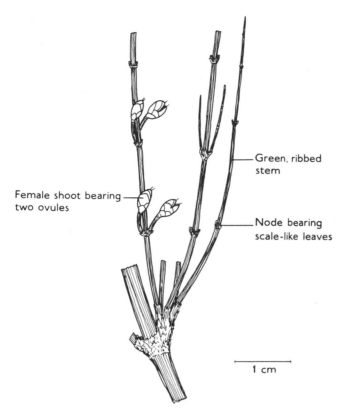

Figure 8.39. *Ephedra* **sp. Portion of shoot system showing the articulated stem, reduced leaves, and female reproductive structures.**

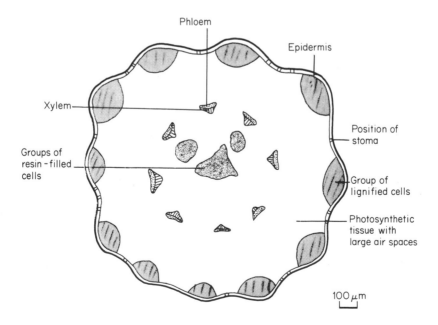

Figure 8.40. *Ephedra* **sp. Transverse section of young stem.**

Ephedra is dioecious. The male reproductive regions are conelike terminations of short shoots which arise in the axils of the scale leaves. The short shoot bears a number of bracts in decussate pairs, and in the axil of each bract is a male flower (Fig. 8.41). This consists of a microsporangiophore, bearing at its summit two to eight microsporangia and enclosed in the basal region by two medianly placed bracteoles. The pollen grains are ellipsoidal and furnished with prominent longitudinal ridges. Nuclear divisions occur immediately after the formation of the grains and when mature they contain four or five nuclei. The first two daughter nuclei, which do not again divide, are regarded as prothallial.

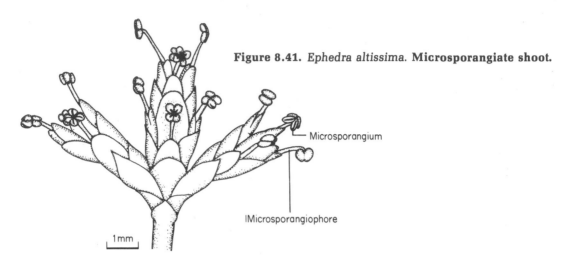

Figure 8.41. *Ephedra altissima.* **Microsporangiate shoot.**

Microsporangium

Microsporangiophore

1mm

The female reproductive organ is similar in structure to the male, but only the uppermost pair of bracts is fertile. Each subtends an upright ovule which is surrounded by a sheath, sometimes two-lipped above and probably homologous with the two bracteoles of the male flower. Hermaphrodite flowers are occasionally found. The radially symmetrical ovule (Fig. 8.42) is bounded by a papery integument, the apex of which is prolonged into a micropylar tube, highly cutinized at maturity. Megasporogenesis is initiated in the nucellus and in the usual way only one megaspore of the tetrad persists. The female gametophyte passes through a period of free nuclear division before becoming cellular. Two, rarely more, archegonia are differentiated at the micropylar end, and they are unusual in being quite deeply sunken into the somatic tissue of the gametophyte. As the eggs mature, the upper part of the nucellus breaks down to form a pollen chamber, and complex cytological phenomena, among them the amoeboid migration of nuclei and endomitosis, occur in the upper cells of the gametophyte.

The pollen is distributed by wind and possibly also by insects attracted by the sugary pollination drop produced by nonfertile ovules associated with the male flowers of some species. Germination of the pollen occurs directly on the surface of the female gametophyte and a pollen tube pushes its way into an archegonium. Two sperm nuclei, produced by division of the spermatogenous cell, enter the egg cell. One fuses with the nucleus of the egg cell and the other, in some species at least, regularly fuses with the ventral canal cell. Only the zygote however has been seen to undergo any further development. The interval between pollination and fertilization, in contrast to the prolonged period in the conifers, may amount to no more than 24 hours in *Ephedra*. The embryology of *Ephedra* is not well known, but it appears established that the zygote first undergoes free nuclear division, the eight nuclei so formed being the initial cells of proembryos. There is thus potential polyembryony, but only one embryo reaches maturity. A suspensor of complex, compound origin drives the embryo into the central region of the female gametophyte, rich in food reserves. The mature embryo has two cotyledons and lies surrounded by the membranous remains of the ovular tissues and the hardened integument. In many species the bracts below the ovules become hard and winglike in fruits, but in the alpine *E. helvetica* they become fleshy and brightly pigmented.

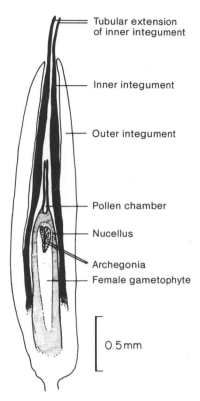

Tubular extension
of inner integument

Inner integument

Outer integument

Pollen chamber

Nucellus

Archegonia
Female gametophyte

0.5 mm

Figure 8.42. *Ephedra* **sp. Longitudinal section of ovule with archegonia. (From a photograph by W. R. Ivimey-Cook in McLean and Ivimey-Cook. 1951.** *Textbook of Theoretical Botany.* **Longmans, London.)**

The genus Gnetum. The genus *Gnetum* is exclusively tropical, occurring in Asia, Africa, and South America. Many species are lianes, but others are small trees. In contrast to *Ephedra* the leaves are well developed (Fig. 8.43) and possess broad, ovate laminae with reticulate venation, some of the veins ending blindly in areolae. Vegetatively, therefore, *Gnetum* has very much the appearance of an angiosperm.

The stem of *Gnetum* usually has a small pith, surrounded by a little primary xylem. Most of the xylem is secondary and is interspersed with broad parenchymatous rays. In the climbing forms the stem is eccentric, and successive cambia give rise to a polycyclic stelar structure, the asymmetry in any particular region depending upon its spatial orientation. In general features, therefore, the stem is closer to that of the cycads and pteridosperms than to that of the conifers. A striking difference, however, is the even closer approach in *Gnetum* than in *Ephedra* to the differentiation of authentic vessels in the secondary xylem. Another peculiarity in *Gnetum* is that parenchymatous cells are closely associated with the sieve cells, recalling the companion cells of angiosperms. The cortex adjacent to the phloem is rich in fibers, occasionally used as cord. In some species the stems contain laticifers, one of the very few instances of these tissue elements occurring outside the flowering plants.

The retention of *Gnetum* in the gymnosperms is justified by the nature of the reproduction. Both male and female reproductive regions are again strobili, usually ter-

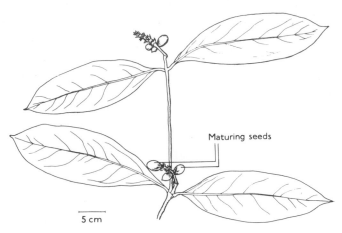

Maturing seeds

5 cm

Figure 8.43. *Gnetum gnemon.* **Portion of shoot bearing female strobili. (After Madhulata, from Maheshwari and Vasil. 1961.** *Gnetum.* **CSIR, Delhi.)**

minating lateral axes, although some species show cauliflory. In the male strobilus (Fig. 8.44a) the axis bears a succession of gallerylike sheaths, usually about eight in number, probably formed from coalesced bracts. In the axil of each sheath are whorls of male flowers, in some species surmounted by whorls of abortive ovules. The male flower (Fig. 8.44b) consists of a single microsporangiophore, terminating in two microsporangia, surrounded at its base by a delicate membranous sheath. The pollen grains which, like those of *Ephedra,* have conspicuous longitudinal striations, are trinucleate when shed, one nucleus possibly being prothallial.

In the female strobilus (Fig. 8.44c) each sheath encloses a whorl of female flowers. Each flower consists of a single, radially symmetrical ovule (Fig. 8.45a) surrounded by three integuments, the outer of which is possibly homologous with the basal sheath of the male flower. The inner integument is extended into a cutinized micropyle, and the nucellus beneath becomes transformed into a pollen chamber. One or more of the tetrad of megaspores formed in the nucellus enters into the formation of the acellular female gametophyte.

Pollination is by wind, and also by insects attracted, as in *Ephedra,* by the sugary pollination drops of abortive ovules. The entry of the pollen into the micropyles of functional ovules initiates renewed growth of the ovule. The germinating grains, trapped by the proliferation of the cells lining the micropyle, send pollen tubes into the nucellus. Meanwhile the female gametophyte (often called in *Gnetum* the embryo sac) completes its development. At maturity, it is shaped like an inverted flask (Fig. 8.45b). Much of the cytoplasm, in which the nuclei are irregularly scattered, lies at the base of the sac, but the remainder is distributed as a thick layer around the periphery, a large vacuole occupying the center. While the embryo sac is completing its development, the pollen tubes, having penetrated the nucellus, approach the sac. One of the nuclei in the male gametophyte moves to the tip of the tube and divides into two sperm nuclei. Meanwhile one or more nuclei in the upper part of the embryo sac in the region adjacent to the closest pollen tube become conspicuously large. The pollen tube, having by now made contact with the sac, discharges the two sperm nuclei into it. They immediately migrate to the nearby large nuclei of the sac, which can thus be identified as egg nuclei. There is no clear segregation of egg cells or nuclei within the sac.

The entry of the male nuclei stimulates general division of the somatic nuclei within the embryo sac. The contents of the sac become cellular, the cells often containing several nuclei, which subsequently fuse. The male nucleus and egg nucleus meanwhile coalesce and form a zygote lying within what can now be regarded as a cellular endosperm.

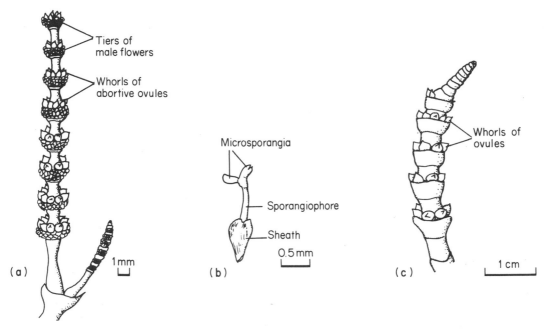

Figure 8.44. *Gnetum gnemon.* (a) Young microsporangiate strobilus with whorls of abortive ovules. (b) Dehiscent microsporangia. (c) Megasporangiate strobilus. (From Maheshwari and Vasil. 1961. *Gnetum.* CSIR, Delhi.)

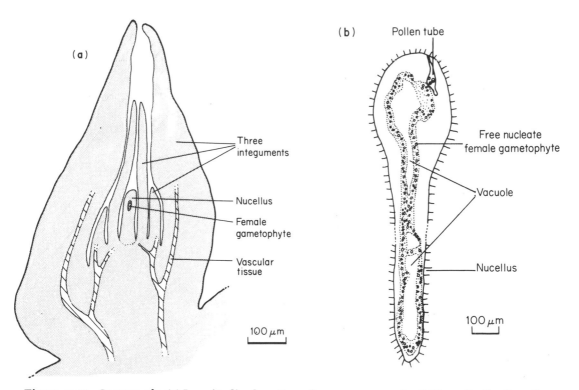

Figure 8.45. *Gnetum ula.* (a) Longitudinal section of very young ovule. (b) Longitudinal section of mature female gametophyte (embryo sac) showing entry of pollen tube. (Both after Vasil, from Maheshwari and Vasil. 1961. *Gnetum. CSIR, Delhi.)*

Development of the zygote proceeds at once, but does not involve any free nuclear division. A complicated suspensor, possibly partly haustorial in function, is formed before the embryo proper. Although, since more than one egg nucleus may be present in the embryo sac and several male nuclei may be discharged into it, there is potential polyembryony, only one embryo usually comes to maturity. The embryo has two cotyledons.

Although the inflorescence is quite different, the seed of *Gnetum* recalls that of some Bennettitales. When ripe, the outer integuments of the *Gnetum* seed become fleshy, and in some species are edible.

The genus Welwitschia. In respect of habit, *Welwitschia* (Fig. 8.46) is one of the most peculiar plants in existence. The genus is monotypic, and the single species is confined to desert regions of southwestern Africa. The stem is short, upright, and mostly below soil level. At the upper end it bears two strap-shaped leaves with indefinite basal growth. Developmentally, these are the first pair of leaves after the cotyledons, and growth soon becomes confined to them. A further pair of decussate leaf primordia is formed in the young plant, but these differentiate into hornlike protuberances. Below, the stem passes into a long taproot which gives rise to an extensive root system. The stem contains much secondary tissue and shows anatomical peculiarities similar to those of *Gnetum* and *Ephedra*.

Welwitschia is dioecious, the conelike inflorescences terminating small branch systems arising in the axils of the leaves. The cones consist of a series of scales, arranged in decussate pairs, in the axils of which are the individual flowers. The male flowers consist of a short axis bearing first two decussate pairs of small scales, and then a ring of trilocular synangia, the filaments supporting them being fused together at the base into a membranous cylinder. In the center of the flower, and terminating its short axis, is an abortive ovule. The pollen grains are binucleate when shed, but neither nucleus can be regarded as belonging to a prothallial cell.

The female flower consists solely of an ovule with two integuments, terminating an axis with occasionally two minute lateral outgrowths. The outer integument, which is broadly winged tangentially to the cone and is traversed by several vascular bundles, may be homologous with the upper pair of bracteoles of the male flower. The membranous inner integument is extended at its apex into a cutinized micropylar tube. The basic symmetry of the ovule seems to be radial rather than bilateral.

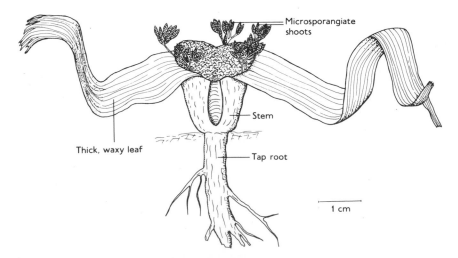

Figure 8.46. *Welwitschia mirabilis.* **Habit of the male plant. (After J. D. Hooker. 1863.** *Transaction of the Linnean Society* **[London] 24.)**

Pollination and the initiation of the embryo sac take place much as in *Gnetum*, but in the later stages, especially at the time of pollination, there are features peculiar to *Welwitschia*. After the initial free nuclear division in the sac, walls are laid down, many of the cells so formed being multinucleate. As the pollen tubes penetrate the nucellus, some of the multinucleate cells in the upper part of the embryo sac give rise to tubular processes. These grow up towards the descending pollen tubes and potential egg nuclei move to their tips. When a pollen tube and process make contact, the separating walls dissolve and the sperm and egg nuclei fuse. The zygote then becomes ensheathed in cytoplasm and a cell membrane forms. There is no free nuclear division in the development of the zygote. A suspensor is formed and below it the embryo proper. Only one zygote yields a mature embryo. There are two cotyledons. In fruit, the outer integument of the seed persists as a broad wing which assists aerial dispersal.

The origins of the Gnetales. Although remains of *Gnetum*-like plants have been found in the Lower Cretaceous of North America, the origins of the Gnetales remain conjectural. Anatomically, especially in the features of the secondary xylem, they recall the pteridosperms rather than the Cordaitales. This view is strengthened by the symmetry of the seeds, which is primarily radial. Nevertheless, the Gnetales are clearly far from the pteridosperms of the Carboniferous and probably the consequence of quite prolonged evolutionary change. Features of the distribution suggest that *Ephedra* originated in Eurasia and *Gnetum* in Gondwana, a vast southern continent which broke up in the Mesozoic era.

Gymnospermy as an Evolutionary Grade

The diversity of the gymnosperms taken as a whole, in both vegetative and reproductive features, indicates that gymnospermy must be regarded as a grade of evolutionary advance, probably first attained in the second half of the Devonian period. The seed habit may, of course, have arisen independently more than once in groups of plants at the progymnospermous level of evolution. This would be in line with the overlapping of the progymnosperms and gymnosperms in geological time, and the diversity present in both. Within the gymnosperms, for example, the affinities of the cycads appear to be pteridospermous, those of *Ginkgo* cordaitalean. Here then are two groups of plants whose evolution has probably been independent for many millions of years, but which are at the same level of advancement in respect of the reproductive process. Siphonogamy may also have arisen independently in the conifers and Gnetales.

The highest grade of gymnospermy is clearly shown by the Gnetales. The peculiarities of the female gametophyte in *Gnetum* and *Welwitschia* indicate the kind of developments which, in some early transitional forms, may have led to the angiospermous embryo sac. Although there are differences in the manner of their formation, the vessels of the Gnetales also foreshadow those of the angiosperms. Further, studies of the nucleotide sequences of the nucleic acids show that those of the Gnetales more closely resemble those of the angiosperms than any other seed plants. It is reasonable on this evidence to suppose that the Gnetales and the angiosperms evolved from a common stock. The relationship between the gymnosperms and angiosperms is analogous to that between the progymnosperms and the gymnosperms.

___ 9 ___
THE SUBKINGDOM TRACHEOPHYTA
Part 4

THE FLOWERING PLANTS

ANGIOSPERMOPHYTA

The angiosperms are the most abundant and widely distributed tracheophytes. They are of outstanding economic importance, being the source of many durable hardwoods, most of our vegetable foodstuffs, and about one-quarter (in monetary value) of commercially marketed drugs. They number some 200,000 species and show remarkable diversity in growth form, morphology, and physiology.

The general features of the Angiospermophyta can be summarized as follows:

Sporophyte herbaceous or arborescent; branching usually axillary. Leaves various, but regarded as megaphyllous in origin. Secondary vascular tissue commonly present. Vascular system usually consisting of vessels and tracheids, and sieve tubes with distinctive companion cells. Heterospory as in the Gymnospermophyta, but the ovules borne within a characteristic structure (carpel), usually closed, the pollen germinating on a specialized region of the exterior (stigma). Female gametophyte always an embryo sac, lacking archegonia. Male cells (sperms) lacking specialized means of locomotion, released into the embryo sac from the filamentous male gametophyte (siphonogamy) (Fertilization characteristically double, yielding in each embryo sac a zygote and a mostly triploid endosperm nucleus. Embryogeny endoscopic. Various forms of asexual reproduction not uncommon.

Growth Forms of Angiosperms

Examination of a tracheophyte flora (frequently consisting principally of angiosperms) will usually reveal that it consists of a number of distinct growth forms, ranging from large woody trees to minute herbs. The extent of the representation of these different morphologies in a given vegetation presents intricate ecological problems, the discussion of which is facilitated by Raunkiaer's concept of life forms. A life form is defined by the length of life of the shoots and the position and protection of the resting buds (Fig. 9.1). Those plants in which the shoots are persistent and the buds are carried well above the soil surface are termed **phanerophytes,** those with resting buds closer to the surface **chamaephytes,** and those with resting buds at the surface **hemicryptophytes.** Familiar examples of these three classes are, respectively, the larger woody plants, small bushes such as *Calluna,* and rosette plants such as *Taraxacum.* The classification, except for certain small specialized categories, is completed by the **cryptophytes** (geophytes) where the resting buds are below the soil surface, and the **therophytes.** The latter are those annuals and ephemerals which tide over unfavorable periods as embryos enclosed in seeds. By use of this classification it is possible to show in a precise statistical manner that, for example, the vegetation of the humid tropics consists predominantly of phanerophytes, and that in the Northern Hemisphere the percentage of hemicryptophytes in general increases with latitude (Table 9.1).

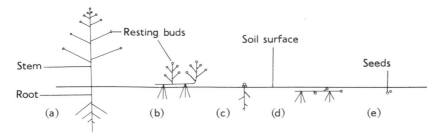

Figure 9.1. Life forms of angiosperms. The circles indicate the positions of the resting buds.
(a) Phanerophyte. (b) Chamaephyte. (c) Hemicryptophyte. (d) Cryptophyte.
(e) Therophyte.

Table 9.1. The relationship between life form and latitude. (Data from Raunkiaer. Succulents, water plants, and specialized epiphytes are omitted.)

Flora	Approx. latitude	Percentage representation				
		Ph	Ch	H	Cr	Th
Amazonian rain forest	0°	95	1	3	1	0
S. Thomas & San Juan (West Indies)	18°N	58	12	9	3	14
Denmark	57°N	7	3	50	22	18
Iceland	65°N	2	13	54	20	11
Spitzbergen	77°N	1	22	60	15	1

Ph: phanerophyte   Ch: chamaephyte   H: hemicryptophyte
Cr: cryptophyte   Th: therophyte

Although there is this evident relationship between life form and distribution, no less important in determining the distribution of individual species are such factors as mean minimum temperatures, and the length of time available for gainful photosynthesis in the yearly cycle.

For convenience of description we can regard the plant body of an angiosperm as consisting of three morphological categories, stem, leaf, and root, although, as we shall see later, there are good reasons for regarding the leaf as a megaphyll and hence of axial origin. The main stem is usually upright, displaying negative gravitropism and positive phototropism. Branching occurs in the axils of leaves. The final orientation of these laterals is probably determined by a combination of complex gravitropic and phototropic responses. Although such branch systems are usually aerial, they may be subterranean (when, of course, light will no longer affect morphogenesis). The genus *Parinarium,* for example, is represented by normal trees of tropical Africa, but in *P. capense* (Fig. 9.2), a species of cooler South Africa, the stem, although of similar woodiness and ramification, is below soil level and only small shoots appear at the surface. This so-called **suffrutescent** habit is also encountered, although less strikingly, in alpine willows.

In the development of the branch system, the main axis may originate in two ways, Either the apical bud continues its vegetative growth indefinitely, or it is extinguished at the end of each season, when either it gives rise to a reproductive system of limited growth or it aborts. Where the apical bud remains active (Fig. 9.3a), the main axis is of simple origin, and growth is monopodial. Where it is extinguished (Fig. 9.3b), growth is continued in the following season by the uppermost lateral. Subsequent positional readjustments often almost wholly obscure the discontinuities in the mature axis. The axis is, nevertheless, of compound origin, and growth is sympodial. These two kinds of growth are represented among plants of all life forms. Among phanerophytes, for example, the growth of the ash (*Fraxinus*) is monopodial and that of the lime (*Tilia*) is sympodial. The

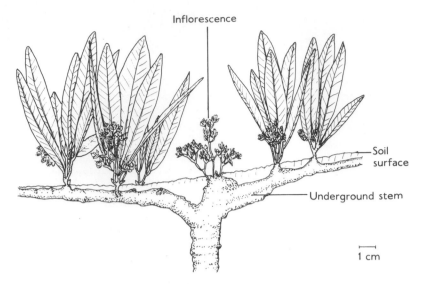

Figure 9.2. *Parinarium capense*. **The suffrutescent habit. (After J. B. Davy. 1922.** *Journal of Ecology* **10.)**

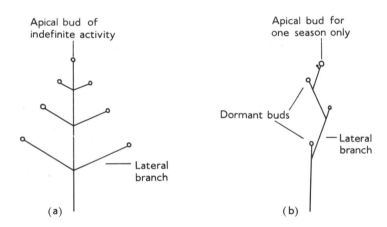

Figure 9.3. The basic types of branching. (a) Monopodial. (b) Sympodial.

factors which cause abortion of terminal buds in plants of sympodial growth are complex, but, in trees, day-length is often of paramount importance. The shortening days of later summer and autumn stimulate the synthesis of a growth inhibitor in the leaves. This is, in turn, transmitted to the terminal buds, causing either dormancy or in some species abortion. In long days the inhibitor is lacking. Experimental alteration of day length can therefore sometimes lead to plants with sympodial growth becoming monopodial.

Aerial stems may sometimes twine in a regular fashion. Familiar examples are the hop (*Humulus lupulus*), where the stem grows with a left-handed screw (clockwise when viewed from above), and bindweed (*Convolvulus*), where the screw is right-handed. The mechanism by which twining growth is achieved remains obscure, but in the hop it has been demonstrated that the curvature of the stem remains the same whatever the diameter of the support. In some flowering plants the main axis takes the form of a horizontal rhizome, a form of growth already encountered in Paleozoic plants. The rhizome of

Aegopodium podagraria (goutweed) and probably of other species is able, by adjusting the orientation of its growing region, to maintain itself at an almost constant depth beneath the surface of the soil. The physiological mechanism underlying this remarkable behavior is still not wholly known. Other highly modified stems are seen in corms, bulbs, and some tubers, in which the stem, which is adapted for the storage of food, either in itself (corms, Fig. 9.4a, or stem tubers, Fig. 9.4b) or in the associated swollen leaf bases or scale leaves (bulbs, Fig. 9.4c), shows very little elongation. A very peculiar stem, resembling a thallose liverwort, is found in the Podostemaceae (Fig. 9.5), a family of small plants growing on rocks by tropical streams.

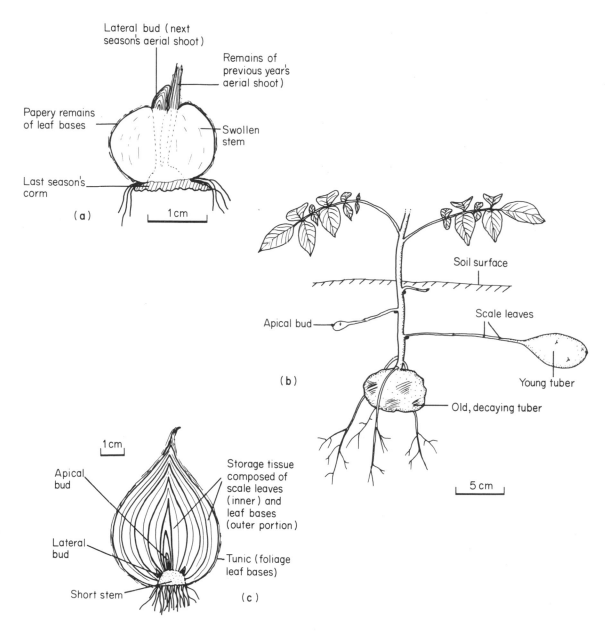

Figure 9.4. Stems modified for storage. (a) Longitudinal section of the corm of *Crocus* (winter condition). (b) Stem tuber formation in *Solanum tuberosum*. (c) Longitudinal section of the bulb of *Allium cepa*.

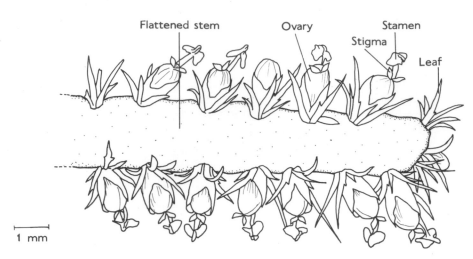

Figure 9.5. *Butumia marginalis* of the Podostemaceae. Flowers are produced at the margin of the flattened stem. (After G. Taylor. 1953. *Bulletin of the British Museum (Natural History), Botany* 1.)

Growth and Anatomy of Stems

The stem of the angiosperm grows from a group of meristematic cells lying near the summit of the apex. The apex itself (Fig. 9.6) is usually organized into two distinct zones recognizable by their geometry and the directions of the divisions of the cells. In the center is the *corpus,* where divisions are in various directions, thus adding to the width of the apex as well as to its length. The corpus is covered by the **tunica,** a layer of cells in which divisions are principally anticlinal, thus increasing the surface of the apex. The tunica is usually two cells thick, but in some plants the thickness may amount to only a single cell and in others to four or five. That these two zones are to some extent independent is shown by the existence of chimaeras in which the cells of the tunica acquire (as a consequence, for example, of treatment with colchicine) aneuploid or polyploid nuclei and retain them through all subsequent divisions, while the nuclei of the cells of the corpus remain diploid. The tunica yields the leaf primordia and ultimately the cortex of the mature stem, and the corpus the vascular tissue and associated parenchyma. Although this kind of apical organization is foreshadowed in the conifers (particularly in *Araucaria*), it is nowhere so distinct as in the angiosperms.

The angiosperms fall into two major groups, referred to as the monocotyledons and dicotyledons, in which the embryos are commonly furnished with one and two cotyledons respectively. These groups also differ in many other features, among them the form of the primary vascular tissue. In the dicotyledons, which are the more numerous, the primary vascular tissue usually consists of a ring of collateral bundles in which the differentiation of the xylem is centrifugal and of the phloem centripetal (Fig. 9.7). The xylem and phloem usually remain separated by a thin layer of undifferentiated cells, later recognizable as the intrafascicular cambium. In the monocotyledons, however, the bundles, although often collateral and mostly oriented with the xylem adaxial, are not in one ring but are irregularly scattered in a parenchymatous matrix, referred to as the ground parenchyma (Fig. 9.8). The bundles of the monocotyledons also lack potentially meristematic tissue between the xylem and the phloem and are consequently said to be closed.

Usually secondary vascular tissue soon appears in a dicotyledonous stem. The undifferentiated layer in the primary bundle becomes meristematic and continuous with a similar layer between the bundles (the interfascicular cambium), and subsequent activity

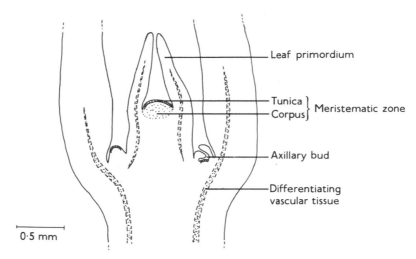

Figure 9.6. *Syringa.* **Longitudinal section of the stem apex.**

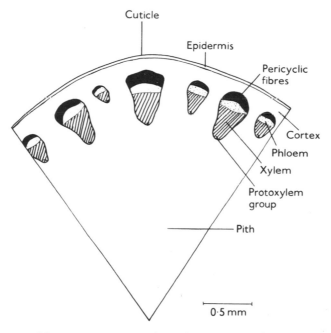

Figure 9.7. *Dahlia.* **Transverse section of a segment of a young stem.**

of the cambial ring resembles that already seen in the gymnosperms (Fig. 9.9). The first cambium does not always continue to function indefinitely; other cambia may arise outside the vascular cylinder, recalling the situation in *Cycas.* In a few arborescent monocotyledons a meristematic zone at the periphery of the stem gives rise to additional collateral vascular bundles as the girth of the stem increases. This activity is, however, limited. In most arborescent monocotyledons, such as the palms, the plant remains for several years as a widening bud bearing a rosette of leaves only a little above soil level. When the bud approaches its mature diameter elongation of the stem takes place very rapidly. For this reason, coconut palms, for example, are not commonly seen with their trunks half extended. The manner of growth of the monocotyledonous trees is thus quite different from that of the dicotyledonous.

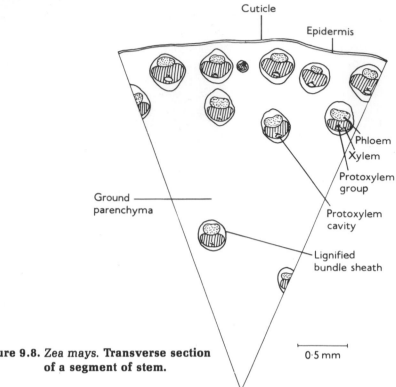

Figure 9.8. *Zea mays.* **Transverse section of a segment of stem.**

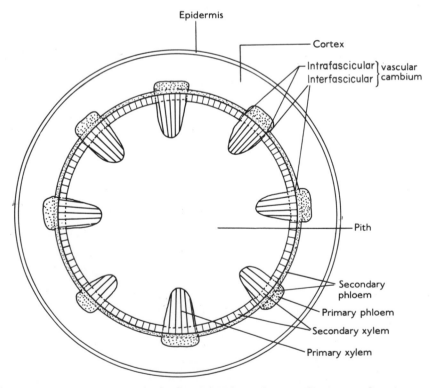

Figure 9.9. Diagram showing the method of production of secondary vascular tissue in a dicotyledonous stem.

The vascular tissue of almost all angiosperms is distinguished from that of all but a few living gymnosperms by the presence of vessels in the xylem and of companion cells in the phloem. Vessels are long tubes, commonly about 10 cm (4 in.) long, but in some plants, such as lianes, reaching or exceeding a length of 5 m (16.5 ft.). These tubes are composed of segments (Fig. 9.10), each of which is derived from a single cell and is equivalent to a single tracheid. During differentiation of a vessel, however, the end walls of the segments are resorbed, leaving either mere rims marking their position or highly perforated diaphragms, referred to as *end plates*. Vessels are often accompanied by tracheids. There are a few angiosperms in which the conducting elements of the xylem remain wholly tracheidal.

The sieve cells of the angiosperms are arranged in longitudinal rows, the whole column being referred to as a sieve tube (Fig. 9.11). The end walls of the sieve cells (or "sieve tube elements") are usually inclined and bear one or more sieve plates. The contents of the sieve cells remain organized, although the cytology is peculiar and the organelles, including the nucleus, are partly degenerate. Deposits of an amorphous polysaccharide, callose, often appear on the sieve plates at the end of the growing season. The companion cell or cells, closely applied to the sieve cell, may be essential for its function. The companion cell has a dense cytoplasm and a prominent nucleus; it arises from the same mother cell as the adjacent sieve cell. The sieve tubes of monocotyledonous trees, in the absence of significant cambial activity, necessarily remain alive and active for many years.

A cork cambium may arise in the outer cortex, particularly in dicotyledonous trees. Any tissue outside it dies and forms a skin covering the first-formed cork. Aeration of the cortex is continued by way of lenticels, passages filled with powdery cork left in the impervious layered tissue.

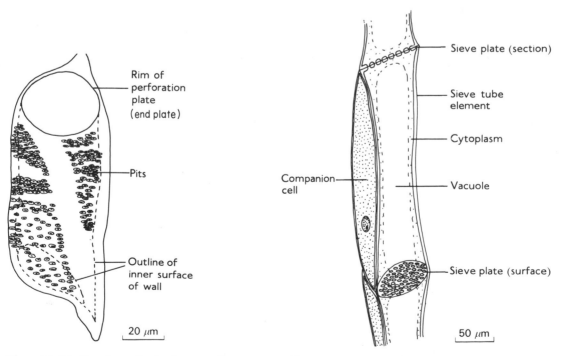

Figure 9.10. *Fraxinus*. **A single vessel segment from a macerate of the secondary wood.**

Figure 9.11. *Cucumis*. **Longitudinal section of a sieve tube and companion cell.**

Economic products of stems. The stems of angiosperms frequently contain abundant fibers, and the formation of periderm in the outer cortex or even closer to the vascular tissue is often extensive. Some trees generate cork very freely, that of *Quercus suber* (cork oak) in the Mediterranean region and Portugal being periodically harvested and finding many uses in commerce. A large part of the economic value of angiosperms, in fact, lies in the stems. Quite apart from the often very valuable timber, stems yield materials as diverse as starch (e.g., sago from the palm *Metroxylon* and arrowroot from the rhizome of South American *Maranta*), sugar (from sugar cane and in the form of syrup from the stem of *Acer saccharinum,* the sugar maple), fibers, spices (e.g., cinnamon, the bark of *Cinnamomum zeylanicum*), rubber (formed from the latex of *Hevea,* and less importantly from that of certain other species), oil (from the trunk of the Amazonian *Copaifera*), and drugs (e.g., quinine from the bark of *Cinchona*).

Leaves

The leaves of angiosperms show a wide range of size and shape, and all forms of phyllotaxy from distichy to decussate, whorled and spiral arrangements are found. The factors determining these different arrangements have not yet been identified. There are, however, theoretical grounds for believing that spiral arrangements (which, with upright stems, result in the minimum amount of shading of a leaf by others above it) are the consequence of a periodicity set up by interacting morphogens at the apex. In most mature leaves a petiole and lamina are usually distinguishable, and two small lateral outgrowths at the base of the petiole, termed stipules, are often present. In many species the leaves are pinnately branched, sometimes even as far as the third order, causing the whole to resemble a lateral branch system. This resemblance is occasionally enhanced (e.g., in *Thalictrum aquilegifolium,* Fig. 9.12) by the presence of small outgrowths, recalling

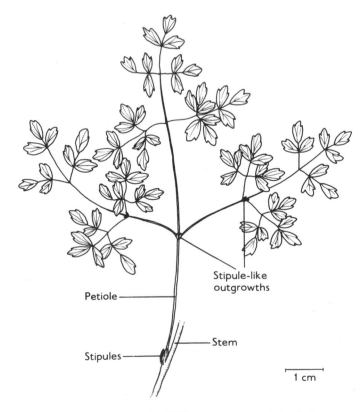

Figure 9.12. *Thalictrum aquilegifolium.* **Compound leaf structure.**

stipules, at the points of branching. At the other extreme, leaves may be little more than scales, photosynthesis being carried out principally if not entirely in the stem. In *Casuarina*, a tree of the tropics and subtropics of the Old World with leaves of this kind, the leaves are arranged in whorls so that the whole shoot comes to bear a striking external likeness to that of *Equisetum*. In a few plants (e.g., *Ruscus aculeatus*, butcher's broom) the reduction of the leaves to scarious scales has been accompanied by the transformation of lateral shoots of limited growth into flattened, leaflike structures called **phylloclades** or *cladodes* (Fig. 9.13).

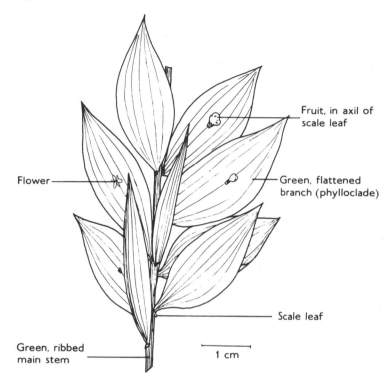

Figure 9.13. *Ruscus aculeatus.* **Habit of shoot.**

An intermediate condition is where photosynthesis is carried on in broadened and flattened petioles, the laminae being absent or rudimentary. These structures, called **phyllodes,** are found in many species of *Acacia* (wattle).

A classification has been developed for leaves, depending upon the area of the mature lamina, and clear relationships have emerged between climate and leaf size. In the tropical rain forest, for example, the leaves of many species tend to be large and of similar area, but, in the dry, scrubby vegetation of the Mediterranean region, the leaves by contrast show a much wider range of considerably smaller areas. Among other relationships, mostly of unknown significance, is the rarity of leaves with toothed margins in tropical vegetation and their abundance in temperate vegetation. Leathery leaves, with their apices extended into "drip tips," are also characteristic of vegetation in regions of high rainfall. Experiments have shown that such leaves do drain faster than comparable leaves lacking such tips. Species growing by streams often tend to have long narrow leaves (**stenophylls**). Species of *Salix* provide familiar examples in north temperate regions.

Although the leaves of most species are differentiated into a lamina and a petiole, sometimes the petiole is lacking. The leaves are then termed sessile. The grasses provide familiar examples of sessile leaves. Most leaves are oriented with the plane of the lamina more or less horizontal, and the insertion into the stem transverse. There are, however, some plants (all with distichous phyllotaxy) in which the plane of the leaf is vertical and the base of the leaf clasps the stem as a rider a horse. This so-called equitant arrangement is well seen in *Iris*. In some species motor cells, usually aggregated into a distinct pulvinus, are able to alter the orientation of the leaf in a striking fashion. In the tropical rain tree (*Pithecolobium saman*), for example, the leaves appear to collapse with the diminishing light of the afternoon (a so-called "sleep movement"), and in *Mimosa pudica* (sensitive plant) similar movements take place if the leaves are mechanically disturbed. The leaves of the "compass plants," of which *Lactuca scariola* (a wild lettuce) is a notable example, move in relation to the sun so that only one edge is fully insolated at any one time. The motive effects of pulvini depend upon changes in turgor and the ability of the cytoplasm in certain conditions to move water actively in and out of vacuoles.

In many monocotyledons the leaf bases become swollen and form a subterranean bulb which overwinters. The cells of the bulb are filled with food reserves and are often also rich in secondary metabolites. The lachrymatory thiopropanol S-oxide of the onion bulb is particularly well known.

Anatomy and development of leaves. Although the leaves of angiosperms are structurally similar to those of gymnospersm, there are a number of new features. The palisade tissue, for example, is frequently sharply differentiated from the rest of the mesophyll and in some plants, especially in those that are shade tolerant (e.g., *Impatiens parviflora*, balsam) the form of the palisade is markedly influenced by the irradiance (Fig. 9.14). The change in cell shape in the shade exposes a greater proportion of the chloroplasts to the incident light with the result that over a wide range there is little variation in net assimilation. Equitant leaves are usually bifacial, stomata and palisade being symmetrically placed on each side. Among other structures found in leaves are oil glands, often giving the leaf a fragrance when crushed, and, usually at the margins or on the

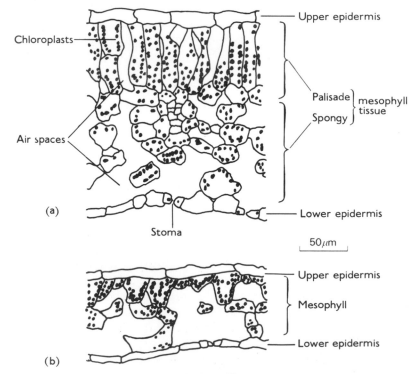

Figure 9.14. *Impatiens parviflora.* **Transverse section of leaf. (a) From plant growing in full sunlight. (b) From plant growing in 7% sunlight. (Both after A. P. Hughes. 1959.** *Journal of the Linnean Society, Botany 56.***)**

petiole, nectaries (as, for example, in cherry). In some species of tropical plants stipular nectaries are associated with symbiotic bacteria ("leaf nodules"). The venation of the lamina is commonly reticulate, patterns of extreme intricacy often being generated by the minor veins, many of which end blindly in the areolae (Fig. 9.15). The symmetry of the vascular supply ascending the petiole is usually clearly dorsiventral, but it becomes almost perfectly radial in the petioles of peltate leaves (e.g., *Tropaeolum majus*, the garden nasturtium).

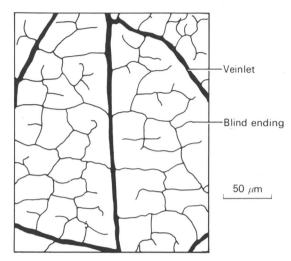

Figure 9.15. *Impatiens parviflora*. **The pattern of leaf venation. (After A. P. Hughes. 1959.** *Journal of the Linnean Society, Botany* **56.)**

The leaf primordium grows from initial cells at its margin, and, at first, cell division predominates over expansion. The pattern of the main veins soon, however, becomes established, and also in dicotyledons that of any branching of the leaf. This is brought about by the marginal meristem becoming discontinuous and its activity confined to definite areas of the periphery. Segmentation in the leaves of monocotyledons (e.g., palms) is a more complicated process and involves the degeneration of tracts of tissue between areas which will subsequently become pinnae.

As growth of a leaf primordium proceeds, cell expansion comes to predominate over division, one of the last products of cell division being the guard cells of the stomata. It is evident that the surface growth of the lamina is closely coordinated with the extent of the vascular framework. In some plants it is possible to disturb this coordination by allowing the primordium to form in one day-length, but to expand in another, leading to deformed leaves. These experiments reveal that the factors controlling the growth of the vascular skeleton are different from those influencing the expansion of the lamina. Standing apart from this general scheme of development are again the leaves of some monocotyledons. The leaf of a grass, for example, grows from a meristem at the base of the lamina, above which there is continuous basipetal differentiation until the leaf reaches its mature length.

In some species leaves are deciduous, being severed from the axis by a distinct abscission layer. Partial digestion and disarticulation of cell walls at this site are brought about by hydrolases, including cellulase. In temperate regions abscission usually occurs during the shortening days of autumn, but some tropical trees also regularly shed their leaves, the periodicity having no evident relation to season.

Leaves of angiosperms show a wider range of surface coverings than those of any other land plants. These may take the form of plates or crystals of wax (as on the upper surface of the leaf of *Zea mays*) or of hairs or scales. Markedly pubescent leaves are often found in desert plants, and such investments may reduce the absorbance of photosynthetically active light by as much as 60%. There is evidence that this is an adaptive response, the degree of hairiness being such that net assimilation is matched to the water available.

The vascular bundles in leaves are frequently surrounded by a chlorophyllous sheath. In some plants this sheath consists of a single layer of large cells containing conspicuous chloroplasts. These can be seen in the light microscope to lack grana, but starch grains are often prominent. This so-called "Kranz" anatomy is associated with a particular kind of photosynthesis in which carbon dioxide is taken first into phosphoenolpyruvate instead of directly into the ribulose bisphosphate of the Calvin cycle. These two kinds of assimilation are termed C_4 and C_3 respectively. C_4 assimilation has less need of water than C_3 and it is able to continue at lower concentrations of carbon dioxide, a feature of significance in conditions of light saturation. It is therefore not surprising that C_4 assimilation is frequently present in plants of dry and sunny habitats, such as deserts, and in plants of saline soils (halophytes).

Juvenile and mature forms of leaves. As in the conifers, the leaves of young plants sometimes differ from those of the mature in both form and arrangements. *Eucalyptus globulus* (Fig. 9.16) provides a striking example of this phenomenon. The juvenile leaves are ovate, decussately inserted, and bifacial in structure, whereas the mature are falcate, spirally inserted, and possess normal dorsiventral structure. A young tree has a cone of juvenile foliage within a crown of mature foliage, showing that the change takes place more or less simultaneously at all the apices when a tree reaches a certain maturity. Sometimes, as in *Hedera helix* (ivy), mature foliage does not appear until the approach of reproduction. Once the production of mature foliage has begun, the system is remarkably stable. Reproductive branches of *Hedera,* for example, can be struck as cuttings, and these yield small bushes quite different in appearance and growth from the scandent vegetative plant. It is very difficult, even by experimental means, to cause the apices of mature plants to revert to juvenile growth. It appears likely that maturity is a consequence of complex and coordinated changes in the ribonucleic acids and proteins in the meristematic cells, and that the original system is only readily re-created during sexual reproduction.

Economic and ecological importance of leaves. Angiospermous leaves are a rich source of foodstuffs and raw materials. Fodder crops often consist largely of leaves, and the leaves of many species are prized as potherbs because of the aromatic oils they contain. The leaves of a wide range of species find medicinal uses, and they are the commer-

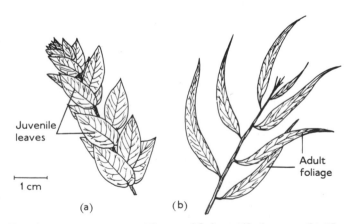

Juvenile
leaves

Adult
foliage

1 cm

(a)   (b)

Figure 9.16. *Eucalyptus globulus.* **(a) Shoot with juvenile leaves. (b) Shoot with adult leaves. (Both after Niedenzu in Engler and Prantl. 1898.** *Die natürlichen Pflanzenfamilien,* 3:7. **Engelmann, Leipzig.)**

cial source of a number of important drugs, among them atropine and hyoscyamine (from *Atropa* and other members of the Solanaceae) and cocaine (from *Erythroxylon*). The hallucinogens of *Cannabis* accumulate within the resinous secretions of the leaves, particularly the bracts of the female inflorescence. The amount produced depends upon temperature and in cool temperate regions is trifling. The leaves of many monocotyledonous species yield valuable fibers, and in tropical regions palm leaves provide a ready and efficient material for thatching. Apart from such direct utilization, angiospermous leaves play a large part in maintaining the fertility of the soil. Since they contain a higher proportion of nitrogenous substances and fewer antiseptic materials, such as tannins and phlobaphene, the leaves of angiosperms decay more rapidly than those of gymnosperms. Angiospermous litter is thus quickly reduced to humus, the organic matter that is the basis of soil fertility. The productivity of much natural vegetation in tropical regions depends upon the steady deposition of litter. Thoughtless removal of this natural cover and destruction of the ecosystem can result in areas of persistent and intractable barrenness.

Roots

The roots of the angiosperms are, in general, organized similarly to those of the gymnosperms. Those of some species are able to develop chlorophyll when illuminated. In *Taeniophyllum* (Fig. 9.17), a peculiar Malaysian epiphytic orchid, photosynthesis is, in fact, confined to bandlike aerial roots, the stem remaining little more than a bud until the production of the inflorescence. A number of angiosperms, especially those of temperate regions that are biennial in habit, develop a swollen taproot which overwinters. These are often used as vegetables, carrots and parsnips being familiar examples. The swollen part of the radish is the transitional region between root and stem.

Remarkable development of roots is characteristic of a number of tropical plants. In some dicotyledonous trees the secondary thickening of the principal roots at the base of the trunk is markedly asymmetrical, leading to the formation of substantial buttresses 1 m (39 in.) or more in height. Some trees develop "prop roots." These leave the trunk at 1 m (39 in.) or more above ground level and descend obliquely towards the substratum. They soon become secondarily thickened. At maturity, the tree is supported by a cone of such outgrowths around its base. Some species of fig (*Ficus*) germinate in humus lodged in the branches of other trees. They then send down branching roots which closely invest and ultimately kill the supporting species ("stranglers"). Roots are not always negatively gravitropic. A number of trees of the Amazonian rain forest, for example, produce auxiliary roots which ascend the trunks of neighboring trees. These probably absorb minerals from the water trickling down from the canopy and augment the poor supply available from the soil. Swamp plants, notably the mangroves (which colonize brackish estuaries in the tropics), often produce negatively gravitropic aerophores from the root system, a development foreshadowed in the conifer. *Taxodium*. The roots produced by the aerial stems of climbers (e.g., *Hedera*) are often of limited growth and modified in a manner which assists adhesion.

Associations between roots and microorganisms are not infrequent. In forest trees the upper rootlets, usually found proliferating in litter, often possess mycorrhizal fungi. The roots of orchids are also mycorrhizal, but these plants, which are all herbaceous, can be grown in pure culture in the absence of the fungus, provided suitable nitrogenous substances are supplied in the medium. In the Leguminosae (the family containing the peas, beans, and several important fodder plants) there is a regular association, involving anatomical modifications, between the roots and certain soil bacteria capable of fixing atmospheric nitrogen. Similar associations with other nitrogen-fixing microorganisms have been confirmed in *Alnus* (alder), *Hippophaë* (sea buckthorn), and *Myrica* (myrtle), and probably exist in a number of other plants.

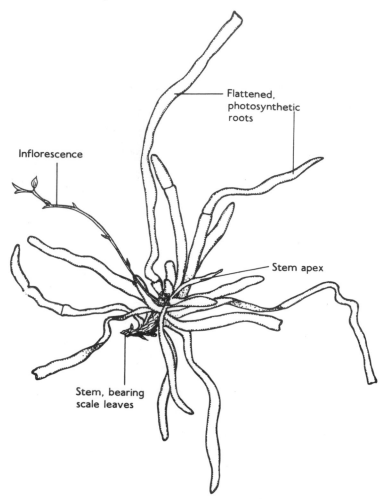

Figure 9.17. *Taeniophyllum* **sp. Habit showing flattened, photosynthetic roots. (After Goebel. 1933.** *Organographie der Pflanzen,* **Fischer, Jena.)**

The manner of growth of roots of angiosperms has been studied in some detail, since it is not complicated by the presence of leaf primordia. Apart from the root cap, which is maintained by a meristem adjacent to the surface of the apex proper, meristematic activity lies principally in the summit of the apex (Fig. 9.18). There is, however, convincing evidence, supported by experiments with radioactive isotopes, that a group of cells at the center of the apex experiences few, if any, divisions. These cells, forming a so-called "quiescent center" can be stimulated into division by wounding and radiation damage, so they are not deficient in essentials. Their normal inactivity must therefore depend upon the physiological organization of the intact apex, but the precise factors involved are still unknown. Behind the meristematic area, the cells elongate, and measurements have shown that there is considerable synthesis of ribonucleic acid and protein in the cells concerned at this time. In some cells, often those which produce root hairs, the nuclei undergo a curious form of mitosis without associated division (endomitosis), leading to a polyploid condition.

A transverse section of a young root of a dicotyledon shows a central core of xylem with radiating arms of protoxylem, between which lies the phloem. In dicotyledons the number of protoxylem groups is usually small (Fig. 9.19), but in the monocotyledons the number is much greater, usually in the region of 15 or 20. An endodermis, separated from

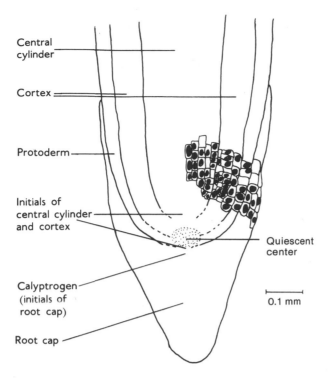

Central
cylinder

Cortex

Protoderm

Initials of
central cylinder
and cortex

Quiescent
center

Calyptrogen
(initials of
root cap)

Root cap

0.1 mm

Figure 9.18. *Trillium* **sp. Longitudinal section of root apex.**

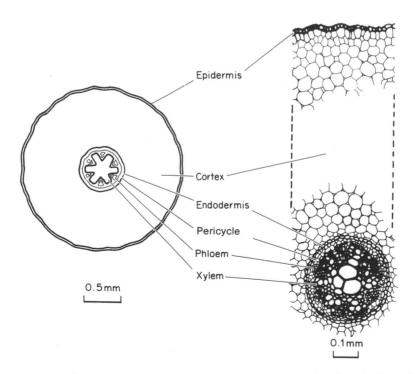

Epidermis

Cortex

Endodermis

Pericycle

Phloem

Xylem

0.5 mm

0.1mm

Figure 9.19. *Ranunculus acris.* **Transverse section of young root, showing the broad cortex and narrow stele. The xylem is exarch. Differentiation proceeds inwards yielding a central core of metaxylem. The phloem lies between the arms of xylem.**

the vascular tissue by a zone of parenchyma (pericycle), is usually present. The root is the only region in the plant body of an angiosperm where an endodermis occurs with any regularity. The presence of the lipoidal Casparian strip in the radial walls may ensure the ordered uptake of mineral nutrients by subjecting the lateral movement of the dissolved salts to the control of the symplast.

The meristems of branch roots arise in the pericycle, an origin termed **endogenous,** and very different from the more superficial (**exogenous**) origin of branches of the stem. Where cork is produced, the cork cambium arises in the pericycle, thus causing the whole of the primary tissue external to it ultimately to die and slough off. In herbaceous plants the parenchymatous tissues of the roots sometimes become locally distended with food materials, occasionally forming distinct tubers (e.g., *Dahlia,* Fig. 9.20). These serve as organs of perennation.

Roots, especially those that are tuberous, are frequent sources of drugs and folk medicines. Aconitine, obtained from the roots of *Aconitum napellus,* frequent in alpine meadows and seen occasionally by shaded streams in Britain, is a well-known example. Derris, a powder made by grinding the roots of *Derris elliptica,* a climbing shrub of Asia, is a powerful insecticide harmless to people.

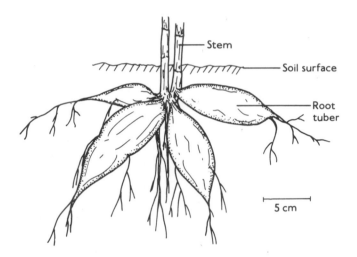

Figure 9.20. *Dahlia* **sp. The production of root tubers.**

Correlation Within the Plant Body

The angiosperms, more than any other division of the Plant Kingdom, have been investigated with the object of discovering the factors responsible for coordinating the growth of the plant as a whole. In the presence of an apical bud, for example, lateral buds commonly remain inactive, an instance of **correlative inhibition.** Correlations of this kind are clearly of great importance in determining the morphology of the mature plant.

Correlation is an example of cell interaction, but the interacting cells are often separated by others not directly concerned. These interactions at a distance are brought about by growth-regulating substances which move through the plant in a manner not yet wholly understood. Research has now revealed a whole series of such substances, which for present purposes can be classified as auxins, kinins, abscisins, gibberellins, and ethylene. Auxins (of which indole-3-acetic acid, IAA, is the most familiar) are those substances which will cause curvature if placed asymmetrically on the tip of a decapitated oat coleoptile. They are usually produced by meristematic cells and are involved (among other effects) in maintaining apical dominance and initiating differentiation. In culture media indole-3-acetic acid (and its nonmetabolizable analogue naphthalene acetic acid) are particularly effective in stimulating the production of roots, a property made use of in horticulture in the rooting of cuttings. Kinins (cytokinins, phytokinins) are substances which promote cell division. They are found in fruits and seeds, but may also be generated in damaged cells and cause proliferation around wounds and their subsequent repair. Abscisins, probably produced in leaves, are effective in causing leaf abscission, and dormancy in apical buds, common phenomena in trees of temperate regions. Gibberellins were first extracted from the fungus *Gibberella* which, when infecting rice, causes the plant to be excessively tall. They are now known to exist in higher plants, and early experiments showed that when administered to dwarf mutants, as, for example, of maize (corn), these would grown to their normal stature. Additionally, they are probably involved in photoperiodic responses and the change from vegetative to reproductive growth. Ethylene has long been known to be produced by ripening fruits, and conversely the process of ripening in stored immature fruits can be controlled by regulation of the amount of ethylene in the atmosphere. Ethylene also produces many other effects, probably as a result of interacting with IAA within the plant.

Although detailed investigation of the growth-regulating substances (which include others less well known in addition to those mentioned here) is the province of physiologists, they are also basic to the causal study of morphology since regulatory substances play an important role in directing the growth of a species in its characteristic and immediately recognizable way. It was inevitable that the flowering plants, because of their ubiquity, familiarity, and economic importance, should figure prominently in research into growth-regulating substances, but the study of simpler systems in algae and the archegoniate plants might more readily reveal how they enter into the molecular biology of the cell and influence morphogenesis.

Reproduction

In the angiosperms the onset of the reproductive phase is frequently (but not always) dependent on the length of day ("photoperiodism"), but short-day and long-day plants being clearly recognizable. The reproductive axes are ordinarily short and of limited growth. The axis itself and its attendant structures, which are often conspicuously colored or shaped, is termed a flower. Flowers may be borne either singly (as in *Anemone nemorosa*) or severally, the branching system bearing the flowers then being called an **inflorescence.** Inflorescences may have complex morphology. Sometimes the inflorescence is contracted and superficially resembles a single flower (as in many Compositae). Inflorescences may also take the form of loose strobili (as the catkins of *Populus*) or (especially in fruit) a coniferous cone (as in *Alnus* of the Northern Hemisphere and *Banksia* of the Southern). The parts of a plant giving rise to flowers or inflorescences are not normally very different from those that are vegetative, but distinct functional separation occurs in some tropical trees. In *Couroupita guianensis* (Cannon ball tree), for example, the uppermost branches are densely leafy and solely vegetative. The reproductive branches, which are almost leafless and bear numerous flowers, arise directly from the trunk and hang down from the crown. In some other tropical trees, flowers are produced only from special regions towards the lower part of the trunk (as in the jack fruit, *Artocarpus*, of tropical gardens), a phenomenon known as **cauliflory.**

In annuals and certain longer-lived plants of warmer regions the production of flowers heralds the end of the life span. Such plants are said to be **monocarpic** or **hapaxanthic.** A spectacular example is the palm *Corypha* of tropical Asia. Growth terminates with the production of an inflorescence reaching a height of 14 m (46 ft.) and a breadth of 12 m (40 ft.), containing more than 100,000 flowers.

Reproduction in angiosperms may be either monecious or dioecious. Monoecious species are further divisible into the monoclinous, where the male and female organs are together in the same flower, and the diclinous, where they are in separate flowers (as in the cucumber, *Cucumis sativus*). Functional separation of the sexes in bisexual flowers is often achieved by the organs of the two sexes maturing at different times, a phenomenon termed **protandry** if the male precede the female, and **protogyny** if the female precede the male. Some orchids show an extreme form of protandry in which the ovules are not formed unless the flower is pollinated. In some species self-pollination is discouraged or prevented by the flowers being in two (as in *Primula*) or even three (as in *Lythrum*) forms.

Male reproductive structures. The male sporangiophore of the angiosperms is termed a stamen (Fig. 9.21). Typically this consists of a stalk (**filament**) terminating in four pollen sacs, the sacs being in two pairs, the pairs lying side by side and joined by the **connective.** The whole of this region is called the anther, and seen in transverse section (Fig. 9.22) resembles the male synangia of some pteridosperms and of *Caytonia*. Apart from congenital fusion of stamens, discussed later, there is some variation in the form of the individual sporangiophore. In *Degeneria*, for example, and in some members of the Magnoliaceae, the "filament" is, in fact, a broadly ovate scale, about 5 mm (0.19 in.) long and 2 mm (0.08 in.) wide, with four pollen sacs partially embedded on the abaxial surface (Fig. 9.23). At the other extreme, the filament may be lacking, one or two anthers being attached directly to a modified floral axis, as in the complex flower of the orchids. Besides simple dehiscence along a longitudinal stomium, other methods of opening, such as the development of pores or the differentiation of distinct valves, are encountered in some species.

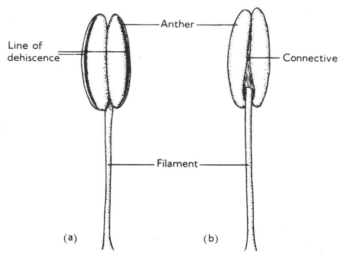

Figure 9.21. Diagram showing the features of a typical angiosperm stamen. (a) Viewed from the front. (b) Viewed from the back. In this anther the pollen sacs face the floral axis (introrse), but the reverse position is also found (extrorse).

In contrast to microsporogenesis in *Pinus*, the development of angiospermous anthers is commonly synchronous. This is probably a consequence of broad cytoplasmic connections between the young spore mother cells, a phenomenon (**cytomixis**) not seen in *Pinus*. Subsequently the pollen mother cells become surrounded by thickened callosic walls and all interconnections are obliterated. After meiosis, the young microspores, while still in the tetrad, begin to be coated with sporopollenin. Concurrently, cellular organization breaks down in the tapetum, metabolites are mobilized by hydrolase activity, and additional sporopollenin is synthesized. The young pollen grains are released into this medium and complete their development at its expense (Fig. 9.24). Much of the tapetal sporopollenin is added to that already coating the grain, probably by selective

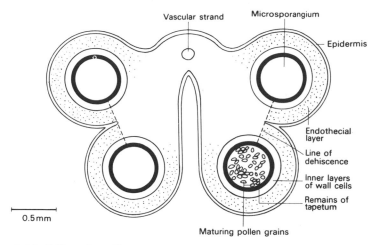

Figure 9.22. *Lilium longiflorum.* **Transverse section of almost mature anther.**

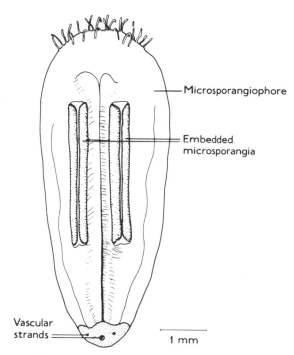

Figure 9.23. *Degeneria vitiensis.* **Adaxial surface of stamen. (After J. E. Canright. 1952.** *American Journal of Botany* **39.)**

accretion. The pattern of deposition and location of the pores (colpi) are established before the grain is liberated from the tetrad, but the manner of the inheritance of this pattern is still unknown (cf. *Sphaerocarpos*). The structure and cytochemistry of the wall of a mature pollen grain are complex. The chambered exine often takes up enzymes and other proteins from the tapetum while the inner cellulosic layer (intine) may contain secretions from the grain itself. Although the tapetal products are commonly regarded as sporophytic, and those of the grain gametophytic, the precise nature of the gene activation in the tapetum is not yet known. In some species part of the tapetal sporopollenin coats an additional membrane delimiting the sac containing the developing microspores (*peritapetal membrane*). The formation of this barrier may change the nature of the gene expression within the sac.

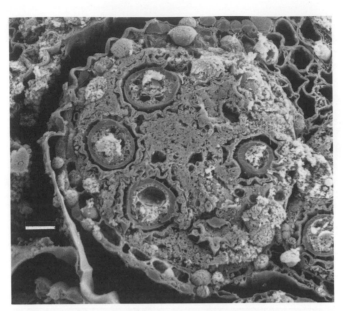

Figure 9.24. Maturing microspores of *Catananche caerulea* (Compositae) lying in the tapetum. Anther transversely freeze-fractured. (Unpublished micrograph provided by S. H. Barnes and S. Blackmore.) Scale bar 1 μm.

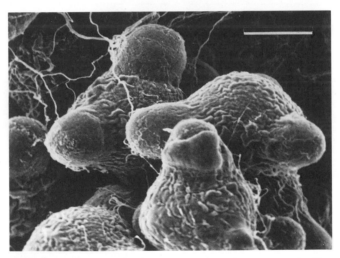

Figure 9.25. *Oenothera organensis.* Pollen grains (each with three projecting colpi), enmeshed in threads of viscin (a polymer produced by the tapetum), lying on the stigma. (Photomicrograph by H. Dickinson.) Scale bar 5 μm.

The patterns of pollen walls, which are established while the grains are still in the tetrad, are often very distinctive (Fig. 9.25). Since sporopollenin, a highly polymerized mixture of esters of complex fatty acids, is extremely resistant to decay, pollen grains partially fossilized in peats and lake muds are often fully identifiable. Most pollen grains are spherical or ellipsoidal and land plants show only few examples of rodlike grains. Marine angiosperms ("sea grasses"), however, possess "confervoid" pollen. These grains lack exine and may reach or exceed 5 mm (0.19 in.) in length. In some species (as in the Ericaceae, or heaths, and the decorative plant *Salpiglossis variabilis*), the pollen remains stuck together in tetrads, a feature facilitating, as in *Sphaerocarpos,* tetrad analysis. Even larger aggregates of pollen (known as "polyads") occur naturally in some genera. Those of *Parkia*, for example, contain 16–32 grains and are not separated by acetolysis (which destroys all structural materials other than sporopollenin). In *Petunia*, where the grains are normally separate, adhesion within the tetrads can be brought about by a recessive gene.

Development of the male gametophyte normally begins while the pollen is still in the anther. Usually two nuclei are present, of which one (vegetative nucleus) becomes large and diffusely staining and the other (the generative nucleus) dense and often transversely elongated (Fig. 9.32a). These nuclei lie on a radius of the original tetrad, and there is evidence that their conspicuously different behavior depends upon gradients set up in the cytoplasm of the cleaving pollen mother cell and persisting in that of the grain. In about one-third of all families (both monocotyledons, including the grasses, and dicotyledons) the generative nucleus divides while still in the grain yielding trinucleate pollen.

Female reproductive structures. The megasporangiophore is the distinguishing feature of the angiosperms for it is normally a closed body (termed a *carpel*), furnished with a distinct stigma (often elevated on a style) on which the pollen germinates (Fig. 9.26). In only a few genera (e.g., *Reseda*) are the carpels open at maturity. A carpel is dorsiventral in symmetry and the fertile region is adaxial. One or several ovules may be present, and if the latter they are commonly borne in two series along the so-called ventral suture of the carpel (Fig. 9.26).

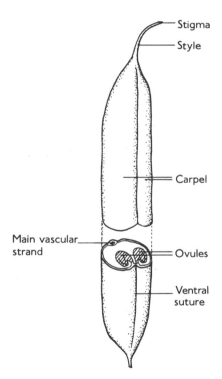

Figure 9.26. The simple carpel characteristic of the family Leguminosae. This arrangement of the ovules is referred to as "marginal placentation".

The ovules are much smaller than those of most gymnosperms, but are structurally similar. They may be upright, inverted, or occasionally more or less horizontal, these three orientations being termed *orthotropous, anatropous,* and *campylotropous* respectively (Fig. 9.27). At least one integument is present, but in many families there are two, and in some as many as four. The last formed integument, irrespective of the total number, sometimes becomes transformed into a fleshy aril in the fruit (as in the durian, *Durio*). The integuments normally develop uniformly, indicating the basic radial symmetry of the angiospermous ovule. A slender vascular trace enters the stalk of the ovule, but in only a few families does this extend into an integumentary vascular system.

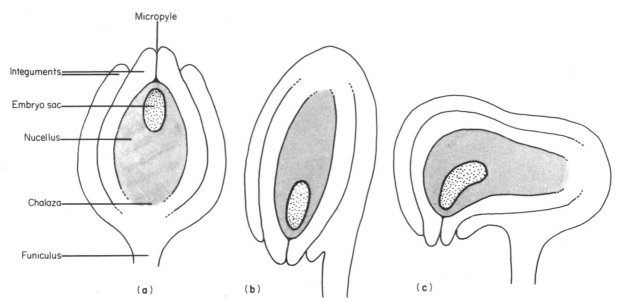

Figure 9.27. Forms of ovule orientation. Diagrammatic. (a) Orthotropous. (b) Anatropous. (c) Campylotropous.

The female gametophyte. The development of the female gametophyte, in the angiosperms termed the embryo sac, begins in a familiar fashion (Fig. 9.28a). A cell in the upper part of the nucellus, immediately below the layer of the cells at its surface, becomes conspicuously large and surrounded by a callosed wall. Meiosis then leads to a tetrad of megaspores (Figs. 9.28b and 9.29a). Although the term "megaspore," because of the evident homology of this cell with the megaspore of the gymnosperms, is retained, the megaspore in the angiosperms is frequently smaller than the microspore. In many instances only the inner (chalazal) megaspore undergoes further development, and the female gametophyte is consequently said to be *monosporic* in origin. The nucleus of the megaspore enters a succession of mitoses, and, in step with the expansion of the ovule as a whole, the multinucleate embryo sac is formed (Figs 9.28e, f, and 9.29b). In a frequent type of embryo sac there are eight nuclei, produced by three successive mitoses, which associate themselves with cytoplasm and arrange themselves in a definite pattern (Fig. 9.28e). Eight cells are usually identifiable at this stage, each bounded by a delicate membrane. At the micropylar end of the sac is a group of three cells, of which one (frequently with a weakly staining nucleus) is the egg cell. The cells accompanying the egg are termed *synergid* cells. At their micropylar ends these cells frequently have conspicuously labyrinthine walls. Together they form the so-called filiform apparatus.

At the other end of the sac (the chalazal end) is another group of three cells, referred to as the *antipodals*. The nuclei of the remaining two cells (termed polar nuclei) come together at the center of the sac and eventually fuse, thus giving rise to a central diploid fusion nucleus (Figs 9.28f, 9.29b). During the development from the megaspore the boundary between the gametophyte and sporophyte remains distinct, and the mature embryo sac is commonly also surrounded by a callosed wall. This is known to be impermeable to large molecules such as nucleotides.

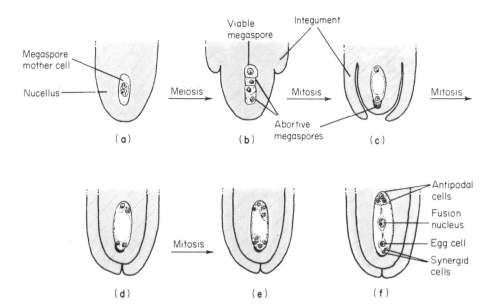

Figure 9.28. The sequence of divisions leading to the production of a monosporic, eight-nucleate embryo sac. From flowers of *Myosurus minimus* (mouse tail). Diagrammatic; not to scale.

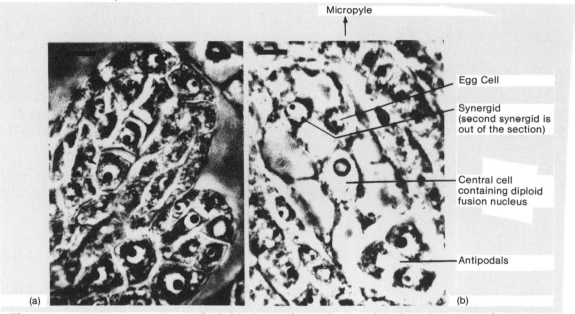

Figure 9.29. *Myosurus minimus*. The initiation and completion of the female gametophyte. (a) The tetrad of spores resulting from meiosis. The innermost (indicated by arrow) is the viable megaspore. (b) The mature embryo sac. Scale bar 10 μm in each instance.

The foregoing is only one of the several kinds of development of the female gametophyte found in the angiosperms. In some families, notably the Onagraceae, the embryo sac develops from the outer and not from the inner megaspore, a feature of unknown cytological and physiological significance. Sometimes, as in *Allium* (onion), the equational division of meiosis is omitted, a dyad of megaspores being produced instead of a tetrad. The embryo sac, which develops directly from the inner spore, is then said to be **bisporic** in origin. Finally, in another kind of development (seen, for example, in

Lilium), termed **tetrasporic,** the nucleus of the megaspore mother cell divides meiotically, but the cell itself does not divide at all. Instead, mitotic divisions follow and the cell expands without interruption to form the sac. Bisporic and tetrasporic embryo sacs occur in a wide range of families (and have already been noted in *Gnetum*), and there is no justification for regarding this kind of origin as abnormal. Another kind of variation lies in the number of nuclei in the mature sac. Although this is commonly eight, irrespective of the mode of origin of the sac, others are known containing four or sixteen. The cytology of development is also variable. Although, where sexual reproduction is normal, the egg nucleus remains haploid, the chalazal polar nucleus may contain more than one set of chromosomes so that the central fusion nucleus becomes triploid or even, as in *Fritillaria*, tetraploid.

Architecture of the flower. The arrangement of the reproductive organs in the flower is often of considerable complexity, and floral morphology is a specialized field with its own terminology. Here we shall use only as much of this terminology as is necessary for general discussion; for further details the reader must consult specialist works. To facilitate illustration of the principles involved in the structure of the flower, we shall first consider a bisexual flower in which the components remain separate (Fig. 9.30), a situation seen, for example, in *Ranunculus*. The flower terminates an axis, and the transition from the vegetative to the reproductive region is marked by the occurrence of one or more whorls of sterile structures resembling rudimentary leaves. These form collectively the **perianth.** Where two whorls are present the segments of the outer whorl are frequently green (and termed **sepals**), whereas those of the inner are brightly colored (and termed **petals**). These two parts of the perianth are termed **calyx** and **corolla** respectively. Following closely upon the perianth are the stamens, forming collectively the **androecium.** The stamens may be either indefinite in number and spirally arranged (as in *Ranunculus*), or definite and arranged in whorls, the positions of individual stamens often bearing a clear relation to each other and to those of the preceding perianth segments.

The termination of the reproductive axis, often termed the **receptacle,** bears the carpels. This female region is called the **gynaecium** or ovary. The carpels may again be spirally arranged, as in *Ranunculus,* or in a single whorl. In the latter event the carpels usually alternate in position with the stamens of the uppermost whorl.

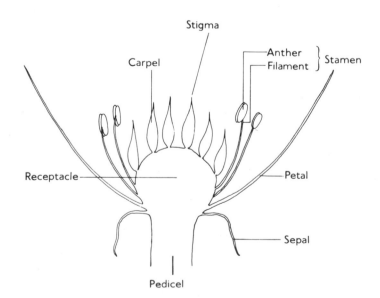

Figure 9.30. Longitudinal section of a flower with separate components. Diagrammatic.

In unisexual flowers the organs of one sex are absent or nonfunctional, although the bisexual condition can sometimes be brought about by treating the very young flower with growth substances such as gibberellic acid. Unisexual flowers may approach the limits of reduction, a feature well seen in catkin-bearing trees. In *Populus* (poplar), for example, the male flower consists of nothing more than a disk bearing up to 20 stamens, and in *Salix* (willow), where the male flower is similar, the number of stamens is commonly no more than two. Congenital fusion of parts of similar nature is widespread in flowers. Both the calyx and corolla independently may become tubular, the composite nature of the tube being indicated by lobes or teeth at its mouth. Filaments of whorled stamens may also fuse for most or only part of their lengths, leading to a cylindrical androecium. Fusion of carpels leads to a syncarpous ovary, in which the individual carpels are often represented by compartments (or loculi), as in the Liliaceae. Externally, the composite nature of the ovary is often indicated by lobing, or by the style, which rises from the center of the ovary, breaking up into branches equivalent in number to the constituent carpels. The ovary may sometimes be sunken in the receptacle (as in *Rosa*) or otherwise surmounted by the perianth and androecium, leading to clearly **epigynous** flowers. Sterilization of one or more stamens may occur, giving rise to staminodes which may either persist as rudimentary structures (as in *Scrophularia*) or occasionally become enlarged and brightly colored, and play a conspicuous part in the organization of the flower (as in *Canna* and *Iris*). Both sterilization and congenital fusion of parts are involved in the complex flowers of the orchids. Nectaries may be associated with almost any part of the flower depending upon the species.

Ontogeny of the flower. Although, in the gymnosperms, the apices giving rise to the reproductive and vegetative axes differ little in organization, in angiosperms they are usually easily distinguishable. In becoming reproductive an apex flattens considerably and may even become concave, regions previously quiescent now showing numerous mitoses. This has given rise to the concept of the vegetative apex containing a *meristème d'attente* active only in the reproductive phase. The perianth whorls and sporangiophores are often produced in acropetal sequence, the members of successive whorls alternating in position, but this is not always so. The direction of initiation may even become locally reversed, and primordia appear between the whorls of those already laid down. This phenomenon affects particularly the androecium, with conspicuous results in the symmetry of the mature flower. In some Caryophyllaceae, for example, a whorl of five stamen primordia appears, each stamen alternating in position with the preceding petals. A second whorl of stamens is then initiated basipetally, the stamens alternating with those of the first whorl, and hence standing opposite the petals. In this way an *obdiplostemonous* flower is formed.

Pollination and growth of the male gametophyte. Pollination in angiosperms is occasionally dependent upon aerial dispersal of the grains (as in catkin-bearing trees), but frequently involves the participation of insects as carriers. Various mechanisms have been evolved which ensure that insects visiting flowers in search of nectar become dusted with pollen, which then is transported to stigmatic surfaces in other flowers. A familiar example of such a pollination mechanism is provided by *Salvia pratensis* (sage), a species pollinated by the bumblebee. The stamens have short filaments, but a greatly elongated connective (Fig. 9.31). The longer upper portion of the connective terminates in an anther, but the shorter lower part in a plate blocking the approach to the nectary. A bee attempting to reach the nectar displaces the plate. Since the connective is hinged about the filament, the anther is in this way forced on to the insect's back, coating it with pollen (Fig. 9.31). A bizarre form of pollination is found in some orchids (e.g., *Ophrys speculum* of the western Mediterranean) where the pattern and conformation of the lip recall the female of the insect concerned, and even an odor may be released resembling the pheromone of the species. The male insect, in attempting to copulate with the lip of the flower, detaches the

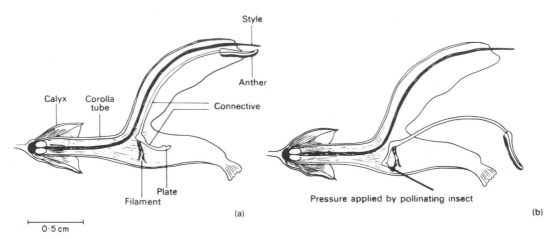

Figure 9.31. *Salvia pratensis.* **(a) Half flower showing the normal position of the stamen. (b) The same after the plate has been depressed. (After Kerner von Marilaun and Oliver. 1902.** *The Natural History of Plants.* **Blackie, London.)**

pollen (which coheres in masses called pollinia) and carries it to another flower. A few tropical flowers are pollinated by birds and bats, and *Aspidistra,* once a favorite pot plant, by snails. The appropriate authorities must be consulted for more detailed accounts of pollination mechanisms.

The pollen, having been brought into contact with the stigma (to which it may be firmly attached as a consequence of electrostatic forces) imbibes water from the surface of the stigma. Germination follows, usually within a few hours. A pollen tube breaks through a colpus in the exine of the grain (Fig. 9.32) and, provided there are no incompatibility barriers, penetrates the style. The growth of the tube is largely confined to the tip, behind which the wall becomes callosed. Its passage towards the ovules often follows a tract of specialized, thin-walled, transmitting tissue, the direction of growth probably being controlled by metabolic gradients of a quite simple kind. In some species (e.g., *Lilium*) the style is hollow. The pollen tube then grows down the surface of the cells lining the canal, the outer walls of which are frequently labyrinthine.

If the generative cell has not already divided into sperm cells, it does so in the extending tube. The two sperm cells are irregular in shape and together with the vegetative cell sometimes form a complex referred to as the "male germ unit." This moves forward with the growth of the tube and remains close to the tip (Fig. 9.32c). Although the mechanism driving the complex forward is not yet clear, actin microfilaments, prominent in the growing tube, together with myosin are probably closely involved. The distribution of the cytoplasmic organelles between the two sperm cells is not always equal. In *Plumbago,* for example, in one of the sperm cells, mitochondria predominate, and in the other plastids. In some species (e.g., *Lycopersicon,* tomato, *Gossypium,* cotton) plastids are either excluded from the generative cell, or degenerate during spermatogenesis, leaving only mitochondria in the cytoplasm. Instances are also known in which the sperm cytoplasm lacks organelles altogether (e.g. *Apium,* Umbelliferae).

Variations in the development of the male gametophyte are not uncommon. In some species, for example, the pollen grains put out two or more tubes, but only one carries the gametes. Some flowers (termed *cleistogamous*) do not open, but are nevertheless fertile. A familiar example is provided by *Viola canina.* The normal (*chasmogamous*) flowers of spring produce little seed, but the budlike cleistogamous flowers of summer are fully fertile. They contain only minute petals and all but the two abaxial stamens are abortive. The pollen of these, however, germinates in the anther and the pollen tubes grow through the

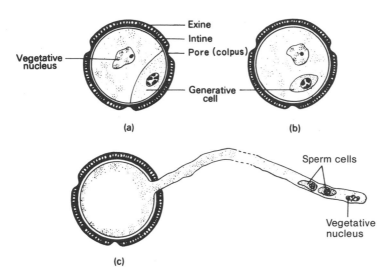

Figure 9.32. Development of a pollen grain in which the final nuclear division occurs in the tube. (a) Condition in the anther during final stages of the growth of the wall. (b) Condition during liberation and pollination. (c) The sperm cells and vegetative nucleus move forward with the pollen tube. Diagrammatic.

wall into the stigma. Even more extensive growth of the pollen tube through vegetative tissue is shown by the aquatic *Callitriche*. Here the flowers are unisexual and those submerged remain closed. The pollen germinates in the anther. The tubes then grow for long distances to female flowers at the same node, penetrating the ovary from below.

Fertilization. A pollen tube enters the ovule either by way of the micropyle or by growth through the chalazal end. Entry into the embryo sac itself is usually at its micropylar end. A region near the tip of the tube, possibly in consequence of enzymes produced by synergid cells, soon breaks open. The contents of the tube are discharged into the sac. The discharge may take place, as in *Gossypium*, into one of the synergids. The sperm cells are then stripped of their cytoplasm and only the male nuclei continue into the sac. Following the entry of the sperm cells or nuclei the "double fertilization," characteristic of the angiosperms, then ensues. One gamete fuses with the egg cell and the other moves through the sac and fuses with the central cell. Fusion of the nuclei gives rise to the diploid zygotic nucleus and the triploid endosperm nucleus respectively. The movements of the sperm cells may not always be random. In *Plumbago*, for example, where the cytoplasms of the gametes differ, the cell richer in plastids fuses preferentially with the egg cell. In general, however, the inheritance of plastids seems to be principally maternal. *Pelargonium* provides one of the few examples of clear biparental inheritance. The inheritance of mitochondria is more difficult to determine and may be more widely biparental. In *Gossypium* and similar instances the inheritance of the cytoplasm and its components is presumably entirely maternal.

Control of mating. In addition to protogyny and protandry various devices are found which discourage or prevent self-pollination and the possible deleterious effects of inbreeding. Conspicuous among these are those involving dimorphic flowers. In *Primula*, for example, the two kinds of flowers are termed "thrum-eyed" with a short style and elevated stamens, and "pin-eyed" with a long style and lower stamens (Fig. 9.33). An insect searching for nectar at the base of the corolla tube will tend to pass pollen from the stamens of the "pins" to the styles of the "thrums" and vice versa. This morphological device is accompanied by a physiological difference which causes the imperfect growth of "pin" and "thrum" pollen in their own styles. The mechanism, which is genetically controlled, depends upon the presence of antibodies in the style which precipitate essential enzymes in the pollen tubes of pollen from the same or like flowers. The pollen of the thrum flowers is some 50% greater in diameter than that of the pin, and this is correlated with a greater rate of extension of the pollen tube of the thrum pollen. *Lythrum salicaria* and some other species have a similar mechanism involving three kinds of flowers, the diameter of the pollen grains again being related to the position of the stamens producing them. Physiological systems of self-incompatibility unaccompanied by structural devices in the flower are widespread. In *Linum,* for example, successful growth of the pollen tube can take place only if the ratio of the osmotic pressure of the style to that of the pollen is the order of 1 : 4, a relationship never present between pollen and style of the same flower. Other mechanisms are seen in *Raphanus* (radish) and *Brassica* (cabbage). Here the pollen germinates following selfing, but the tube is prevented from penetrating the stigma by the rapid development of pads of callose in the stigmatic cells beneath its tip. In *Oenothera* the incompatible tube penetrates, but in the style it becomes plugged and growth ceases.

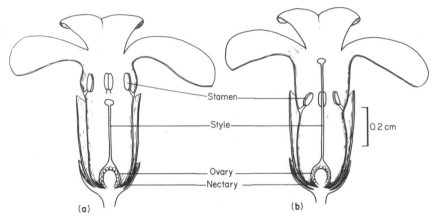

Figure 9.33. *Primula* **sp. Half flowers showing (a) thrum arrangement and (b) pin arrangement.**

In general, two kinds of incompatibility can be distinguished, sporophytic and gametophytic. In the former, the failure of pollination is a consequence of interaction between products of tapetal origin (called collectively **tryphine**) carried by the exine of the pollen grain and the papillae at the surface of the stigma. In gametophytic incompatibility, the interaction is directly between the male gametophyte itself, represented by the pollen tube with its thin permeable wall, and the tissue of the style. The distinction between sporophytic and gametophytic systems is not, however, absolute. In *Raphanus,* for example, in addition to an initial sporophytic system at the surface of the stigma are further interactions of a gametophytic kind in the style. It seems likely that basic to both sporophytic and gametophytic systems of incompatibility is an interaction of the kind seen throughout land plants generally wherever the cells of gametophyte and sporophyte are intimately apposed.

Other, less studied forms of incompatibility also exist. In species of *Castanea* (chestnut), for example, incompatible pollen tubes liberate gametes but fertilization fails. In *Gasteria* incompatible pollination leads to the formation of embryos, but they subsequently abort.

Embryogeny. Following fertilization, the contents of the embryo sac are commonly resorbed, the zygote and endosperm being the only sites of further growth. As in other seed plants, the first dividing wall of the zygote and endosperm being the only sites of further growth. As in other seed plants, the first dividing wall of the zygote lies transverse to the longitudinal axis of the ovule (Fig. 9.34), and the subsequent embryogeny is endoscopic. Free nuclear division is not, however, a characteristic of the development of

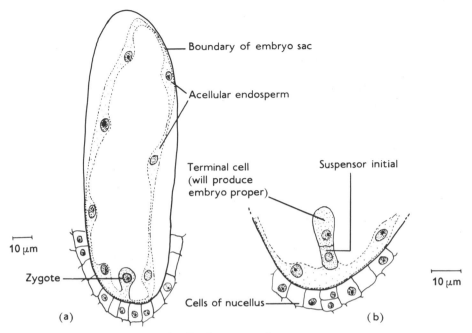

Figure 9.34. *Myosurus minimus.* Longitudinal section of embryo sac following fertilization. (a) After several divisions of the primary endosperm nucleus. (b) After the first division of the zygote. The zygote is at the micropylar end of the sac.

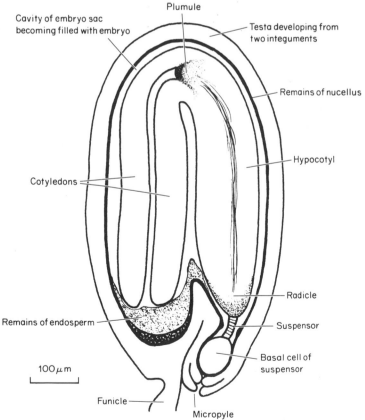

Figure 9.35. An almost mature seed of *Capsella bursa-pastoris* sectioned longitudinally and showing the orientation of the embryo.

the proembryo, as it is in many gymnosperms. Instead, embryogenesis begins with a series of precise cell divisions, showing little variation between species, and meristematic activity does not pause until the embryo differentiates and the seed is formed. The mature embryo (except in the orchids where it remains undifferentiated) possesses a stem apex (*plumule*), one or two cotyledons, and a root apex (*radicle*). Even though the mature embryo may be folded (as in *Capsella bursa-pastoris* (shepherd's purse), Fig. 9.35), the radicle is always directed towards the micropyle. The extent to which the fully formed embryo has drawn upon the food reserves of the endosperm varies widely with the species. In endospermous seeds this reserve remains considerable, but in the nonendospermous little remains, and much of the food materials in these seeds often becomes transferred to the cotyledons. Examples of such seeds are those of the legumes where the swollen cotyledons entirely replace the endosperm in both space and function (Fig. 9.36a).

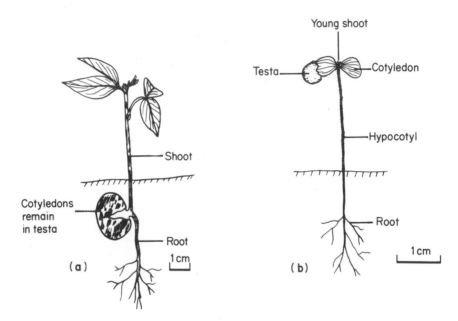

Figure 9.36. (a) *Vicia faba.* **Hypogeal germination. (b)** *Sinapis* **sp. Epigeal germination.**

Features of the endosperm. The formation of the endosperm usually begins before the division of the zygote. The primary nucleus and its daughters undergo successive mitoses, giving rise to an endosperm that is either cellular from the first, or initially acellular (fig. 9.37a) and only later partly or wholly cellular. Cytologically the endosperm is a remarkable tissue, with unique cytoplasmic and nuclear properties, significantly different from the haploid "pseudoendosperm" of the gymnosperms. It is a rich source of growth-regulating substances or their immediate precursors, many of them not yet chemically identified. The successful culture of young embryos and other plant tissues, for example, is often impossible in fully defined media, but is facilitated by the addition of coconut milk (the liquid endosperm of the coconut seed). Endosperm itself can sometimes be obtained in pure culture and, since the cell walls of young endosperm often have little thickening, such cultures have been used to study the details of mitosis *in vivo*. In some endosperms the nuclei reach high and irregular levels of endopolyploidy, and amitotic nuclear division has been reported in a number of instances.

In keeping with the function of providing for the nutrition of the embryo and young plant, the cells of the mature endosperm are often filled with food materials. Carbohydrates, fats, and proteins are present in various labile forms. The occurrence of these materials is sometimes so abundant that the seeds concerned acquire vast economic importance. The cereal grains, where the endosperm yields starch and protein in a form readily palatable to humans, and the oil seeds, such as *Ricinus* (castor bean) and *Linum* (linseed), are familiar examples. Polymers of mannose (mannans) also occur as reserve products, a peculiar example being provided by *Phytelephas*, a palm. The large endosperm of this species becomes so heavily indurated with a mannan that it enters commerce under the name of "vegetable ivory", once used for the manufacture of buttons and billiard balls. In a few plants, notably the orchids, the endosperm undergoes only trifling development and in others it is remarkably aggressive. In *Pedicularis*, for example, the endosperm produces haustoria which invade the integumentary tissue, leading eventually to its complete resorption.

In relation to the gymnospermous cycle the endosperm is a new feature in the life cycle of the seed plants. As a consequence of its polyploid genome, its gene products are likely to be sporophytic in nature rather than gametophytic, and directly compatible with the metabolism of the embryo. Further, the juxtaposition of embryo and endosperm provide little impediment to the transfer of metabolites. This, no doubt, makes for the generally rapid development of the angiosperm embryo compared with the gymnosperms, conferring upon them a significant selective advantage.

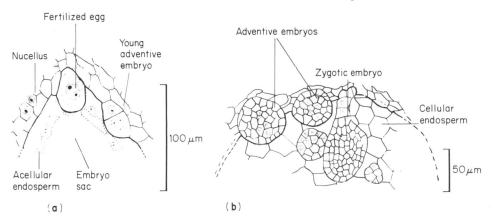

Figure 9.37. *Citrus trifoliata.* **Longitudinal section of micropylar end of embryo sac in which adventive embryony is occuring. (a) Immediately after the fertilization of the egg. (b) After the endosperm has become cellular. Only the true embryo has a suspensor. (Both after Osawa, from Maheshwari. 1950.** *An Introduction to the Embryology of Angiosperms.* **McGraw-Hill, New York.)**

Maturation of the seed and fruit. Associated with the maturation of the embryo are considerable changes in the nucellus and integuments. In some seeds, food reserves are laid down in the nucellus, and in the mature seed this then forms a distinct tissue known as the perisperm. Frequently, however, only a little of the nucellus remains, and this, together with the integument, becomes transformed into the outer covering of the seed. This often involves the deposition of much cutin and lignin, and the resulting seed coat (**testa**) is consequently often remarkably impervious.

As the seeds mature, the ovary becomes a fruit. There are numerous forms of fruits, but the wall is formed either from the carpel (**pericarp**), or by the fusion of the pericarp with surrounding tissues. The pericarp may become hard, as in nuts, or dry and brittle, as in the capsules of the Caryophyllaceae, or partly or wholly fleshy, as in common edible fruits. In berries (e.g., tomato, *Lycopersicon*) the pericarp is wholly fleshy; in drupes (e.g., cherry, *Prunus*) the outer part becomes fleshy while the inner hardens to form a stone; and in pomes (e.g., apple, *Malus*) a thin layer of pericarp hardens while the remainder fuses with the surrounding receptacle and forms the flesh. Where fruits are palatable to humans or animals the seeds commonly have a testa able to survive passage through the digestive tract.

Seeds are often distributed while still in the fruit, and the fruit in this instance commonly bears hooks or spines (as in the burdock, *Arctium*), or wings (ash, *Fraxinus*), or parachutelike fringes of hairs (as in the pappus of many Compositae), which assists dispersal by attachment to animals or by wind. Other angiosperms have evolved mechanisms whereby tensions are set up in the carpel wall as it dries, causing eventually an explosive dehiscence which scatters the seeds far and wide. This brief summary of a wealth of morphological detail shows how the angiosperms have evolved a great variety of ways which ensure wide dissemination, a feature in which they differ sharply from the gymnosperms, again contributing to their present success.

Dormancy and germination. The cause of the cessation of growth of the embryo as the seed matures is not altogether clear, but the partial dehydration of the interior of the seed at this time is probably an important factor. Certainly the imbibition of water is essential for renewed growth, but this by itself is often insufficient for the production of viable seedlings. A period of dormancy ("after ripening") must often ensue before successful germination will occur. The seeds of some species also need to be chilled, a treatment which apparently activates the enzymes which make available to the embryo the food reserves of the endosperm. This cold requirement also provides a biological advantage, since the seeds in a seasonal climate will not germinate until after the winter, and the seedlings thus avoid prolonged freezing. Other seeds, for example, those of lettuce, germinate only after illumination, and the phytochrome system (which depends upon a protein sensitive to red light) is here involved in the renewal of growth. In *Banksia* and some other Australian Proteaceae the fruits open only after bush fires (cf. "closed cone" pines) and the seeds germinate in the ash layer. Not only is this a more favorable substratum than loose humus, but the seeds of at least one species need heat treatment before they will germinate. The seeds of many tropical plants, on the other hand, germinate immediately in moist conditions and soon lose their viability if stored. The tropical mangroves, plants of coasts and estuaries, are outstanding in that, in some species, the embryo continues to develop in the ovule, and a swollen radicle protrudes from the withered flower. The young plant eventually falls and, with the radicle correctly oriented for penetration of the mud beneath, it rapidly establishes itself.

As a germinating seed imbibes water, there is usually considerable swelling and eventually rupture of the testa. The radicle is the first organ to emerge, followed closely by the plumule. The cotyledons may remain within the seed at, or below, soil level (**hypogeal** germination), as in *Vicia faba* (broad bean) (Fig. 9.36a) or become elevated (**epigeal** germination) and photosynthetic, as in *Sinapis* (mustard) (Fig. 9.36b). Germination is basically similar in monocotyledons and dicotyledons, but in the grasses the plumule and radicle are first enclosed in sheaths called the **coleoptile** and **coleorhiza** respectively. Coleoptiles show marked phototropic and gravitropic responses and, being devoid of appendages, are particularly suitable for the experimental investigation of these phenomena.

Asexual reproduction. In addition to sexual reproduction, the angiosperms show many forms of asexual reproduction. One form of such reproduction, termed **apomictic,** superficially resembles sexual reproduction. Although its presence can be readily inferred from the genetics of the species concerned, the elucidation of the accompanying cytology has sometimes demanded very careful investigation. In one kind of apomixis, parthenogenesis, the egg develops without fertilization. Where this occurs regularly, the nuclei involved in the formation of the embryo sac and the nucleus of the egg contain an unreduced number of chromosomes. Such embryo sacs may result from a modification of the meiosis which yields the megaspores (as in *Taraxacum*, dandelion), or from the spatial and functional replacement of the megaspore by an adjacent nucellar cell (as in *Hieracium*, hawkweed). Although the formation of the embryo occurs without fertilization, the stimulus of pollination is often needed before the process will begin (a phenomenon known as **pseudogamy**), and an endosperm may even be formed with the participation of one of the sperm nuclei. The pollen, although without genetic effect in the species producing it, is sometimes able to form hybrids with related sexual species. This parallels the situation in apogamous ferns, such as *Dryopteris affinis*.

Another form of apomixis involves the formation of embryos directly from the cells of the nucellus, often in addition to the normal zygotic embryo. A classical example of this phenomenon, known as **adventive embryony**, is provided by *Citrus* (Fig. 9.37), where the mature seeds may contain several viable embryos. Although in some species adventive embryony is independent of pollination, there are others (probably including *Citrus*) in which the embryos do not mature unless it occurs. In these instances the germinating pollen probably provides a chemical factor essential for continued growth. It is significant in this connection that in one orchid, *Zygopetalum machaii*, adventive embryony, although dependent upon pollination, is quite as effectively stimulated by pollen from another genus as by that from the species itself. Multiple seedlings from normally one-seeded fruits are a feature of many tropical trees, and this suggests that apomixis may be widespread in the tropical rain forest.

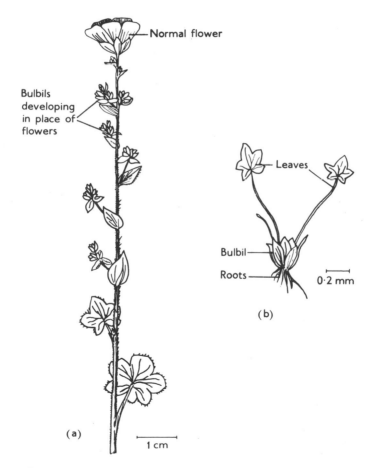

Figure 9.38. *Saxifraga cernua.* **(a) Inflorescence with one normal flower and the rest developing as bulbils. (b) Germinating bulbil. (Both after Kerner, von Marilaun, and Oliver. 1902.** *The Natural History of Plants,* **Blackie, London.)**

The only remaining form of apomixis is more conspicuous since it involves the formation of bulbils or dwarf shoots in place of flowers. *Saxifraga cernua* and *Festuca vivipara* provide examples. Some species (e.g., *Saxifrage cernua*, Fig. 9.38) producing these structures are also able to reproduce in a normal sexual manner, bulbils being borne in the lower part of the inflorescence and flowers in the upper. In some grasses, and probably in other plants displaying this phenomenon, the way the plant reproduces can be influenced by the length of day in which it is grown.

Other forms of asexual reproduction are purely vegetative. Stems may arch over and root themselves, as in *Rubus*, and leaves of a number of species produce plantlets either at the summit of the petiole, as in *Tolmiea* (a frequent house plant), or marginally, as in *Kalanchoë*. Species which grow from bulbs usually reproduce themselves by the production of axillary buds at the base of the axis which develop into daughter bulbs. Corms multiply themselves by a similar process. Fleshy roots which are able to give rise to buds largely account for the success with which plants such as *Convolvulus arvensis* (bindweed) and *Lepidium draba* (hoary cress) are able to multiply themselves. In addition to these various forms of vegetative reproduction·occurring in nature, layering, budding, and grafting are frequently used in horticulture, and are the only means of propagating many valuable varieties whose sexual reproduction is defective.

Most remarkable of all, perhaps, is that in experimental conditions isolated cells and even pollen grains will give rise to whole plants of normal growth and form. Those raised from pollen grains are particularly useful in breeding. They initially contain the reduced number of chromosomes, but this can be doubled by colchicine so that a pure homozygote is obtained which shows normal sexual reproduction.

Evolutionary Aspects of Angiosperms

The evolution of the angiosperms has many different aspects. Of particular interest to systematists is the evolution of individual species, genera, and families. Another approach, of greater appeal to morphologists, is to consider the evolution of certain morphological or anatomical features—such as the form of leaves and the nature of vascular tissue—in the angiosperms as a whole. Unfortunately the amount of information the fossil record can offer directly on these points is limited. By the time the angiosperms become firmly identifiable (mid-Cretaceous), they are referable, to a surprising extent, to modern families and even genera. The inception of the angiosperms was evidently followed by a burst of radiative evolution which had a profound effect upon the earth's flora and fauna.

The Emergence of the Angiosperms

The first angiosperms. There are no simple criteria by which a fossil can be accepted as an angiosperm. Vessels, for example, are found in the Gnetales, and archegonia are absent from *Gnetum* and *Welwitschia*. The presence of a closed carpel and pollen tubes in stylar tissue are not readily demonstrated in fossil material. With the earliest fossils angiospermous affinity has largely to be assessed by the extent of the resemblances to living forms. The position with regard to fossil pollen is fortunately clearer. Pollen with columellar exine (Fig. 9.39) is unknown outside the angiosperms, and there is no convincing evidence that fossil gymnosperms ever produced pollen of this kind.

Some have envisaged the angiosperms appearing in upland regions (and hence away from sites of fossilization) as early as the late Paleozoic. The first specimens of angiosperm pollen are not, however, found until the Lower Cretaceous (Table 9.2). In view of the long distances over which pollen can be transported by wind and water, an origin of the angiosperms before the Mesozoic seems unlikely. The oldest reproductive parts assigned to the angiosperms also come from the Lower Cretaceous. They are mostly

Ectexine

Endexine

Tectum

Columellar layer

Foot layer

Figure 9.39. Diagram of a representative section of the exine of the angiospermous pollen grain, showing the thin endexine and the three layers of the ectexine. The exine covering the colpus is both thinner and simpler.

Table 9.2. Stages of the Cretaceous.

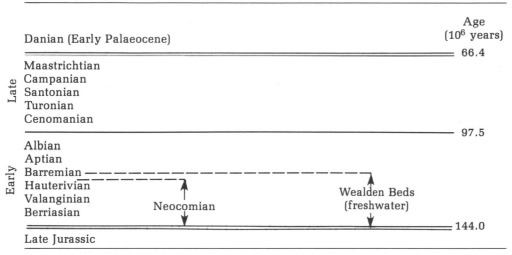

Danian (Early Palaeocene)

Age (10⁶ years)

66.4

Late
Maastrichtian
Campanian
Santonian
Turonian
Cenomanian

97.5

Early
Albian
Aptian
Barremian
Hauterivian
Valanginian
Berriasian

Neocomian

Wealden Beds (freshwater)

144.0

Late Jurassic

seeds and fruits lacking structural preservation. Their resemblances to those of any living species are too superficial to permit firm identification of their affinities. Later in the Lower Cretaceous (in rocks of Aptian and Albian age) are found flowers with free carpels and fruits with small seeds evidently produced from anatropous ovules. These can be confidently accepted as angiosperms. Both bisexual flowers of a magnoliaceous kind (with their parts in a spiral arrangement), and unisexual flowers in catkinlike inflorescences, resembling those of living *Platanus* (plane), can be recognized by the Albian stage. Leaves with angiospermlike venation, already known from pre-Albian rocks, become more numerous and varied at this time.

At the beginning of the Upper Cretaceous (Cenomanian) the frequency and diversity of angiosperm remains increase substantially. Many of the leaves and flowers begin to resemble those of modern genera (Fig. 9.40). Flowers with a cyclic arrangement of the floral parts also appear in the fossil record, some with syncarpous gynaecia. By the late Cenomanian, flowers with inferior ovaries are also found. Although many of the mid-Cretaceous flowers remain unfamiliar, a number of living families, among them the Magnoliaceae, Platanaceae, and Winteraceae, can now be clearly recognized.

Features of the ancestral angiosperms. The immediate ancestors of the angiosperms should therefore probably be sought in late Jurassic and the earliest Cretaceous rocks. No fossils of this age have yet been generally accepted as intermediate

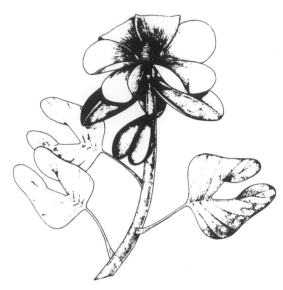

Figure 9.40. Reconstruction of a flowering shoot of the magnoliaceous Cenomanian (mid-Cretaceous) plant *Archaeanthus*. The deeply divided leaves are unlike those of any living magnoliaceous plant, but there is an approach to this condition in *Liriodendron* ("tulip tree") of North America and China. (After D. L. Dilcher and P. R. Crane. 1985. *Annals of the Missouri Botanical Garden 71*.)

between gymnosperms and angiosperms. Despite this absence of direct evidence, however, the features to be expected in the ancestral angiosperms can be inferred from a comparative examination of living species, making use of the "principle of correlation." This principle maintains that, in allied organisms undergoing evolution, the primitive states of the evolving features, even though these are evolving independently, will tend to remain associated in the group as a whole. Conversely, the most advanced states of the features will also tend to be associated. The degree of association and the probability of its occurring by chance can be assessed statistically. The recognition of a significant association as primitive must rest upon evidence external to the analysis, preferably provided by the fossil record.

Using this procedure it is found that, among living angiosperms, there is, for example, a very significant correlation between tracheidlike xylem, free petals, separate carpels, binucleate pollen, and woody habit. Since free petals and separate carpels are characteristic of early Cretaceous flowers, and tracheidlike xylem would be expected in plants emerging from the gymnosperms, the features listed can be taken as primitive, and were probably characteristic of the ancestral forms. Also, since they are still overrepresented in tropical families there is the implication that the angiosperms may, indeed, have arisen in warm and humid conditions.

Origin of the carpel. There is little direct evidence of the origin of the carpel. It is however significant that in *Drimys* (Fig. 9.41) and *Degeneria* (Fig. 9.42), representative of families in which the features identified as primitive in the foregoing section are well represented, the carpel is a clearly conduplicate structure with a lateral stigma lying along part or whole of the ventral suture. In *Degeneria* the ventral margins of the carpel do not even fuse (Fig. 9.42), but become interlocked by hairs between which the pollen tubes force their way.

If *Drimys* and *Degeneria* indicate the form of the ancestral carpel, it is conceivable that it was derived from the kind of ovule-bearing structures present in the late pteridosperms. Fossils of possibly great significance in the evolution of the angiosperms

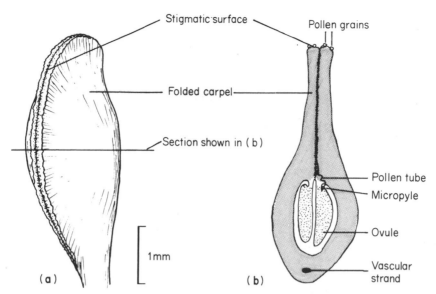

Figure 9.41. *Drimys piperita.* **(a) Carpel with stigmatic surface along the ventral suture. (b) Transverse section of carpel. (Both after Bailey and Swamy, from Foster and Gifford. 1949.** *Comparative Morphology of Vascular Plants.* **Freeman, San Francisco. © 1949.)**

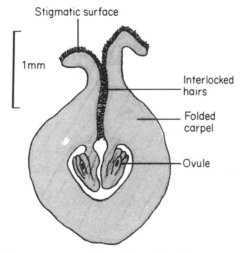

Figure 9.42. *Degeneria* **sp. Transverse section of conduplicate carpel. The ventral lobes, although appressed, do not fuse, but become interlocked by hairs. The flared ventral margin provides a stigmatic surface. (After Swamy, from Foster and Gifford. 1949.** *Comparative Morphology of Vascular Plants.* **Freeman, San Francisco. © 1949.)**

have come from the Permian of the Southern Hemisphere. Curious reproductive organs are found associated with a reticulately veined leaf and coniferlike wood. All are now thought to be parts of one plant, *Glossopteris.* The megasporangiophores, although leaflike, were inserted in the axils of foliage leaves (Fig. 9.43). They bore several ovules on one side (initially thought to have been abaxial, but this has been disputed), each about 1.5 mm (0.06 in.) long. The fertile area was largely covered by the margins of the sporangiophore. The male sporangiophores were also leaflike with lateral tassels of pollen sacs. The pollen was winged in the manner of that of *Pinus. Glossopteris* was probably a bush or small tree.

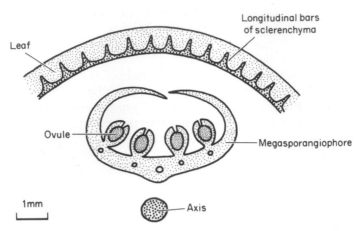

Figure 9.43. *Glossopteris.* **Diagrammatic representation of the female reproductive region. The leaflike megasporangiophore was borne in the axil of the foliage leaf. (From information in R. E. Gould and T. Delevoryas. 1977.** *Alcheringa* **1.)**

The importance of *Glossopteris* is this: it reveals that, already by the end of the Paleozoic era, there was a clear tendency for ovules to become enclosed in carpellike structures. This tendency is seen again in the Caytoniales, frequent in the Yorkshire Jurrassic, and in Greenland. Although the growth form of these plants is not clear (since the stems rarely exceed a few centimeters in diameter they were probably bushy), the reproductive organs and leaves are well known. The female organ consisted of an axis, 5 cm (2 in.) or more in length, and possibly dorsiventral in symmetry. This axis bore several short pinnae, arranged in more or less opposite pairs, each pinna terminating in a hollow, spherical body about 0.5 cm (0.2 in.) in diameter (Fig. 9.44a). This sphere contained the ovules, and, because of its analogies with the structures partially enclosing the ovules of some Carboniferous pteridosperms, it is termed a cupule. At the base of the cupule, on the upper side, and concealed by a small flap of tissue, was a pore communicating with the interior. Up to 32 upright ovules, each about 2.5 mm (0.1 in.) in length, were arranged in lines on the inner surface of the cupule, more or less opposite the basal pore (Fig. 9.44b).

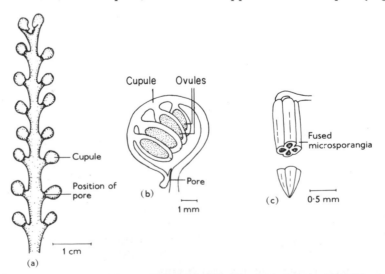

Figure 9.44. *Caytonia nathorsti.* **(a) Reconstruction of fertile shoot. (b) Longitudinal section of cupule. ([a], [b] After Thomas, from Andrews. 1961.** *Studies in Paleobotany.* **Wiley, New York.) (c)** *Caytonanthus kochi.* **Reconstruction. (After Harris, from Andrews. 1961.** *Studies in Paleobotany.* **Wiley, New York.)**

The symmetry of the male reproductive organ was similar to that of the female, but the pinnae terminated in several short irregular branches, each of which bore one or more synangia (Fig. 9.44c). Each synangium was divided longitudinally into four pollen sacs (and consequently, although symmetrical, bore a striking resemblance to the anther of a modern flowering plant, Fig. 9.22), the loculi containing winged pollen grains. The leaves consisted of a rachis ending in three to six leaflets, the leaflets palmately arranged and reticulately veined. Plants with similar ovulate cupules also occurred as late as the Lower Cretaceous, coincident with the earliest flowering plants.

Since pollen has been found in the micropyles of the ovules of *Caytonia*, there is no possibility of *Caytonia* having been an early angiosperm. Nevertheless, if the cupule in related plants had become closed, and the basal flap stigmatic, the structure would have resembled a carpel. To have functioned as such, there would, of course, have needed to be an absence of any adverse interaction which prevented the growth of pollen tubes through sporophytic tissue to the ovules.

Evolution of Morphological Features Within the Angiosperms

The herbaceous habit. Although comparative analysis of living floras indicates that woody habit is correlated with primitive reproductive features, it should be noted that woody plants generally reproduce less rapidly than herbaceous, thus limiting the rate of evolutionary advance. Nevertheless, there is no evidence as yet that herbaceous forms occurred in the pteridosperms of the late Paleozoic and Mesozoic, the most likely source of the angiosperms. It seems reasonable to conclude from present knowledge that the earliest angiosperms were woody, but possibly not large trees. Herbaceous forms probably first evolved as the angiosperms diversified in the Aptian and subsequent stages of the Lower Cretaceous, but massive trees, such as found in the tropical rain forest, may not have appeared until the Upper Cretaceous (Fig. 9.45).

Xylem. Although the secondary xylem of a particular species of angiosperm has a characteristic anatomy, within the angiosperms as a whole there is considerable variation in such features as the length of vessel segments and the form of perforation and orientation of the end-plate. These features vary independently of each other, but statistical analysis has revealed that two combinations appear significantly frequently. Long vessel segments (1.3–2 mm/0.05–0.08 in.), for example, tend to be associated with obliquely placed end-plates with scalariform perforations, and short segments with transverse end-plates with single large pores (see Fig. 9.10). Long vessel segments are also associated with a number of other features, such as the angularity of the outline in transverse section, small cross-sectional area and uniform thickening of the walls. Since vessel segments of this kind closely resemble the tracheids of gymnosperms, this combination of characters, invoking the "principle of correlation" utilized earlier, can be taken as primitive. Conversely, it seems likely that the features associated with short vessel segments are more recently evolved.

As we have seen, tracheidlike xylem is associated with floral features appearing early in the fossil record. The wood of a few angiosperms (e.g., *Drimys*) lacks vessels altogether and consists wholly of tracheids. These species also tend to be primitive reproductively.

Leaves. Although little can be said with certainty concerning the evolution of leaves, there are many indications of stemlike properties in the leaves of angiosperms, indicating their affinity with megaphylls rather than micropylls. Apart from the stemlike branching of some leaves and the readiness of others to yield shoot buds, many pinnate leaves are hardly distinguishable from lateral shoots of limited growth, bearing leaves in a single plane. In *Chisocheton*, a tree of southeastern Asia with a large pinnate leaf, the apex of the main rachis remains active for several seasons, producing a pair of pinnae annually. The leaves of the related tropical shrub *Guarea* behave in a similar way and the

Figure 9.45. Diagram showing the changing composition and form of the Earth's vegetation during the Cretaceous period. The current climax vegetation at middle latitudes is included for comparison. (a) Araucarian conifer. (b) Taxodiaceous conifer. (c) Cycadalean shrub. (d) *Cycadeoidea*-like bennettitalean. (e) Herbaceous lycopod. (f) Fern. (g) Angiosperm shrub. (h) Angiosperm herb. (i) Gnetalean herb or small shrub. (j) Angiosperm tree. (After Crane, from Friis, Chaloner, and Crane. 1987. *The Origins of Angiosperms.* Cambridge University Press.)

leaf may continue to grow for as long as ten years. Inbreeding may also reveal the developmental similarity of stems and leaves. In some improved varieties of tomato, for example, buds arise freely along the rachis of the leaf, and the distal part often continues growth as a shoot.

Comparatively recent evolution in the form of leaves has evidently been concerned in some instances with modification for a particular function. The insect traps of *Drosera* (sundew, Fig. 9.46), *Nepenthes* (pitcher plant, Fig. 9.47), and *Dionaea* (Fig. 9.48) are examples of this kind of development. Sometimes, evolution has been in relation to, and perhaps coupled with, that of a particular insect, usually a species of ant, which lives in and is nourished by the plant concerned. In several tropical species of *Acacia*, for

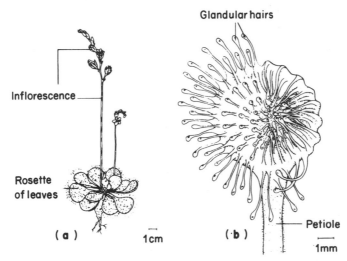

Figure 9.46. *Drosera rotundifolia.* **(a) Habit. (b) Single leaf in which some of the glandular hairs have closed on an entrapped insect. (After Kerner, von Marilaun, and Oliver. 1902.** *The Natural History of Plants.* **Blackie, London.)**

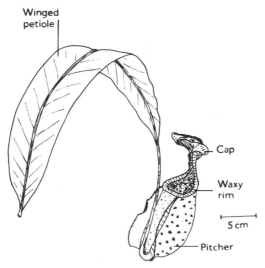

Figure 9.47. *Nepenthes.* **Leaf modified to form a pitcherlike insect-trap. (After** *Botanical Magazine,* **from Wettstein. 1935.** *Handbuch der systematischen Botanik.* **Deuticke, Leipzig.) The inner surface of the pitcher is covered with tilelike plates of wax making it impossible for the insect to climb out.**

example, the long thornlike stipules are hollow and provide shelter, while the leaflets terminate in small globules of parenchyma (Belt's corpuscles) which are devoured by the ants (Fig. 9.49). Succulence of leaves (the anatomical and physiological modification of the mesophyll enabling it to retain large quantities of water) also reaches its highest development in the angiosperms and is a feature of many desert and maritime species.

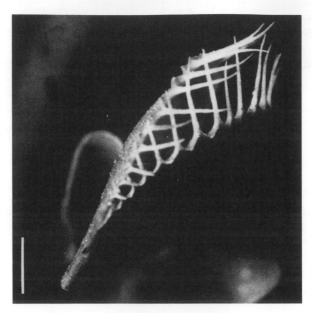

Figure 9.48. *Dionaea muscipula* **(Venus fly trap): an insectivorous plant in which the leaf is highly modified. The terminal lamina of the leaf is in two halves which rapidly close together when the sensitive hairs on the surface are touched, the marginal teeth interlocking. Entrapped insects are digested by proteolytic enzymes coming from glandular hairs on the inner (adaxial) surface of the lamina. Scale bar 2 mm.**

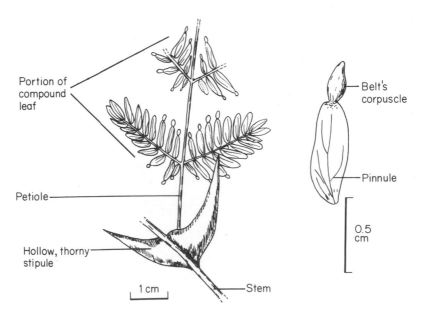

Figure 9.49. *Acacia sphaerocephala.* **Portion of leaf showing the hollow stipules and Belt's corpuscles.**

Flowers and inflorescences. The fossil evidence points to the first angiosperm flowers having had the members of both the sterile and fertile parts of the flower separate from each other. In present-day floras such *chorichlamydeous* flowers are found not only in primitive woody species, but are also characteristic of such families as the herbaceous Ranunculaceae and the monocotyledonous Alismataceae. Nevertheless, congenital fusion of parts (which involves no actual fusion, but is a consequence of the delayed appearance of separated meristems in development), both of the perianth and reproductive region, can be looked upon with fair certainty as a later feature in the evolution of the flower. As we have seen with the algae (Table 3.1) simple and presumably earlier features can continue to exist together with the more complex provided the organisms concerned continue to fill an ecological niche and their attributes are not selectively disadvantageous.

Quite apart from fusion of parts within the flower, another tendency, clearly evident in some alliances, has been towards the evolution of compact inflorescences. The ultimate form of such an inflorescence is the dense capitulum, occurring in a number of families, but characteristic of the Compositae. In many instances, owing to the differentiation of disk and ray florets, capitula have come superficially to resemble simple flowers, a feature well shown by the common daisy, *Bellis perennis*. In *Syncephalantha*, a Central American composite, we even find a racemose aggregate of capitula which, taken as a whole, also has a flowerlike form owing to the asymmetry of the marginal capitula. The sporadic occurrence of these flowerlike inflorescences in a number of families indicates that the radiate form of the reproductive region, whatever its composition, has biological significance. Natural selection would account for the repeated emergence of this pattern despite the increasing morphological complexity of the reproductive region.

The evolution of spurred and otherwise zygomorphic petals, distinctive scents, nectaries, and peculiar stamens, and of striking pigmentation of the perianth are all usually associated with insect (and rarely bird) pollination. Sometimes the life cycle of both plant and insect are so closely interwoven that one could not exist without the other. Reproduction in the figs (*Ficus*) provides a particularly striking example. Here evolution has led to an urn-shaped receptacle, almost closed at the mouth. The female flowers, the first to open in the cavity of the receptacle, are pollinated by particular wasps which force their way through the narrow entrance and bring in pollen from elsewhere. Eggs are laid in some of the ovaries and the emergence of the new generation of wasp coincides with the opening of the male flowers. In escaping from the fruit they take pollen with them and the cycle is repeated in another developing fruit. In some species the cycle is even more complicated, two species of wasp being involved. These situations raise profound problems of evolution, since, at some stage, flower and insect must have begun to evolve together, leading ultimately to complete mutual dependence.

The inflorescences of the Araceae, in which the small individual flowers are crowded on a rodlike spadix surrounded by a tentlike spathe, are remarkable for the heat generated as the flowers mature. This is brought about by vigorous respiratory activity in the spadix. The high temperature (as much as 30°C/54°F above that outside the spathe) vaporizes unpleasantly smelling compounds which attract pollinators.

Physiology. It is evident that morphological diversification in the angiosperms has been accompanied by physiological. Apart from the wide range of secondary metabolites in angiosperms are many instances of physiological adaptation to extreme habitats. In the salt marsh plant, *Limonium,* for example, glands on the leaves actively excrete the excess ions taken up with the soil water. In a number of angiosperms evolution has led to the abandonment of autotrophy, the species concerned becoming, in consequence, obligate heterotrophs, a transition already seen in other divisions of the Plant Kingdom. Examples among the angiosperms are provided by *Epipogium,* an orchid which lacks chlorophyll and derives its nutrition from a highly developed mycorrhiza, and such parasites as *Lathraea* (toothwort), *Orobanche* (broomrape), and *Cuscuta* (dodder, Fig. 9.50). A number of angiosperms, such as *Viscum* (mistletoe), *Euphrasia* (eyebright) and *Striga* (a pest of *Sorghum* in the semi-arid Tropics), have become parasitic without losing the ability to photosynthesize, so they are not entirely dependent upon their hosts.

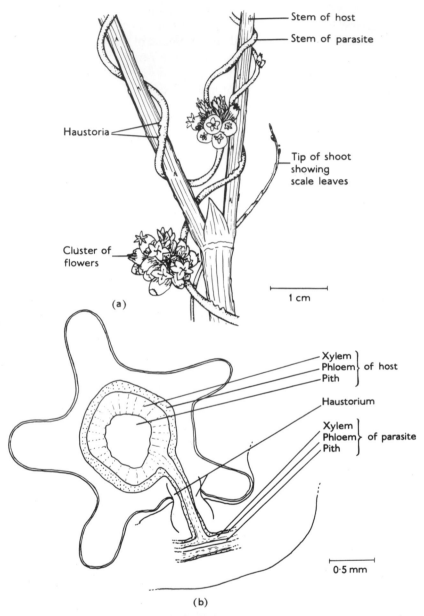

Figure 9.50. *Cuscuta*. **(a) Habit. (After Kerner, von Marilaun, and Oliver. 1902.** *The Natural History of Plants.* **Blackie, London.) (b) Transverse section of host stem showing the penetration of the vascular tissue of the parasite into that of the host.**

Recent Evolution Within Families and Genera.

Knowledge of recent evolution in the angiosperms has come principally, as with the ferns, from studies of chromosome pairing in hybrids, often produced under experimental conditions. Hybridization of diploid species, followed by allopolyploidy, appears to have been the origin of a number of well-known plants, among them *Spartina anglica* (cord-grass) and *Galeopsis tetrahit* (hemp nettle). Observations of chromosome size may assist in assessing evolutionary relationships of a wider character. The Commelinaceae, for example, include some genera in which the chromosomes are small, others in which they are strikingly large, and yet others in which the nuclei contain chromosomes of both sizes. Fossils have contributed little to knowledge of this kind of evolution. A few fossil series of recent age are known, mostly of angiospermous seeds, but they indicate little other than minor changes in structure and shape. Seeds of the water plant *Stratiotes,* for example, are found in successive strata throughout the Tertiary era. There was evidently a slight, but progressive lengthening and narrowing of the seed during this period, together with minor changes in the relative development of parts.

Comparative biochemistry can also reveal affinities of living plants and indicate possible lines of evolution. The grouping of species in *Aesculus,* for example, is assisted by the occurrence in some species of distinctive nonprotein amino acids and their absence in others. Isozyme analysis can also indicate evolutionary pathways within genera. The constitution of chloroplast DNA has also been subject to evolutionary change in recent time. Restriction enzyme analysis of chloroplast DNAs from both monocotyledons and dicotyledons has assisted in elucidating phylogenetic relationships within genera and families.

The lead- and copper-tolerant forms of the grasses *Agrostis tenuis* and *Festuca ovina,* found on spoil heaps of old mines, are splendid examples of recent evolution. This is generally considered to have occurred by selection of resistant mutants appearing spontaneously in natural populations. Species tolerant of normally toxic concentrations of various heavy metals are found in a wide range of angiosperm families. The metal ions are usually removed from the general metabolism by complexing with specific proteins. Plants of this kind often serve as "indicator plants" to those prospecting for ores of copper, nickel, and other metals.

The Main Trends of Evolution in the Angiosperms

To summarize, we can envisage the main evolutionary trends in the angiosperms to have been as follows. They emerged from the gymnosperms, probably, in view of the radial symmetry of the ovules and certain anatomical similarities, from a group evolving from the Pteridospermopsida of the late Paleozoic. The angiosperms thus inherited a seed habit of long standing, and a serviceable form of axial structure in which lateral branch systems had long ago become megaphylls, and in which megaphylls, in turn, were becoming differentiated into petioles and laminae with reticulate venation. They improved upon this structure, replacing coarse and angular ramification by elegant branching. Anatomically, strengthening tissues tended to become reduced to those required by mechanical laws, and stems and branches consequently acquired a pliancy that is a protection against storms. Vessels were also developed from tracheids, and a more highly organized phloem containing sieve tubes and companion cells from simple sieve cells and parenchyma, both developments probably having advantageous physiological consequences. The reproductive regions became highly specialized, their evolution often being correlated with that of pollinating organisms.

Some have suggested that the angiosperms had more than one origin (i.e., that they, like the gymnosperms, are polyphyletic). This, however, seems unlikely in view of the general occurrence of the characteristic "double fertilization" in the embryo sac. Nevertheless, the relationship between some monocotyledons, especially the palms, and dicotyledons certainly appears remote, and we must assume that dicotyledons and monocotyledons have followed more or less parallel lines of evolution for a considerable period.

We have considered the primitive angiosperms to be woody, but diversification soon led to herbaceous forms. The subsequent evolution of the herbs was undoubtedly influenced by their interaction with grazing animals and the establishment of angiospermous forests. In tropical regions many herbaceous angiosperms are, in fact, epiphytes upon angiospermous trees and shrubs, and it seems clear that these followed rather than preceded the arboreal vegetation. Because of the rapidity with which the relatively succulent leaves of angiosperms decay in tropical conditions, humus is soon formed wherever they accumulate. Consequently, fertile substrata are found in abundance on horizontal surfaces and in the crotches of branches. This leads to an ecological situation different from that in gymnospermous forests, where the leaves mostly reach the ground and the humus, owing to the slow decay of the the leaves, is covered by harsh and unconsolidated material unfavorable for the establishment of delicate herbaceous forms. It thus seems reasonable to suppose that the epiphytic angiosperms (and ferns) emerged in response to, and are in the process of, exploiting the new ecological habitat created by the rise of the angiosperm forests. Their evolution would then have been, in geological time, a relatively recent event.

Today, the angiosperms resemble successful individuals reaching affluent middle age, determined to let no human experience escape them. They experiment freely with metabolic novelties, such as fragrant oils, peculiar alkaloids, and unusual carbohydrates (such as inulin). In sexual reproduction, both as regards pollination and the cytology of the gametophytes, they show sometimes bizarre variations. We probably witness the angiosperms at the height of their success. What they will be like in their senescence we can safely leave to our successors.

So long as the photosynthetic membrane, solar radiation, and the atmosphere continue their existence, there is no reason why the evolution of plants should cease. Many more remarkable forms may yet appear.

GLOSSARY

Terms relating to special features defined in the text should also be traced through the Index.

Abaxial, of lateral organs. The side away from the main axis (the same as DORSAL).

Abscission. The separation of structures from the stem of a plant after the formation of a layer of specialized cells (abscission layer).

Achene, of flowering plants. A dry, indehiscent, one-seeded FRUIT.

Acropetal. Developing from the base towards the apex (i.e., the youngest structures are adjacent to the apex) or, with regard to the movement of materials, towards the apex.

Actinomorphic. Radially symmetrical.

Adaxial, of lateral organs. Said of the side towards the main axis (the same as VENTRAL).

Adventitious. Said of plant organs which appear in unusual positions (e.g., roots from stems, or buds not in the AXILS of leaves).

Adventive embryony. The development of EMBRYOS from cells other than EGG cells, and without FERTILIZATION.

Aerenchyma. A plant tissue composed of unthickened, often irregularly shaped, cells surrounding large air spaces.

Aerophore. A negatively GRAVITROPIC root produced by certain plants growing in waterlogged conditions and believed to assist the aeration of the root system.

Akinete, of algae. A single-celled, nonmotile, resting SPORE in which the original cell wall forms part of the spore wall.

Allopolyploidy. The type of POLYPLOIDY found in organisms of hybrid origin, at least two complete sets of CHROMOSOMES from each of the original parents being present in the SPOROPHYTIC phase.

Alternation of generations. An outmoded term referring to the manner in which in a life cycle a SPOROPHYTE is succeeded by a GAMETOPHYTE and this in turn by a sporophyte.

Amphiphloic. Having PHLOEM tissue on each side of the XYLEM.

Amyloplast. A PLASTID without CHLOROPHYLL, involved in the synthesis and storage of starch.

Anemophily. Pollination by wind.

Aneuploid. Possessing a chromosome number not an exact multiple of the HAPLOID number.

Anisogamy. The production or fusion of GAMETES which are visibly different.

Annual. A plant which completes its life cycle in one growing season and then dies.

Antheridium. A general term for a male GAMETANGIUM.

Anthesis, of flowering plants. The time of opening of the FLOWER and exposure of the mature reproductive organs.

Anticlinal. Of cell divisions in which the new cell wall is perpendicular to the outer surface of the region in question.

Antithetic (relating to life cycles). The view that the SPOROPHYTE has evolved by elaboration of a transient ZYGOTIC phase interpolated between successive GAMETOPHYTES.

Aplanospore, of algae. A single-celled, nonmotile SPORE which is liberated from the parent cell.

Apogamy. The ASEXUAL production of a SPOROPHYTE from a GAMETOPHYTE.

Apomixis, of seed plants. Reproduction by SEEDS which have not developed as a consequence of sexual process.

Apoplast. The space formed by the hydrated framework of cell walls outside the protoplasts in which soluble substances and ions may freely move.

Apospory. The production of a GAMETOPHYTE directly from the SPOROPHYTE without the intervention of MEIOSIS.

Archesporium. The group of cells giving rise to the SPORE mother cells (MEIOCYTES).

Areola (Areole). Area surrounded by VEINS in a RETICULATELY veined leaf.

_____ , of some cacti. The tuft of spines indicating a lateral BUD.

Aril. A fleshy outer covering of the OVULE of certain SEED plants.

Asexual. Reproduction not involving the fusion of GAMETES.

Autogamous, of FLOWERS. Normally self-pollinated.

Autopolyploidy. The occurrence of three or more identical sets of CHROMOSOMES in the nucleus.

Autosome. A CHROMOSOME that is not a sex chromosome.

Auxin. A general term for organic substances (naturally occurring or synthetic) which promote or regulate plant growth.

Axil. The upper angle which a leaf makes with a stem.

Basipetal. Developing from the apex towards the base (i.e., the youngest structures are the most distant from the apex) or, with regard to the movement of materials, towards the base.

Berry, of flowering plants. A many-seeded FRUIT with a fleshy PERICARP, produced from a single flower.

Biennial. A plant in which the life cycle requires two growing seasons for completion, reproduction being followed by death.

Bifid. Divided deeply into two parts.

Bifurcate. To divide into two branches.

Bilateral symmetry. Symmetry with respect to a single plane; when divided by the plane, the two halves are mirror-images.

Blade. The broad, flattened portion of a leaf (the lamina).

Bordered pit. A PIT in which the thin area of primary wall is overhung by lignified secondary wall. The center of the pit membrane is often thickened, forming the torus.

Bract, of seed plants. A reduced leaf with a reproductive structure in its AXIL.

Bract scale, of gymnosperms. The scale (in some species becoming woody) in the female CONE in the AXIL of which arises the ovuliferous scale.

Bracteole, of flowering plants. A reduced leaf produced on a FLOWER stalk.

Bud. The apical MERISTEM of a shoot, surrounded by leaf primordia and sometimes enclosed in scale leaves.

Bulb, of flowering plants. An underground PERENNATING and storage organ, consisting of a reduced stem surrounded by fleshy leaves.

Caducous. Said of an appendage shed as the cell or organ bearing it becomes mature.

Callus. A PARENCHYMATOUS tissue formed over wounds or in culture, often capable of being sub-cultured indefinitely in nutrient media.

Cambium. A secondary MERISTEMATIC tissue: the VASCULAR CAMBIUM produces secondary XYLEM and PHLOEM; the CORK CAMBIUM produces CORK and secondary CORTEX.

Capitulum, of flowering plants. An INFLORESCENCE in which the FLOWERS are closely grouped on a disk.

Carotenoids. A group of yellow pigments occurring in chloroplasts, CHROMOPLASTS and elsewhere.

Carpel, of flowering plants. The OVULE-bearing structure, usually regarded as a MEGASPOROPHYLL.

Casparian band (strip). A girdle of SUBERIZED thickening occurring on the radial and upper and lower walls of the cells of the ENDODERMIS.

Cauliflory, of flowering plants. The continued lateral production of FLOWERS on old woody stems.

Cauline. Appertaining to the stem.

Cell sap. Watery contents of the large VACUOLE usually present in plant cells.

Centrifugal. Developing from the center towards the outside (i.e., the oldest structures are at the center, and the youngest at the outside).

Centripetal. Developing towards the center from the outside (i.e., the oldest structures are at the outside, and the youngest at the center).

Chemotaxis. The movement of a motile organism, SPORE, or GAMETE in response to a chemical stimulus.

Chimaera. An organism with tissues of more than one GENOTYPE. It may be produced by mutation or by grafting.

Chlorophyll. A group of structurally similar green or blue-green PHOTOSYNTHETIC pigments common to all phototrophic plants.

Chromatin. A complex of deoxyribonucleic acid and basic proteins present in CHROMOSOMES and defined by its staining properties.

Chromatophore. An intracellular body containing pigments; hence a general class containing CHROMOPLASTS and chloroplasts, but not often used for the latter.

Chromoplast. A PLASTID deficient in or lacking CHLOROPHYLL, but accumulating other pigments, principally CAROTENOIDS.

Chromosomes. The organized structure in the NUCLEUS, carrying the Mendelian genes.

Cilium. A short, threadlike, locomotory ORGANELLE which beats in a regular manner. Usually numerous.

Circumnutation. The sweeping movements made by the tip of a stem, particularly marked in twining plants, caused by unequal rates of growth around the axis.

Cladode. A short, flattened shoot of limited growth, replacing the leaves as the site of photosynthesis.

Coccoid, of algae. A unicellular, nonmotile, spherical condition of the adult organism.

Coenobium. A COLONY of cells firmly attached to each other in which there is some degree of coordination.

Coenocyte. A multinucleate plant body or cell.

Coleoptile, of flowering plants. The sheath around the PLUMULE of certain MONOCOTYLEDONOUS plants.

Coleorhiza, of flowering plants. The sheath around the RADICLE of certain MONOCOTYLEDONOUS plants.

Collateral bundle. A VASCULAR bundle in which the XYLEM and PHLOEM are CORADIAL, the xylem lying towards the center of the axis and the phloem away from it.

Collenchyma. A strengthening tissue composed of columnar cells with heavy deposits of cellulose at the angles of the primary walls.

Colloid. A system of minute, charged particles suspended in a liquid. The colloid itself maybe in the form of a liquid (sol) or solid (gel).

Colony, of algae. A group of adhering unicellular organisms, all of which are equivalent and independent of one another.

Colpus. Each of the thin areas in the EXINE of a POLLEN grain, through one of which the pollen tube usually emerges.

Columella. A central, sterile tissue found in certain SPORANGIA. Also used in relation to the structure of the EXINE of POLLEN.

Companion cells, of flowering plants. The relatively small, nucleate cells with dense

PROTOPLASM associated with SIEVE TUBES in the PHLOEM tissue.

Complanate. Arranged in a single plane.

Conceptacle, of brown algae. The flask-shaped cavity containing the sex organs in certain genera.

Concrescent. Grown together, coalesced.

Cone. An axis bearing SPOROPHYLLS (usually closely packed).

Congenital (as used in **Congenital fusion**). Said of an inherited characteristic which becomes apparent during growth.

Coplanar. In the same plane.

Coradial. Lying on the same radius.

Coralloid. Corallike as a consequence of repeated branching.

Cordate. Heart-shaped.

Cork. The outer tissue formed by the activity of a special CAMBIUM and consisting of cells which become SUBERIZED and impervious to water.

Corm, of flowering plants. A subterranean storage and PERENNATING organ consisting of a short, swollen stem base.

Corolla, of flowering plants. A collective term for the PETALS of a FLOWER.

Corpus. The central part of the apex of a stem in which cell divisions occur in a variety of directions.

Cortex. The outer tissue of a multicellular axis, often little differentiated.

Cotyledons, of seed plants. The first leaf or leaves of the EMBRYO.

Crenulate. Toothed, but the teeth small and rounded.

Cupule, of seed plants. An extra cup-shaped organ surrounding the OVULE of certain fossil plants, or a cup formed by concrescent bracts at the base of the fruit of certain living genera (e.g., acorn).

Cuticle. The noncellular waxy coating of the EPIDERMIS of all land plants and certain internal surfaces adjacent to air spaces.

Cymose branching. The type of branching in which the apices abort successively, growth being continued by laterals in each instance.

Cytoplasm. The PROTOPLAST of a cell, apart from the NUCLEUS.

Deciduous plant. One which sheds its leaves at intervals, in temperate regions usually at the end of a growing season, by means of an ABSCISSION layer.

Decumbent, of a shoot. Either initially, or soon becoming, prostrate.

Decussate. Arranged in opposite pairs, alternate pairs being at right angles to each other.

Dehiscence. Opening at maturity, by splitting, especially of SPORANGIA and FRUITS.

Diarch xylem. XYLEM tissue consisting of two strands, or having two PROTOXYLEM groups.

Dichogamy. The maturing of male and female organs of a FLOWER at different times.

Dichotomous branching. Branching into two equal parts.

Diclinous, of flowering plants. Having STAMENS and CARPELS in separate FLOWERS, but the flowers borne by the same individual. Also used correspondingly of archegoniate plants.

Dicotyledonous. Having two COTYLEDONS.

Differentiation. The development of cells, tissues, and organs during which differences arise in structure and function.

Dimorphic. Existing in two separate forms.

Dioecious. Bearing male and female sex organs on different individuals.

Diploid (2n). Having twice the basic (HAPLOID) number of CHROMOSOMES.

Distal. Distant from a point of reference or symmetry (antonym: PROXIMAL).

Distichous. Arranged in two ranks, diametrically opposed.

Dorsal, of lateral organs. The same as ABAXIAL.

_____ , of prostrate plants. The side away from the substratum.

Dorsiventral. Having distinct upper and lower sides.

Drupe, of flowering plants. A one-SEEDED FRUIT, formed from a single flower, in which the PERICARP is part fleshy and part stony.

Egg. A nonmotile, female GAMETE.

Elaters. Sterile cells or parts of cells interspersed with spores and assisting dispersal by hygroscopic movements.

Emarginate, of leaves. Having a small depression at the apex.

Embryo. The young, partly developed SPOROPHYTE.

Embryo sac, chiefly of angiosperms. The mature female GAMETOPHYTE.

Embryogeny. The development of the EMBRYO from the ZYGOTE.

Endarch, of xylem. XYLEM tissue which differentiates CENTRIFUGALLY, leaving the PROTOXYLEM nearest the center.

Endodermis. The inner layer of cells of the CORTEX of a stem or root, distinguished by the presence of CASPARIAN BANDS, and often by starch grains.

Endogenous. Developing or emanating from the inside.

Endolithic. Living within pores or narrow channels in rocks.

Endomitosis. Duplication of the CHROMOSOMES without nuclear division, resulting in POLYPLOIDY.

Endophyte. A plant which lives inside another organism, but is not PARASITIC upon it.

Endopolyploidy. POLYPLOIDY which is the result of ENDOMITOSIS.

Endoscopic embryogeny. EMBRYOGENY in which the first division of the ZYGOTE is such that the apex of the EMBRYO develops from the inner cell.

Endosperm, of flowering plants. The nutritive tissue in SEEDS derived from the primary endosperm cell of the EMBRYO SAC. Notable for the frequent presence of TRIPLOID nuclei and as a source of AUXINS.

Endosporic gametophyte. A GAMETOPHYTE which develops within the SPORE wall.

Endosymbiosis. SYMBIOSIS between two organisms, one of which lives inside the other.

Entomophily. Pollination by insects.

Enzyme. A soluble protein, or protein complex, which catalyses a biochemical reaction.

Epicotyl. That part of the axis of a seedling immediately above the COTYLEDONS.

Epidermis. The primary outer layer of cells of all plant organs.

Epiphyte. A plant which grows attached to another, but is not PARASITIC upon it.

Etiolation. The abnormal type of growth occurring in heavy shade or darkness, involving particularly reduction in the amount of CHLOROPHYLL, and an exaggerated elongation of the INTERNODES.

Eustele. A STELE consisting of a ring of COLLATERAL bundles.

Exarch. XYLEM tissue which differentiates CENTRIPETALLY, leaving the PROTOXYLEM furthest from the center.

Exine. The outer layer of the wall of a SPORE or POLLEN grain.

Exogenous. Developing or emanating from the outside.

Exoscopic embryogeny. EMBRYOGENY in which the first division of the ZYGOTE is transverse, the apex of the EMBRYO developing from the outer cell.

Exoskeleton, of algae. A firm outer layer, especially calcium carbonate, secreted by certain species.

Facultative. Of a response which depends upon conditions; implying that the organism in other conditions is able to respond in another way.

Falcate. Sickle-shaped.

Fascicular. Relating to a cluster or bundle, usually implying association with VASCULAR bundles. Hence FASCICULAR CAMBIUM.

Fertilization. The fusion of GAMETES.

Fiber cell. An elongated cell with a thick lignified wall, tapering at each end.

Filiform. Threadlike.

Fission. The ASEXUAL division of a unicellular organism into two similar organisms.

Flagellum. A whiplike organ of locomotion, single, paired, or many, found in motile algae and GAMETES.

Flower. A reproductive region of a nonarchegoniate SEED-bearing plant, usually consisting of a PERIANTH and one or more STAMENS or PISTILS, or both, and terminating the PEDUNCLE or PEDICEL.

Foot. The portion of the EMBRYO in the bryophytes and lower tracheophytes which anchors the embryo in the GAMETOPHYTE and may absorb some nutrients from it.

Frond. The leaf of a fern, or of a cycad.

Fruit. The SEED-containing structure derived from the OVARY, sometimes associated with other parts.

Funicle. The delicate stalk of an OVULE of a flowering plant, sometimes (as with inverted ovules) wholly or partly concrescent with the outer INTEGUMENT.

Fusion, in relation to organs. Used figuratively, implying CONCRESCENCE either during development or in the course of evolution.

Gametangium. An organ producing GAMETES.

Gamete. A mature sex cell, capable of fusing with another to form a ZYGOTE.

Gametophyte. The (normally) HAPLOID phase producing GAMETES which fuse to form the ZYGOTE from which develops the DIPLOID SPOROPHYTE.

Gemma, chiefly of bryophytes. A multicellular, ASEXUAL, reproductive structure which becomes detached from the parent and is capable of growing into a new plant.

Generative cell, of seed plants. The cell of the male GAMETOPHYTE which by division gives rise to the two male GAMETES.

Genotype. The genetic constitution of an organism.

Gravitropism (Geotropism). A growth movement of a plant dependent upon the direction of gravitation.

Guard cell. One of a pair of cells, containing chloroplasts, which occur in the EPIDERMIS and form a STOMA.

Halophyte. A plant living in a habitat of high sodium chloride concentration (e.g., a saltmarsh).

Haploid (n). Possessing a single set of CHROMOSOMES.

Hapteron. Same as HOLDFAST.

_____ , in *Equisetum*. A synonym for ELATER.

Haustorium. An absorptive organ of PARASITES which penetrates the HOST tissue.

Heterogamy. The condition in which GAMETES are morphologically distinct.

Heteromorphic. Having more than one growth form.

Heterophyllous. Having more than one leaf form.

Heterospory. The production of spores of two sizes (MEGASPORES and MICROSPORES).

Heterothallic, of algae. The condition in which GAMETES from the same individual or strain of individuals either cannot fuse or, if they fuse, yield an inviable ZYGOTE.

Heterotrichous, of algae. Having two kinds of FILAMENTOUS growth, one prostrate and the other erect, in the same individual. Also used in relation to the PROTONEMA of mosses.

Hilum. Region of attachment of FUNICLE to OVULE.

Histone. A highly basic protein often associated with deoxyribonucleic acid in the CHROMOSOMES.

Holdfast, of algae. The organ or cell which anchors the plant to the substratum.

Homologous. Said of organs believed to be identical in nature (but not necessarily in form), often (but not always) implying an evolutionary relationship.

Homospory. The production of only one type of SPORE.

Homothallic, of algae. The condition in which GAMETES from the same individual commonly fuse and yield viable ZYGOTES (antonym: HETEROTHALLIC).

Host. The organism from which a PARASITE obtains its food.

Hyaline. Clear, transparent.

Hydathode. A water-secreting gland or pore found in the leaves of many plants.

Hygroscopic. Water absorbing.

Hypocotyl. The stem of a seedling between the ROOT and the COTYLEDONS.

Hypogynous, of flowering plants. A FLOWER in which the perianth and androecium are inserted below the gynaecium.

Incompatibility. The inability of conspecific GAMETES to fuse and form a ZYGOTE, resulting from a number of different mechanisms in different kinds of plants.

Inflorescence. A group of FLOWERS with a common stalk.

Integuments, of seed plants. The envelopes surrounding the NUCELLUS.

Intercalary. Of MERISTEMS that lie between two apices.

Internode. The portion of a stem occurring between two NODES.

Intine. The inner layer of the wall of a SPORE or POLLEN grain.

Involucre. A whorl of sterile structures beneath a terminal reproductive region.

Isodiametric. Of equal height, length, and breadth.

Isogamy. The condition in which GAMETES are morphologically identical.

Isomorphic. Of the same form.

Lacuna. A cavity.

Lamina. See BLADE.

Lanceolate. Spear-shaped; flattened, and tapering to a point, with the widest part near the center.

Leaf gap. A gap in the VASCULAR cylinder of a stem immediately above the departure of a LEAF TRACE.

Leaf trace. The VASCULAR supply to a leaf, consisting of one or more strands.

Lenticel. A channel filled with loosely packed CORK cells permitting the diffusion of gases in stems (and occasionally in roots) formed by a characteristically shaped CORK CAMBIUM.

Leucoplast. A PLASTID lacking CHLOROPHYLL and other pigments, but often containing starch.

Lignin. A complex macromolecule formed by dehydrogenative polymerization of p-hydroxycinnamyl alcohols and deposited in the walls of certain cells.

Ligule, of grasses. A collarlike outgrowth at the inner junction of the leaf-sheath and BLADE.

——— , of lycopods. A scale inserted into the upper surface of the leaf or SPOROPHYLL.

Littoral zone. The zone on the sea shore between the high- and low-tide marks.

Loculus, of flowering plants. The cavities (loculi) in an OVARY, each containing one or more OVULES.

Lumen. The space bounded by a cell wall, more often used in relation to dead cells than living.

Lysigenous (of spaces). Formed by the dissolution of a cell or cells (cf. SCHIZOGENOUS).

Medulla. The central region of an axis or organ.

Megasporangiophore. A structure bearing MEGASPORANGIA.

Megasporangium. A sporangium producing only MEGASPORES.

Megaspore. A SPORE which on germination produces a female GAMETOPHYTE.

Megasporocyte. The cell whose nucleus undergoes MEIOSIS and which yields MEGASPORES.

Megasporophyll. A structure bearing MEGASPORANGIA, and believed to be equivalent to a leaf.

Meiocyte. A cell in which MEIOSIS occurs.

Meiosis. The two consecutive nuclear divisions, in the first of which the CHROMOSOME number is halved, so that each NUCLEUS ultimately contains only one set of chromosomes. The pairing of the chromosomes in the first division provides the opportunity for genetic recombination.

Meristele. The unit of a DICTYOSTELE, composed of a central strand of XYLEM completely surrounded by PHLOEM and delimited by an ENDODERMIS.

Meristem. A group of undifferentiated cells retaining the capacity to divide indefinitely.

Mesarch. XYLEM tissue which differentiates both CENTRIFUGALLY and CENTRIPETALLY, the PROTOXYLEM consequently lying within the METAXYLEM.

Mesophyll. The PARENCHYMATOUS tissue of leaves lying between the upper and lower epidermis; the site of most of the chloroplasts.

Metaxylem. The portion of XYLEM tissue which is derived from the PROCAMBIUM (and is hence PRIMARY) and which DIFFERENTIATES after the PROTOXYLEM.

Micropyle, of seed plants. The minute pore in the distal end of the INTEGUMENTS of the OVULE.

Microsporangiophore. A structure bearing MICROSPORANGIA.

Microsporangium. A sporangium producing only MICROSPORES.

Microspore. A SPORE which on germination produces a male GAMETOPHYTE.

Microsporophyll. A structure bearing MICROSPORANGIA and believed to be equivalent to a leaf.

Midrib. The central VEIN of a leaf.

Mitochondrion. An ORGANELLE of variable shape bounded by a distinct membrane, and occurring in all eukaryotic cells. The site of the major part of respiration.

Mitosis. Nuclear division in which two identical NUCLEI are formed.

Monoclinous, of flowering plants. Having functional STAMENS and CARPELS in the same FLOWER. Also used correspondingly of archegoniate plants.

Monocolpate. Said of a SPORE or POLLEN grain which possesses only one COLPUS.

Monocotyledonous, of flowering plants. An EMBRYO possessing single COTYLEDON.

Monoecious. The condition in which both male and female sex organs are produced on one plant.

Monolete spore. The kind of SPORE which, having been formed in a tetrad without tetrahedral symmetry, bears a linear scar on its PROXIMAL face.

Monopodial branching. The type of branching in which the main apex remains active, the shoots produced from AXILLARY BUDS remaining clearly lateral.

Morphology. The study of the form and related anatomy of living organisms.

Mycorrhiza. A SYMBIOSIS between the roots of a plant and a fungus.

Myrmecophily. The association between certain plants and ants, possibly of advantage to both organisms.

Nectar, of flowering plants. A sugary fluid produced by many species, often in FLOWERS, and collected by insects.

Nectary. A gland which secretes nectar.

Nerve, of mosses. A bundle of elongated cells normally at the center of the leaf.

Node. The position on a stem at which leaves and branches are attached.

Nucellus, of seed plants. The tissue in which the MEGASPOROCYTE arises.

Nucleolus. A body lying within the NUCLEUS, rich in ribonucleic acid.

Nucleus. A conspicuous body in the protoplasm containing most of the deoxyribonucleic acid of the cell, and the site of the Mendelian genes.

Organelle. A part of a cell with certain definite functions and characteristic morphological features.

Organism. An individual animal or plant.

Ovary, of flowering plants. The female region of the FLOWER.

Ovule, of seed plants. The NUCELLUS (containing the EMBRYO SAC) and INTEGUMENTS yielding the SEED after FERTILIZATION.

Palisade tissue. The portion of the leaf MESOPHYLL (usually the uppermost) which consists of regular, columnar cells, containing most of the CHLOROPLASTS.

Palmate. Shaped like a hand with fingerlike lobes.

Palmelloid state, of algae. The aggregation of normally separate individuals into a gelatinous, more or less PALMATE COLONY.

Paraphyses, of algae and bryophytes. Sterile hairs found among the reproductive organs in certain genera.

Parasite. An organism which lives at the expense of another, the HOST, upon which it confers no benefits.

Parenchyma. A tissue composed of VACUOLATED, thin-walled, more or less ISODIAMETRIC cells, functioning chiefly as a ground and storage tissue, the cells often retaining the ability to divide.

Parthenogenesis. The development of a female GAMETE into a new individual without FERTILIZATION.

Parthenospore, of algae. A female GAMETE which develops into a resting SPORE without FERTILIZATION.

Pedicel, of flowering plants. The stalk of a single FLOWER in an INFLORESCENCE.

Peduncle, of flowering plants. The main axis of the INFLORESCENCE, or the stalk of a single FLOWER if it is solitary.

Pellicle, of algae. An elastic membrane surrounding a unicellular organism, found chiefly in the Euglenophyta.

Peltate. Umbrella-shaped.

Perennating organ. A specialized part of a plant ensuring its survival from one season to the next.

Perennial. A plant with an indefinite life span.

Perianth, of flowering plants. Collective term for the PETALS and SEPALS of a FLOWER.

Pericycle. The tissue lying between the vascular tissue and the ENDODERMIS.

Periderm. The collective term for the PHELLEM, PHELLOGEN, and PHELLODERM.

Perigynous, of flowering plants. A FLOWER in which the PERIANTH and androecium are borne on the rim of a concave receptacle, separate from the central gynaecium.

Perispore, of ferns. An irregular accretion around the spore, derived from the TAPETUM.

Petals, of flowering plants. The upper PERIANTH segments, especially when these are distinct by form or pigmentation from the lower.

Petiole. A leaf stalk.

Phellem. See CORK.

Phelloderm. The layer of cells produced towards the inside by the PHELLOGEN (secondary CORTEX).

Phellogen. The CAMBIUM producing the CORK and PHELLODERM.

Phenotype. The visible, or chemically or biologically detectable, manifestation of the GENOTYPE produced as a consequence of growth and development.

Phloem. VASCULAR tissue, composed of SIEVE CELLS or SIEVE TUBES, and associated PARENCHYMATOUS and FIBROUS cells.

Phototaxis. The movement of a whole organism in response to light.

Phototropism. The change in the direction of a plant's growth in response to unequal illumination.

Phyllode. A PETIOLE which is broad and leaflike, the LAMINA often being little developed.

Phyllotaxis. The arrangement of the leaves on a stem.

Phylogeny. The evolutionary history of an organism or group of organisms.

Physiology. The study of chemical and physical processes which occur in living organisms.

Pinna. The primary division of a PINNATE leaf.

Pinnate, of leaves. Having a series of leaflets on each side of a common RACHIS.

Pinnule. The ultimate division of a PINNATE leaf.

Pistil, of flowering plants. The OVARY, STYLE, and stigma.

Pit. A thin area of cell wall traversed by PLASMODESMATA through which substances may pass from cell to cell.

Pith. The core of PARENCHYMA which may occur at the center of a STELE.

Placenta, of flowering plants. The region of attachment of the OVULE to the OVARY wall.

Plagiotropic. Tending to take up a position at a definite angle to an orientating influence (e.g., gravity).

Plankton. Collective term for the microscopic organisms, both plant (phytoplankton) and animal (zooplankton), which occur at the surface of fresh and salt water.

Plasma membrane. Same as PLASMALEMMA.

Plasmalemma. The membrane which bounds the PROTOPLASM of a cell and lies next to the cell wall.

Plasmodesma. A tubular connection between PROTOPLASTS traversing the cell wall.

Plastid. Collective term for chloroplasts, LEUCOPLASTS, AMYLOPLASTS, and CHROMOPLASTS.

Platyspermic, of seed plants. Having flattened, BILATERALLY SYMMETRICAL SEEDS.

Plumule. The embryonic shoot.

Polarity. DIFFERENTIATION in structure or function between two ends of an axis.

Pollen, of seed plants. The MICROSPORES, often showing ENDOSPORIC germination.

Pollination, of seed plants. The process by which POLLEN is transferred from the ANTHER to the CARPELS or OVARY.

Pollinium, of certain genera of flowering plants. A mass of POLLEN grains held together by a sticky secretion.

Polycyclic. Arranged in concentric circles.

Polyembryony, of seed plants. The development of more than one EMBRYO in an OVULE.

Polyploidy. The possession of three or more sets of CHROMOSOMES per NUCLEUS.

Polystelic. Having several STELES.

Primary tissues. Tissues produced as a direct consequence of the activity of apical MERISTEMS.

Procambium. A strand of relatively UNDIFFERENTIATED cells which later gives rise to the PRIMARY XYLEM and PRIMARY PHLOEM tissues.

Proembryo, of gymnosperms. The structure developing immediately from the zygote.

Prothallus. A little DIFFERENTIATED, free-living GAMETOPHYTE.

Protocorm, chiefly of lycopods and orchids. A tuberous structure developing from a portion of the young EMBRYO.

Protonema, chiefly of mosses. The FILAMENT or small plate of cells produced by the germinating SPORE.

Protoplasm. The contents of a living cell.

Protoplast. That part of a living cell remaining after removal of the wall.

Protoxylem. The portion of the PRIMARY XYLEM which DIFFERENTIATES first.

Proximal. Adjacent to a point of reference or symmetry (antonym: DISTAL).

Pseudoendosperm, of gymnosperms. The nutritive tissue in SEEDS derived from the female GAMETOPHYTE (and hence HAPLOID).

Pseudogamy. The development of a female GAMETE into a new individual after stimulation by a male GAMETE, but without actual FERTILIZATION.

Pseudoparenchyma, in algae and fungi. A tissue consisting of closely appressed, often inter-woven FILAMENTS, giving the impression of PARENCHYMA.

Raceme, of flowering plants. An INFLORESCENCE in which FLOWERS are borne on branches of the main axis, those at the base maturing first.

Rachis. The main axis of a pinnate leaf.

Radial symmetry. Symmetry about a central axis; when divided longitudinally along any diameter, the two halves are mirror-images.

Radicle. The embryonic root.

Radiospermic, of seed plants. Bearing RADIALLY SYMMETRICAL SEEDS.

Raphe, of diatoms. The longitudinal fissure in the wall of pennate diatoms.

_____ , of flowering plants. A line marking the position of the FUNICLE in a seed developed from an anatropous OVULE.

Ray. A vertical sheet of PARENCHYMATOUS cells traversing the STELE radially.

Reniform. Kidney-shaped.

Reticulate. In the form of a network.

Rhizoids. Threadlike anchoring and absorbing organs produced by plants lacking ROOTS.

Rhizome. A PLAGIOTROPIC underground stem.

Root. That part of a VASCULAR plant which in branching produces only simple axes like itself, usually subterranean.

Root cap. A cap of tissue over the ROOT apex.

Root hair. Hairlike outgrowths of the EPIDERMAL cells of the young ROOT which absorb water and minerals.

Rootstock. A short, vertical, underground stem, bearing ROOTS; in horticulture the plant providing the root system on the stem of which another kind of plant is grafted.

Saprophyte. An organism obtaining its food in the form of complex molecules from dead organic matter.

Scalariform. In the form of a ladder.

Scandent. Climbing.

Scarious. Thin, dry, and membranous.

Schizogenous (of spaces). Formed by the separation of cells (cf. LYSIGENOUS).

Sclerenchyma. A strengthening tissue composed of FIBERS or STONE CELLS.

Secondary tissues. Produced from MERISTEMS which arise after the DIFFERENTIATION of the PRIMARY TISSUES.

Seed, of seed plants. The product of the OVULE after fertilization, comprising the EMBRYO, with its surrounding food reserves and protective coverings.

Sepals, of flowering plants. The lowermost PERIANTH segments, especially when green, and thus distinguished from PETALS.

Septum. A partition.

Sessile. Without a stalk.

Shrub. A woody plant with no distinct trunk, the main branches arising near ground level.

Sieve cell. A cell with a PROTOPLAST but lacking a distinct NUCLEUS, concerned with the transport of organic materials. A principal component of PHLOEM.

Sieve plate. The perforate area of the cell wall between two SIEVE CELLS or SIEVE TUBE elements.

Sieve tube, of flowering plants. A column of SIEVE CELLS (here usually called SIEVE TUBE elements) with more or less transverse end walls bearing SIEVE PLATES.

Sinuose. Undulating.

Siphon(ac)eous. Tubular.

Spike, chiefly of flowering plants. A RACEMOSE INFLORESCENCE in which the FLOWERS are SESSILE.

Spongy tissue. The lower portion of the MESOPHYLL of a leaf, consisting of irregular cells with large air spaces.

Sporangiophore. A structure bearing one or more SPORANGIA.

Spore. A unicellular, ASEXUAL reproductive cell, usually uninucleate.

Sporophyll. A structure bearing SPORANGIA, and believed to be equivalent to a leaf.

Sporophyte. The (normally) DIPLOID asexual phase of the life cycle producing HAPLOID spores which give rise to the sexual GAMETOPHYTE.

Stamen, of flowering plants. The MICROSPORANGIA (anther) together with their common stalk (filament), usually regarded as a MICROSPOROPHYLL.

Staminode, of flowering plants. An infertile STAMEN often highly modified or reduced.

Stele. The VASCULAR tissue of a root or stem, extending to the ENDODERMIS if present.

Stipule, of flowering plants. Outgrowths (usually two) at the base of the PETIOLE.

Stolon. A short, PLAGIOTROPIC shoot which develops a new plant at the tip, eventually severing connection with the parent.

Stoma A pore in the EPIDERMIS bounded by a pair of GUARD CELLS.

Stone cells. More or less isodiametric, heavily LIGNIFIED cells.

Strain. A mating group within a species, or a variety within a species with distinctive PHYSIOLOGICAL or MORPHOLOGICAL features.

Strobilus. See CONE.

Style, of flowering plants. An elongation of the distal end of the CARPEL, bearing the stigma.

Suberin. An impervious waxy substance with which CORK and other cells are impregnated.

Subsidiary cells. Cells occurring adjacent to the GUARD CELLS or STOMATA, which are usually of constant shape and position, and distinguishable from other EPIDERMAL cells.

Suspensor, chiefly of seed plants. The cell or group of cells arising with the EMBRYO proper from the ZYGOTE, and pushing the young EMBRYO into the nutritive material of the SEED.

Swarmer. See ZOOSPORE.

Symbiosis. An association between two organisms in which there is usually mutual benefit.

Symplast. The continuum formed by PROTOPLASTS linked together by the PLASMODESMATA.

Sympodial branching. The type of branching in which the terminal bud ceases to grow, growth being continued in each instance by the uppermost lateral branch. A form of CYMOSE BRANCHING.

Synangium. A composite structure formed by the CONCRESCENCE of SPORANGIA.

Syngamy. The fusion of GAMETES.

Tangential. Perpendicular to a radius. Used principally to describe the orientation of longitudinal sections of axes, or of cell division.

Tapetum. A layer of nutritive cells surrounding the SPORE mother cells in the SPORANGIUM. Conspicuous only in plants with VASCULAR TISSUE.

Taproot. A persistent primary ROOT, often swollen with food reserves.

Tendril, of flowering plants. A modified leaf, leaflet, or stem, often sensitive to contact and able to coil round objects it touches, thus supporting and anchoring the plant.

Testa, of flowering plants. The covering of the SEED derived from the INTEGUMENTS.

Tetrad. The group of four cells produced by a MEIOTIC division, frequently genetically dissimilar.

Thallus. A relatively undifferentiated plant body.

Tonoplast. The membrane surrounding the VACUOLE, often also called the vacuolar membrane.

Trabeculum. A bar, or elongated cell, traversing a cavity.

Tracheid. An element of the XYLEM tissue consisting of an elongated, elaborately pitted, lignified cell with oblique end walls.

Tree. A woody plant with a definite trunk; the main branching well above ground level.

Trichothallic growth, of algae. The type of growth in which cell divisions are confined to a region at the base of the filament.

Trilete spore. The kind of SPORE which, having been formed in a tetrahedral TETRAD, bears a triradiate scar on its PROXIMAL face.

Triploid. Having three times the basic (HAPLOID) number of chromosomes.

Trisomy. The presence of one chromosome in triplicate in the diploid nucleus. A form of ANEUPLOIDY.

Tuber. A rounded storage organ formed from a ROOT (root tuber) or a stem (stem tuber).

Tunica. The outer layer or layers of cells at the apex of a stem in which the divisions are principally ANTICLINAL.

Vacuole. The portion of the cell which contains the CELL SAP, bounded by the TONOPLAST.

Vascular tissue. A collective term for the XYLEM and PHLOEM.

Veins. The VASCULAR strands of leaves.

Venation. The pattern of VEINS in a leaf.

Ventral, of lateral organs. The same as ADAXIAL.

————, of prostrate plants. The side towards the substratum.

Vessels, of flowering plants. Tubes occurring in the XYLEM and concerned with the conduction of water. Formed from columns of broad, short cells (vessel segments), the end walls of which break down during DIFFERENTIATION.

Xanthophyll. A class of yellow CAROTENOID pigments associated with CHLOROPHYLL in the CHLOROPLAST.

Xeromorph. A plant possessing the features often found in XEROPHYTES, but not necessarily confined to dry places.

Xerophyte. A plant growing characteristically in dry situations, and often possessing anatomical and physiological features enabling it to withstand prolonged drought.

Xylem. VASCULAR tissue which may comprise VESSELS, TRACHEIDS, and FIBERS, together with some PARENCHYMA.

Zoosporangium. An organ producing ZOOSPORES.

Zoospore. A motile SPORE.

Zygomorphic. BILATERALLY SYMMETRICAL.

Zygote. The cell formed by the fusion of two GAMETES.

SUGGESTIONS FOR FURTHER READING

CHAPTER 1

Physiological and biochemical aspects of the evolution of the algae and the colonization of the land:

Raven, J. A. 1986. Evolution of plant life forms. In *On the Economy of Plant Form and Function.* Ed. T. J. Givnish. New York: Cambridge University Press. 421–492.

Swain, T., and G. Cooper-Driver. 1981. Biochemical evolution of early land plants. In *Paleobotany, Paleoecology and Evolution,* Vol. 1. Ed. K. J. Niklas. New York: Praeger. 103–134.

The morphological nature of the colonizers of the land:

Church, A. H. 1981. *Revolutionary Botany, "Thalassiophyta" and Other Essays.* London: Oxford University Press.

Fritsch, F. E. 1945. Studies on the comparative morphology of algae. IV. Algae and archegoniate plants. *Annals of Botany (New Series)* 9:1–30.

The causal aspects of life cycles:

Bell, P. R. 1989. The alternation of generations. *Advances in Botanical Research* 16:55–93.

CHAPTER 2

General accounts of algae:

Bold, H. C., and M. J. Wynne. 1985. *Introduction to the Algae.* 2nd ed. Englewood Cliffs: Prentice Hall.

Cosper, E. M., V. M. Bricelj, and E. J. Carpenter, eds. 1989. *Novel Phytoplankton Blooms.* Berlin: Springer-Verlag.

Fritsch, F. E. 1935, 1945. *The Structure and Reproduction of the Algae.* Vols. 1 and 2. Cambridge: University Press.

Lee, R. E. 1989. *Phycology.* 2nd ed. Cambridge: University Press.

Round, F. E. 1981. *The Ecology of the Algae.* Cambridge: University Press.

Rowan, K. S. 1989. *Photosynthetic Pigments of Algae.* Cambridge: University Press.

Cyanophyta:

Carr, N. G., and B. A. Whitton, eds. 1982. *The Biology of Cyanobacteria.* 2nd ed. Oxford: Blackwell.

Fay, P. 1983. *The Blue-Greens.* Institute of Biology's Studies in Biology No. 160. London: Arnold.

Fay, P., and C. Van Baalen, eds. 1987. *The Cyanobacteria.* Amsterdam: Elsevier.

Rhodophyta:
>Cole, K. M., and R. G. Sheath, eds. 1990. *Biology of Red Algae.* Cambridge: University Press.
>Dixon, P. S. 1973. *Biology of the Rhodophyta.* Edinburgh: Oliver and Boyd.
>West, J. A., and M. H. Hommersand. 1981. Rhodophyta: Life histories. In *The Biology of Seaweeds.* Eds. C. S. Lobban and M. J. Wynne. Berkeley and Los Angeles: University of California Press. 133–193.
>Woelkerling, W. J. 1988. *The Coralline Red Algae.* London: British Museum (Natural History).

CHAPTER 3

General considerations of the chlorophyll a and b algae:
>Raven, J. A. 1987. Biochemistry, biophysics and physiology of chlorophyll b-containing algae. In *Progress in Phycological Research.* Eds. F. E. Round and D. J. Chapman. Bristol: Biopress. 5:1–122.

Prochlorophyta:
>Lewin, R. A. 1989. *Prochloron.* London: Chapman and Hall. (This book also contains information about *Prochlorothrix.*)
>Chisholm, S. W., R. J. Olson, E. R. Zettler, R. Goericke, J. B. Waterbury, and N. A. Welschmeyer. 1988. A novel free-living prochlorophyte abundant in the oceanic euphotic zone. *Nature* (London) 334:340–343.

Extensive treatment of the Chlorophyta will be found in general accounts of the algae. Complementary works are:
>Brook, A. J. 1981. *The Biology of the Desmids.* Oxford: Blackwell.
>Irvine, D. E. G., and D. M. John, eds. 1984. *The Systematics of the Green Algae.* London: Academic Press.
>Pickett-Heaps, J. D. 1975. *Green Algae.* Sunderland, MA: Sinauer.

Euglenophyta:
>Buetow, D. E. 1968, 1982. *The Biology of* Euglena. Vols. 1, 2, 3. New York: Academic Press.

CHAPTER 4

>Green, J. C., B. S. C. Leadbeater, and W. L. Diver. 1989. *The Chromophyte Algae.* Systematics Association Special Volume No. 38. Oxford: Clarendon Press.
>Kristiansen, J., and R. A. Andersen. 1986. *Chrysophytes: Aspects and Problems.* Cambridge: University Press.
>Round, F. E., R. M. Crawford, and D. G. Mann. 1990. *The Diatoms.* Cambridge: University Press.
>Taylor, F. J. R., ed. 1987. *The Biology of Dinoflagellates.* Oxford: Blackwell.
>Werner, D. 1977. *The Biology of Diatoms.* Oxford: Blackwell.

CHAPTER 5

Articles concerning sporogenesis, spermatogenesis, and the sporophyte-gametophyte junction in bryophytes will be found in *Advances in Bryology*. 1988. Vol. 3. Other works of reference are:

Chopra, R. N., and P. K. Kumra. 1988. *Biology of Bryophytes*. New Delhi: Wiley Eastern.

Dyer, A. F., and J. G. Duckett. eds. 1984. *The Experimental Biology of Bryophytes*. London: Academic Press.

Schofield, W. B. 1985. *Introduction to Bryology*. New York: Collier Macmillan.

Schuster, R. M. 1981. Paleoecology, origin, distribution through time, and evolution of Hepaticae and Anthocerotae. In *Paleobotany, Paleoecology and Evolution, Vol 2*. Ed. K. J. Niklas. New York: Praeger. 129–191.

Watson, E. V. 1978. *The Structure and Life of Bryophytes*. London: Hutchinson.

CHAPTER 6

Physiological features of early land plants:

Raven, J. A. 1984. Physiological correlates of the morphology of early vascular plants. *Botanical Journal of the Linnean Society* 88:105–126.

General accounts of living and fossil forms:

Gifford, E. M., and A. S. Foster. 1989. *Morphology and Evolution of Vascular Plants*. 3rd ed. New York: Freeman.

Stewart, W. N. 1983. *Paleobotany and the Evolution of Plants*. Cambridge: University Press.

CHAPTER 7

Bower, F. O. 1923–28. *The Ferns*. Vols. 1, 2, 3. Cambridge: University Press.

Dyer, A. F., ed. 1979. *The Experimental Biology of Ferns*. London: Academic Press.

Dyer, A. F., and C. N. Page, eds. 1985. Biology of Pteridophytes. In *Proceedings of the Royal Society of Edinburgh*. 86B:1–474.

Raghavan, V. 1989. *Developmental biology of fern gametophytes*. Cambridge: University Press.

Sheffield, E., and P. R. Bell. 1987. Current studies of the pteridophyte life cycle. *The Botanical Review*. 53:442–490.

CHAPTER 8

Progymnosperms:

Beck, C. B., and D. C. Wright. 1988. Progymnosperms. In *Origin and Evolution of Gymnosperms*. Ed. C. B. Beck. New York: Columbia University Press. 1–84.

Gensel, P. G., and H. N. Andrews. 1984. *Plant Life in the Devonian*. New York: Praeger.

Stubblefield, S. P., and G. W. Rothwell. 1989. Evidence for heterosporous progymnosperms in the Upper Pennsylvanian. *American Journal of Botany*. 76:1415–1428.

Gymnosperms:

Beck, C. B., ed. 1988. *Origin and Evolution of Gymnosperms.* New York: Columbia University Press.

Friedman, W. E. 1987. Growth and development of the male gametophyte of *Ginkgo biloba. American Journal of Botany* 74:1816–1830.

Konar, R. N., and A. Moitra. 1980. Ultrastructure, cyto-and histochemistry of female gametophyte of gymnosperms. *Gamete Research* 3:67–97.

Moitra, A., and S. P. Bhatnagar. 1982. Ultrastructure, cytochemical and histochemical studies on pollen and male gametophyte in gymnosperms. *Gamete Research* 5:71–112.

Rothwell, G. W. 1987. New interpretations of the earliest conifers. *Review of Palaeobotany and Palynology* 37:7–28.

Singh, H. 1978. *Embryology of Gymnosperms.* Berlin: Borntraeger.

Stewart, W. N. 1983. *Paleobotany and the Evolution of Plants.* Cambridge: University Press.

CHAPTER 9

Crane, P. R. 1985. Phylogenetic analysis of seed plants and the origin of the angiosperms. *Annals of the Missouri Botanical Garden* 72:716–793.

Eichler, A. W. 1954. *Blüthendiagramme.* 2nd ed. Eppenheim: Koeltz.

Esau, K. 1965. *Plant Anatomy.* 2nd ed. New York: Wiley.

Friis, E. M., W. G. Chaloner, and P. R. Crane. 1987. *The Origins of Angiosperms and Their Biological Consequences.* Cambridge: University Press.

Friis, E. M., and P. K. Endress. 1990. Origin and evolution of angiosperm flowers. *Advances in Botanical Research* 17:100–162.

Gifford, E. M., and A. S. Foster. 1989. *Morphology and Evolution of Vascular Plants.* 3rd ed. New York: Freeman.

Hughes, N. 1976. *Palaeobiology of Angiosperm Origins.* Cambridge: University Press.

Johri, B. M., ed. 1984. *Embryology of Angiosperms.* Berlin: Springer-Verlag.

Knuth, P. 1889–1905. *Handbuch der Blütenbiologie.* Leipzig: Engelmann.

Lewis, D. 1979. *Sexual Incompatibility in Plants.* Institute of Biology's Studies in Biology No. 110. London: Arnold.

Proctor, M., and P. Yeo. 1973. *The Pollination of Flowers.* London: Collins.

Raunkiaer, C. 1934. *The Life Forms of Plants.* London: Oxford University Press.

Schultes, R. E., and A. Hoffman. 1980. *The Botany and Chemistry of Hallucinogens.* 2nd ed. Springfield: Thomas.

Steeves, T. A., and I. M. Sussex. 1989. *Patterns in Plant Development.* 2nd ed. Cambridge: University Press.

Weberling, F. 1989. *Morphology of Flowers and Inflorescences.* Cambridge: University Press.

INDEX

Bold page numbers indicate definitions.